A Guide To Successful Construction

Effective Contract Administration

Arthur F. O'Leary

Fellow, American Institute of Architects
Member, Royal Institute of the Architects of Ireland

BNI Building News
1612 South Clementine Street
Anaheim, California 92802

BNi® Building News

Editor-In-Chief
William D. Mahoney, P.E.

Design
Robert O. Wright

BNI PUBLICATIONS, INC.

LOS ANGELES	ANAHEIM
10801 National Blvd.	1612 S. Clementine St.
Los Angeles, CA 90064	Anaheim CA 92802

BOSTON	WASHINGTON, D.C.
629 Highland Ave.	502 Maple Ave. West
Needham, MA 02194	Vienna, VA 22180

1-800-873-6397

Copyright © 1997 by BNI Publications, Inc. All rights reserved. Printed in the United States of America. Except as permitted under the United States Copyright Act of 1976, no part of this publication may be reproduced or distributed in any form or by any means, or stored in a data base or retrieval system, without the prior written permission of the publisher.

While diligent effort is made to provide reliable, accurate and up-to-date information, neither BNI Publications Inc., nor its authors or editors, can place a guarantee on the correctness of the data or information contained in this book. BNI Publications Inc., and its authors and editors, do hereby disclaim any responsibility or liability in connection with the use of this book or of any data or other information contained therein.

Revised edition of **Construction Administration in Architectural Practice**
by Arthur F. O'Leary (McGraw-Hill, Inc. 1992).

ISBN 1-55701-173-7

Preface

Arthur F. O'Leary is universally recognized as the dean of forensic architecture in the State of California: indeed, he might almost be said to have invented this branch of the profession. His forensic activities do not stand alone. The forensic work grew naturally out of a busy and wide-ranging architectural practice and his service on the office practice and the ethics committees of the American Institute of Architects.

The final third of the twentieth century has seen a melancholy increase in litigation and arbitration of disputes relating to construction projects. Such disputes are time consuming, expensive, and demoralizing to architects, engineers, owners, prime contractors, and specialty contractors. Many such disputes could have been avoided by strong supervision and administration. Ironically, strong supervision and administration by architects and engineers has been discouraged by errors and omissions insurers in an attempt to avoid the types of liability that flow from negligent or otherwise inadequate supervision. The very word "supervision" has been banished from the AIA contract documents. The campaign to eliminate supervision from the architect's professional responsibilities was shortsighted and has increased the risk of litigation. The potential for litigation is strikingly reduced when the architect or the engineer performs strong contract administration, as well as design services.

The reason is not difficult to find. It is a commonplace observation that no set of drawings is perfect. When an imperfection comes to light during construction, the problem should be exposed, discussed, studied and resolved immediately. The architect or engineer who is closely associated with the administrations of the project is in a position promptly to discover and resolve problems of design, interpretation, and performance. There is no person in the profession better suited to teach the topic of construction contract administration other than Arthur F. O'Leary.

James Acret
Pacific Palisades, California
April 1996

Acknowledgments

I respectfully acknowledge the early educational influence and inspiration exerted by Professor Henry Charles Burge, FAIA, at the School of Architecture, University of Southern California. He also instilled the superiority of objectivity, thorough analysis, and the superiority of valid basic principles over rote and arbitrary rules.

Those who provided consultation and encouragement include my partners Toshikazu Terasawa, FAIA, Edward K. Takahashi, AIA, CCS, Rudolph V. DeChellis, FAIA, Lawrence Chaffin Jr, FAIA, and Takashi Shida, FAIA; my professional associates Justin J. Gershuny, AIA, and Keith J. Randall; and my associates in forensic architecture Herbert A. Wiedoeft II, AIA, and Russell W. Hobbs, AIA, CCS.

My sincere belief in the value and effectiveness of the arbitration process was developed through service as a construction arbitrator in over 150 arbitrations through the Los Angeles office of the American Arbitration Association since 1957.

My limited command of legal principles and construction law has been absorbed vicariously from the scores of construction industry lawyers I have consulted with through the years and especially from my close professional association with James Acret of the Los Angeles bar and his extensive writings in construction law.

Skilled assistance in preparation of the manuscript of the original book, *Construction Administration in Architectural Practice*, was provided by Felicity Matthews. In revising and expanding the material to produce this Second Edition, which was renamed *A Guide to Successful Construction: Effective Contract Administration*, my task was made immeasurably easier by my acquiring and learning to use a word processor. I am indebted to Raymond McEnaney, computer technologist, who advised me on which equipment to acquire and taught me how to use it.

To Inny, my patient, beloved wife, this book is dedicated.

Arthur F. O'Leary, FAIA, MRIAI

About the Author

Arthur F. O'Leary is a leading expert on construction industry arbitration and architectural practice. He founded the Los Angeles architectural firm of O'Leary Terasawa Partners in 1949, has served as arbitrator and consultant on more than 300 industry litigations, and has written, lectured, and published extensively in over 40 years of professional activity. He received his Bachelor of Architecture from the University of Southern California, where he served on the faculty for 10 years. Mr. O'Leary is a member of the National Panel of Arbitrators of the American Arbitration association. He is also recipient of the Distinguished Achievement Award from the American Institute of Architects. He currently resides near Dublin in Drogheda, County Louth, Ireland.

VI

Contents

1 Introduction 1

2 The Contract Documents 11

3 Architectural Services Agreements 23

4 Construction Contracts 39

5 Project Delivery Systems 65

6 Selecting the Contractor 81

7 The Preconstruction Jobsite Conference 95

8 Consultants and Advisors 107

9 Construction Insurance 113

10 Site Observation and Administration of Construction 117

11 Construction Administration When the Contract Is Bonded .. 133

12 Shop Drawing Procedures 141

13 Payment Certifications 155

14 Change Orders 161

15 Responsibilities of the Owner in a Construction Contract 169

16 The Contractor's Responsibilities 183

17 Role of the Construction Superintendent 197

18 The Architect's Relationship with Subcontractors 209

19	Closing Out the Job	217
20	Termination of the Construction Contract	233
21	Architect's Decisions Based on Design Concept, Aesthetic Effect, and Intent of the Contract Documents	247
22	Resolution of Construction Disputes	259
23	Preventing Time and Delay Disputes in Construction Contracts	269
24	Professional Standard of Care	277
25	Owner's and Contractor's Legal Claims Against Architects	297
26	Analyzing Liability for Construction Defects	317
27	Obtaining the Owner's Instructions	321
28	Designing to a Program and a Budget	325
29	Written Communication	335
30	Graphic Communication	349
	Appendix	357
	Index	475

Detailed Contents

Acknowledgments

1 Introduction 1

    Standard Construction Documents
    Architects as Contract Administrators
    Architect's Versus Contractor's Spheres of Responsibility
    Construction Contract Benefits Both Parties
    Contract Administration by Contractors
    Contract Administration in the Owner's Behalf
    The Contract Documents
    Contractual Relationships
    Figure 1.1, Contractual Relationships
    General Principles

2 The Contract Documents 11

    Contract Documents Defined
    Which Are the Contract Documents
    ...And Which Are Not
    More Non-Contract Documents
    Two More Contract Documents
    The Project Manual
    Oral Agreements Are Contracts But Not Documents
    Enumeration of the Documents in a Contract
    Document Review
    Errors of Inconsistency
    Precedence of the Contract Documents
    Establishing Priority of Documents
    Architect's Interpretations
    Identifying the Contract Documents in Controversies Involving the Correct Drawings
        Controversial Construction Drawings
        Causes of Confusion
    Change Procedures
        Dates on drawings
        Drafting and Reproduction
        Reproductions
    Record of Drawing Revisions
    Issuing Changed Drawings

3 Architectural Services Agreements ... 23

    Architect as Construction Administrator
    Standard Form of Agreement Between Owner and Architect, Fourteenth Edition, AIA Document B141, 1987 (Appendix A)
    Alternative Standard AIA Contract Forms
        A Short Form of Agreement
        A Shorter Form of Agreement
        A Long Form
        Specialized Agreement Forms
        Using Standard Forms
    Oral Agreements and Letter Agreements
        Oral Agreements
        Letter Agreements
    Architectural Fees. How Architects Charge for their Services
        Fee Schedules, Historically
        Private Fee Schedules
        Lack of Fee Study Committees
        Methods of Charging
        Percentage Method
        Multiple of Direct Personnel Expense
        Stipulated Hourly Rates
        Guaranteed Maximum Fee
        Agreed Lump Sum
        Engineering Consultants and Reimbursable Expenses
    Termination of an Architectural Contract
        Ending the Contractual Relationship
        Meaning of Termination, Suspension, and Abandonment
        Reasons for Termination
        Termination Expenses
        Suspension of the Project by the Owner
        Suspension of Services by the Architect
        Additional Reasons for Termination
        Resolution of Disputes
    Construction Administration by the Architect's Consultants

4 Construction Contracts 39

    Types of Standard AIA Contract Forms
        The Two Main Forms
    Alternative Standard AIA Contract Forms
        Short Forms of Agreement
        A Shorter Form of Agreement
        Construction Management (CM) Agreements
        Design/Build Agreements
    Negotiated Contracts. What is Negotiable?
        Everything Is Negotiable
        AIA Standard Contracts
        Who Does the Negotiation?
        Terms and Conditions to Be Negotiated
        The Scope of Work
        The Contract Price
        A Guaranteed Maximum Price (GMP)
        Shared Savings
        The Contract Time
        Liquidated Damages
        Retainage

Change Orders
Surety Bond
Interest on Past Due Amounts
Progress Payments
The Warranty Period
Types of Construction Contracts
Lump Sum Contracts
Cost Plus Fee Contracts
The Contractor's Fee
Unit Prices and Allowances
Agreement Forms
Retainage: Protection for Owner and Surety?
Retainage Must Be Agreed To
Retainage is Contractor's Burden
Reduction, Elimination, or Release of Retainage
Amount of Retainage
Retainage Conditions in Contract
Defeating the Retainage
Manipulating the Schedule of Values
Table 4.1, Schedule of Values and Payment Request # 3
Practical Recommendations
Liquidated Damages and Bonus Clauses
Establishing the Construction Time
Contract Clauses
Deciding the Amount
Administering the Contract Provisions
Determining Completion
Cost Plus Contracts
Effectiveness of Financial Incentives
Allowances in Construction Contracts
Purpose of Allowances
Contract Provisions for Allowances
Owner's Selections
Contractor's Purchase Cost and Related Costs
Specifying Cash Allowances

5 Project Delivery Systems 65

The Traditional System
The Search for an Ideal Contracting System
Indispensable Entities
Some Alternative Construction Delivery Systems
Package Deal
Design/Build
A Closer Look at Construction Management
What Does It Really Mean? And Who Does It?
Qualification Standards for CMs
Diverse Background of CMs
Function of the CM -- What CMs Do
Early CM Appointment
Later CM Appointment
CM's Expanded Services
Why Owners Find CM Attractive
Some Problems with CM
Standard CM Contracts

Two Types of CM
When the CM Is the General Contractor
The CM as the Owner's Impartial Agent
Fast Track Construction
What Is Fast Track Construction?
Figure 5.1, Fast Track Schedules
Understanding Fast Track
Faster Track
Selecting the Contractor
The Fast Track Construction Contract Form
The Guaranteed Maximum Price (GMP)
The Architect's Problems
Will It Save Time or Money?
Will the Conventional System Work?

6 Selecting the Contractor 81

Negotiated Contract or Competitive Bidding
Establishing the Bidding Conditions
Bidding Practices
Recommended Bidding Procedures
Selection of Bidders
Invitations to Bid
Instructions to Bidders
The Bid Package
Bidding Period
Base Bid, Alternate Bids, and Unit Prices
Bid Bond and Proposal Form
Opening of Bids and Determining the Low Bidder
Award of Contract
Considerations in the General Contractor's Decision to Bid or Not to Bid
Probability of Getting the Job
Availability of the Firm
Size of the Project
Location of the Project
Type of Project
Available Personnel
Contractor's Capital Requirements
The Competitive Environment
Reputation of the Architect
Construction Funding
A Fair Bidding Procedure
Award to the Low Bidder
The Most Important Factors
Construction Industry Folklore -- Unlikely Bidding Systems
The Low Bidder

7 The Preconstruction Jobsite Conference 95

Commencement of the Construction Phase
What Is a Successful Building Project?
Preconstruction Jobsite Conference: A Communication Tool
Errors in the Contract Documents
Matters Pertaining to Use of the Site
Contract Documents
Design Intent

Index   XI

Specification Substitutions
Progress Schedules
Record Keeping on the Jobsite
Communications
Architect's Job Visits
Construction Methods and Safety Procedures
Contractor's Requests for Payment
Testing and Inspections
Notice to Proceed
Private Meeting
Owner's Financial Capacity
Unit Prices and Allowances
Contractor's Overhead and Profit
Separate Contractors
Preconstruction Submittals
Owner's Insurance
Liquidated Damages
Architect's Decisions

8 Consultants and Advisors 107

Liability for Mistakes of Outside Consultants
Limiting Exposure to Liability
Limiting Liability for Consultants'
    Shortcomings
    AIA Agreement Forms
    Fair Liability Apportionment
Owners' Versus Architects' Consultants and
    Advisors

9 Construction Insurance 113

Financial Responsibility for Accidental Losses
AIA General Conditions
Owner's Instructions Regarding Insurance and
    Bonds
Certificates of Insurance

10 Site Observation and Administration
of Construction 117

Construction Observation: How much Is
    Enough?
Architectural Service Agreements
Construction Phase
Construction Phase Services
    Advice and Consultation
    Site Visitation
    Reporting to the Owner
    Providing Communication Channel
    Certifying Contractor's Payments
    Reviewing Submittals
    Processing Change Orders
    Resolving Disputes
Budgeting the Fee for Construction Phase
    Services
Construction Observation
    Architect's Field Report
    Pending Claims
    Architect's Versus Contractor's Field
        Function

Architect's Personnel in the Field
Record Keeping During Construction
    Administration
Keeping the Owner Informed
    Construction Period Conditions
    Advising and Informing the Owner
    Certificate for Contractor's Payment
    Architect's Field Report
    Architect's Behind the Scenes Activities
Architect's Fiduciary Duty
Advantage of Effective Communications
    Written Communications
Limitations of Architect's Authority
The Architect's Project Representative
    Historical Background
    Line of Communication
    What Governs the APR's Activities
    APR's Duties and Responsibilities
    APR's Limitations of Authority

11 Construction Administration
When the Contract Is Bonded 133

Construction Bonds
Obligations of Surety
Cost of a Bond
Contractor Prequalification
Surety as Adversary
Surety as Ally
Owner's Right to Approve Contractor's Surety
AIA Standard Bond Forms
Transmitting Bond to Owner
Commencing Work Before Issuance of Bond
Keeping Surety Informed
Termination of the Construction Contract
Consent of Surety
Overpaying the Contractor
Owner's Claim against Bond

12 Shop Drawing Procedures 141

Shop Drawings: Friend or Foe?
Shop Drawing Procedures in AIA Documents
Submittals Defined
Are Shop Drawings Really Needed?
    Unspecified Shop Drawings
Specifying Unneeded Shop Drawings
Specifying Shop Drawings
Figure 12.1, Shop Drawing Procedure
Scheduling Shop Drawings
The Normal Submission Routine
Contractor's Review of Shop Drawings
Consultants' Review of Shop Drawings
Monitoring Progress of Submissions
Keeping the Client Informed
Qualified Personnel
Shop Drawing Stamps
Architect's Approval of a Shop Drawing
Improper Use of Shop Drawings
    Architect Misuse
    Contractor Misuse

Index

13 Payment Certifications 155

    Contractor's Application for Payment
    Representations and Limitations of Certificates
    Decisions to Withhold Certificate
    Overcertification
    Certificate Must Be Fair
    Final Payment
    Accord and Satisfaction
    Substantiation for Payment Requests

14 Change Orders 161

    Who Benefits from Changes During Construction?
    Owner's Right to Make Changes
    Change Orders and Construction Change Directives
    Change Order Procedure
    Pricing and Billing of Change Orders
    Changes in Time
    Acceleration and Impact Claims
    Architect and Surety
    Contractor's Claims for Extra Compensation
    Architect's Minor Changes
    Use Change Orders for All Contract Changes

15 Responsibilities of the Owner in a Construction Contract 169

    The Owner as a Member of the Construction Team
    AIA General Conditions
    Owner's Responsibilities
    Signing the Contract
    After the Contract Is Signed
    Owner Must Furnish Information
    Insurance and Bonds
    Payments
    Owner's Failure to Pay
    Owner's Right to stop the Work
    Owner's Right to Carry Out the Work
    Separate Contractors
    Owner's Claims against the Contractor
    Defective Work
    Owner and Architect
    The Ideal Owner
    Custom Construction Is a Complex Venture
    Owners are Unique
    Owner's Rejection of Work
    Owner's Untimely Decision Making
    Effects of Delayed Completion
    Unexpected Charges for Extra Work
    Completion Formalities
    Solving the Problem with a Tough Contract

16 The Contractor's Responsibilities . 183

    What Is Expected of the General Contractor
    Signing of the Contract
    Drawings and Specifications
    Contractor Responsible for Results
    Construction Quality
    Taxes and Licenses
    Safety and Accident Prevention
    Separate Contractors
    Communications
    Subcontractors and Suppliers
    Hazardous Waste Materials
    Contractor's Warranty
    Warranty Period
    Architect's Services During Warranty Period
    Contractor's Indemnification of Owner and Architect
    Limitation on Owner's Rights to Appoint New Architect or Change Architect's Duties
    The Building Contractor's Dilemma When Confronted with Defective Construction Documents
    The Contractor's Legal Responsibility
    Correction of Errors Can Effect Contract Price and Time
    Contractor's Voluntary Correction of Errors
    Deliberate Breach of the Contract
    Contractor's Expertise
    Errors Caused by Owner's Faulty Directions to Contractor
    Contractors' License Laws

17 Role of Construction Superintendent 197

    Contractor's Superintendent
    Preliminary Site Visit
    Review of Contract Documents and Field Conditions
    Preconstruction Jobsite Conference
    Daily Construction Log
    Construction of the Work
    Supervision and Construction Procedures
    Progress Schedule
    Documents and Samples at the Site
    Shop Drawings, Product Data, and Samples
    Cutting and Patching
    Communications Facilitating Contract Administration
    Continuing Contract Performance
    Claims for Concealed or Unknown Conditions
    Claims for Additional Time
    Cooperation with Owner and Architect and with Separate Contractors
    Changes in the Work
    Time of Construction
    Substantial Completion
    Protection of Persons and Property
    Asbestos and PCB
    Uncovering and Correction of Work
    Tests and Inspections
    Replacement of the Superintendent

Index

18 The Architect's Relationship with Subcontractors 209

    Contractual Relationships
    Relationships among the Parties Governed by Contracts
    Communication
    Shop Drawings
    Contractor's Right to Subcontract
    Subcontractor List
    Subcontract Agreement
    Unwritten Expectations
    Architect's Authority to Reject Work
    Architect's Administration of Payments

19 Closing Out the Job 217

    Orderly Conclusion of the Construction Contract
    When Is the Construction Completed?
    The Contract Time
    Determination of Substantial Completion and Final Completion
    Certificate of Substantial Completion
    Notice of Completion (as Distinguished from Certificate of Substantial Completion)
    Mechanics Lien Laws
    Completion as Seen by Judges
    Certificate of Occupancy
    Lender's Final Disbursement
    Public Work
    Leased Premises
    So, What Is Completion?
    Final Submissions
    Final Completion
    Determination of Final Completion
    Architect's Final Certificate and Final Payment
    Decisions to Withhold Certificate
    Owner's Partial Use or Occupancy
    Retainage
    Liquidated Damages
    Final Reconciliation of Cash Allowances
    Owner's Final Payment to Contractor
    Termination of the Contract
    Architect's Decisions
    Owner-Architect Relationship
    Amicable Conclusion of the Contract

20 Termination of the Construction Contract 233

    Construction Is a Unique Manufacturing Method
    Termination Provisions in the Contract
    Termination of the Contract
        What Do Contractors Expect?
    When the Owner's Behavior Deviates from the Norm
        Reluctant Architects
    Termination by the Contractor

    The Architect's Position when the Contractor Terminates
    When the Owner Ejects the Contractor and Terminates the Contract
        Honorable Intentions
        When the Bubble Bursts
        Considering Termination
    Termination by the Owner for Cause
    The Architect's Approval
    Owner's Rights After Termination
    Balance of the Contract Sum
        A Practical Alternative
    Suspension by the Owner for Convenience
    Position of the Surety
    Position of the Architect
    The Architect's Certificate
    Practical Considerations in the Decision to Terminate
        Architect's Independent Evaluation
        Protecting and Completing the Work
        Late Completion
        Economic Conservation
        Bankrupt Contractor or Owner
    Warranty Responsibility

21 Architect's Decisions Based on Design Concept, Aesthetic Effect, and Intent of the Contract Documents 247

    Imprecise Standards for Architect's Decisions
    Intent of the Contract Documents
    Design Concept
    Architect's Decisions
    Aesthetic Effect Decisions
    Architect's Authority to Reject Defective Work
    Standards for Judging
    Owner Can Accept Defective Work
    Substantial Performance
    Consequential Damages
    Architect's Decision to Reject Work
    Architect's Minor Changes
        Concerns of the Parties
        Clarifying the Contract Documents
    Architect's Interpretation of the Contract Documents
    Limitations on Architect's Authority to Order Minor Changes
    Abuse of the Minor Change Process
    Obtaining Contractor's and Owner's Viewpoints

22 Resolution of Construction Disputes 259

    Misunderstandings During Construction
    General Conditions of the Contract
    Claims and Disputes
    Architect's Aesthetic Effect Decisions
    Architect's Minor Changes
    Assistance from Surety

Architect's Failure to Render a Final Decision
Alternative Dispute Resolution Methods
    Negotiation
    Mediation
    Conciliation
    Arbitration
Filing an Arbitration Demand
    American Arbitration Association (AAA)
    AAA Fees
Use of Lawyers
Participants' Control Over the Outcome
Continuing Contract Performance
Arbitration Award
Conclusion

23 Preventing Time and Delay Disputes in Construction Contracts 269

Changed Conditions
AIA General Conditions
Liquidated Damages
Working Days versus Calendar Days
Unforeseen or Differing Conditions
Delay Caused by Owner or Architect
Construction Schedule
Weather Delays
Critical Path
Substantial Completion
Delay Damages
The Moment of Truth

24 Professional Standard of Care 277

Professional Responsibility
Professional Negligence
Regional Variations in Practice
Determining Professional Standards
Failure of Materials or Procedures
The Hazards of Innovation
Unsuccessful Innovation
Responsible Innovation
Innovation by Engineering Consultants
Other Forms of Innovation
Obtaining the Client's Concurrence
Reviewing One's Own Work
Research by Architects
Others' Reliance on The Architect's Skill
Expert Witnesses
Is Perfection Possible?
Selling Architectural Services
When Should the Architect Pay for Errors?
    Negotiated Compromise
Measuring the Architect's Liability
Architects Do Not Guarantee their Work
Standard of Care
Insuring a Guarantee of Perfection
Disclosure to Owner
Practical Recommendations
Handling Mistakes Realistically
Errors Stemming from Ignorance
Errors of Inadvertence
Budget Problems
    Undershooting
    Overshooting
Finding and Eliminating Errors
Repercussions from Errors
Owners' Attitudes
Contractors' Attitudes
Architects' Attitudes
Purpose of Professional Liability Insurance
An Open Discussion Could Be Revealing
Architects' Decisions

25 Owner's and Contractor's Legal Claims Against Architects 297

When the Architect Gets Sued
Arbitration Demand or Lawsuit
Responding to the Complaint
Professional Liability Insurance
Legal Counsel
Selecting a Lawyer
Preparations to Assist the Lawyer
How Long Should Records Be Kept?
Statutes of Limitations
Settlement
Appearing as a Witness
Third Party Legal Claims Against Architects
    Pitfalls for the Architect who Prepares Plans and Specifications for the Owner-Builder-Developer
    Complete Architectural Services
    Architectural Services for Owner-Builders
    Third-Party Claims
    The Legal Claim
    Developer's Autonomy
    Abbreviated Documents
    Expensive Legal Process
    Indemnity: A Possible Solution
How to Attract a Professional Liability Lawsuit
    Sources of Professional Liability Claims
    Overselling: The Empty Promise of Perfection
    Exhibiting Professional Liability Insurance
    Lack of Communication with the Client
    Failure to Answer Letters and Return Telephone Calls
    Sitting on Complaints
    Not Checking the Work Product
    Violation of Sound Business Principles
    Frequent Payments
    Establishing an Appropriate Fee
    Maintaining High Quality Office Standards
    In Summary

26 Analyzing Liability for Construction Defects 317

Types of Defects
Analyzing the Situation
Identifying the Source of Responsibility
Mediation, a Possible Solution

27 Obtaining the Owner's Instructions 321

An Efficient Way of Obtaining the Owner's Instructions
Part A, Owner's Instructions Regarding the Construction Contract
Part B, Owner's Instructions for Insurance and Bonds
Part C, Owner's Instructions Regarding Bidding Procedures

28 Designing to a Program and a Budget 325

Designing Within the Budget, A Shared Responsibility
The Overall Budget
    The Owner's Program
        Figure 28.1, Owner's Overall Budget
    The Overall Time Schedule
Complying with the Client's Program
Designing to a Budget
Relating the Program to the Budget Price
Measuring Quantity and Quality
The Building Portion of the Budget
Schematic Design Phase
Design Development Phase
Construction Documents Phase
Bidding or Negotiation Phase
Cost Estimates Not Guaranteed
Fixed Limit of Construction Cost
Elimination of Cost Estimating from Architectural Services

29 Written Communication 335

What Must Be in Writing
Architect's Certifications
Certificates Required by the AIA General Conditions
Contractor's Certificate for Payment
Architect's Certificate for Contractor's Payment
Architect's Certificate of Substantial Completion
Architect's Certificate of Final Completion
Architect's Certificate of Approval of Owner's Correction of Deficiencies
Architect's Certificate of Sufficient Cause

Others' Reliance on Certificates
Notices
Submittals
Additional Agreements
Orders, Authorizations, Approvals, and Objections
Additional Written Communications

30 Graphic Communication 349

The Graphic Part of a Construction Contract
Understanding Technical Drawings
People Who Cannot Read Graphics
Client Approval of Drawings
Drafting Conventions
    Orthographic Projection
    Scale
    Drafting Standards
    Drafter's Prerogative
    Legends
Measuring Systems in Construction
Engineering and Architectural Scales
Conversion from One Scale to Another
Referencing
Drawings versus Specifications
Graphically Illiterate Lawyers, Judges, and Arbitrators

Appendix 357

Appendix A: AIA Document B141 -- Standard Form of Agreement Between Owner and Architect, 1987 Edition

Appendix B: AIA Document A101 -- Standard Form of Agreement Between Owner and Contractor (where the basis of payment is a Stipulated Sum), 1987 Edition

Appendix C: AIA Document A111 -- Standard Form of Agreement Between Owner and Contractor (where the basis of payment is the Cost of the Work Plus a Fee with or without a Guaranteed Maximum Price), 1987 Edition

Appendix D: AIA Document A201 -- General Conditions of the Contract for Construction, 1987 Edition

Appendix E: AIA Document B163 --Standard Form of Agreement Between Owner and Architect with Descriptions of Designated Services and Terms and Conditions, 1993 Edition

Appendix F: AIA Document G612 -- Owner's Instructions Regarding the Construction Contract, Insurance and Bonds, and Bidding Procedures, 1987 Edition

Appendix G: Synopses of AIA Standard Form Documents, 1994 Edition

Index

1

Introduction

Standard Construction Documents

The continual evolution of the construction process in the United States has been propelled by remarkable improvements in the technology of materials, building systems, contracting methods, architectural design, and environmental and structural engineering. Concurrent with technological advancement is the continuing development of administrative procedures which must always progress to reflect the latest conditions of accountability, contractual considerations, liability apportionment, governmental regulation, and legal requirements.

Various construction industry organizations have been diligent in providing standard form documents which are consistent with the demands of changing requirements. For more than a century, the American Institute of Architects (AIA), through its national study committees, has responded to the needs of the industry by issuing and updating a complete range of architectural and construction forms and contracts. These documents have provided a measure of uniformity to a decentralized industry. The contents and structure of the AIA documents, and the general principles on which they are based, are widely known and accepted by the industry. Members of the recognized national contracting, subcontracting, and engineering societies consult with architects and the legal counsel of the AIA document committee in developing and revising the standard documents.

Architects and other contract administrators across the nation do not always conform to a precisely defined, single invariable approach to construction administration, but the widespread use of standardized form documents has unquestionably resulted in a fairly high degree of uniformity. Members of the construction industry tend to conduct themselves in accordance with the standard documents, while in turn the standard documents reflect the current professional and industry practices. Thus, the continual change in the standard form documents is closely parallel to the evolution in design and construction practice.

Architects as Contract Administrators

Although much has been written to assist contractors in appropriate procedures needed for conducting a construction business, very little has been made available to aid construction professionals in understanding their professional functions and responsibilities during the construction period.

This book is written primarily for professionals who are faced with the necessity of performing those services which they have contracted to do during, and in anticipation of, the construction phase of the typical contract.

The book will also aid those who represent the owner's interests in a construction contract. This includes facility managers and construction managers. It is a necessity in their job to know what the architect should be doing during the administration phase. Contractors who use this book for reference will better understand their relationship with the owner and architect. Construction lawyers will obtain an overview of the duties of architects, contractors, and owners when the standard AIA documents form the contract.

Large architectural and engineering firms generally have several projects in the construction phase at any one time and, depending on their organizational structure, may have one or more full-time architectural construction administrators on staff. These architects are in the construction phase of the firm's contracts on a more or less continuous basis and are well qualified and experienced. Some larger firms, however, if organized in separate teams for each project, may not have full-time construction administrators. This function will be fulfilled as needed by the project architect or administrator, much like principals of smaller firms.

The principals and key personnel of small and medium-sized firms must be capable of rendering the construction phase services as and when called upon, even though this might represent only a small and infrequent part of their total involvement in the firm's obligations. Ninety percent of their time may be used up by various other pressing professional activities such as seeking new assignments; conferring with consultants, clients, and employees; designing; drafting construction documents; writing specifications; and office administration, along with other equally necessary, diverse, and demanding pursuits. It is these diversified and conscientious practitioners who will find this book most useful as a practical reference for guidance when carrying out their contracted responsibilities during the construction phase.

Introduction

Administration of the contract provides the architect the opportunity to observe the construction as it progresses and the means to suggest adjustments when necessary to preserve the design integrity of the project. The architect will also be able to quickly recognize when contractors and suppliers have misunderstood the intent of the contract drawings and specifications. This will be beneficial to all involved, as it will often reduce the economic waste incurred in the correction of errors. It has been a tradition during the past century for architects to administer the construction contracts under which their projects were built. However, in the last two or three decades some liability insurance carriers and legal advisors to the architectural profession have reasoned that there would be considerably less exposure to professional liability claims if architects had less direct involvement in the construction process. Accordingly, many architects and firms declined to render services beyond completion of the construction documents. Although this approach eliminated the possibility of some types of architectural malpractice occurrences during construction, it also resulted in a greater possibility of construction errors caused by contractors' faulty interpretation or misunderstanding of the construction documents. If the architect was not available during construction, errors or omissions in the documents were not discovered in time to be rectified in an economical manner.

In recent years, informed opinion has reverted once again to favor the architect's traditional role of monitoring the construction contract. This has several advantages which outweigh the disadvantages. With the architect more intimately involved in the conversion from drawings and specifications to physical reality, there is a greater chance of preventing contractor misconceptions and misinterpretations in a timely manner. It also affords the architect an opportunity to correct errors and anomalies in the documents before the construction progress makes them impossible, impractical, or too costly to rectify.

Architect's Versus Contractor's Spheres of Responsibility

A significant twofold outcome of this evolution is that (1) architects now recognize that the physical construction is the contractor's sole responsibility and architects no longer interfere with the contractor's determination and control of means, methods, techniques, sequences, and procedure of construction; and (2) the architect's role during the construction period is limited to counseling, administering, monitoring, observing, and reporting. It is of extreme importance that these two principles not be violated by architects in their preparation of construction documents or in construction administration.

Another way of distinguishing the differences in contractors' and architects' duties is that the contractor is responsible for carrying out the construction contract, while the architect's role is to monitor it. The monitoring process includes interpreting the contract, providing technical assistance to the contractor, observing the work to determine its compliance with the contract, reporting all significant and relevant information to the owner, certifying payments to the contractor, resolving disputes, facilitating communication, and consulting with the owner. An interesting aspect of this evolution is that the architect's construction period role used to be termed *supervision* of construction, whereas now it would be referred to as *observation* of construction. A subtle but significant difference.

Construction Contract Benefits Both Parties

Owners and contractors mutually share the benefits and burdens of their construction contracts. The failure of either to realize their expected legitimate objectives is often caused by shortcomings in carrying out the physical work of the contract but is just as often caused by inadequate or inefficient administration of the contract.

The contract, usually arrived at through negotiation, is the medium of expression by which each party delineates the goals it seeks to obtain. The contractor expects to secure desirable construction employment for its forces, which can be performed under favorable physical and economic conditions, with the reasonable expectation of recovering its costs and a fair profit. The owner aspires to receiving its required construction project built according to the contract drawings and specifications, available for use on or before the agreed delivery date, and to pay no more than the stipulated price.

Contract Administration by Contractors

The largest part of the operation of a construction business is concerned with administration of the contracts which form that company's portfolio of business. Numerous authoritative reference books are available to the construction industry explaining the management and operation of contracting businesses of all sizes, organizational types, and building specialties.

The procedures referred to in this book will apply primarily to construction work contracted by private parties. The general principles also apply generally to public work. However, when contracting with governmental entities, additional statutory requirements will always apply. For the protection of public funds, public works contracting is subject to a large body of specialized laws and regulations which apply to bidding procedures, forms of contracts, specifications, change orders, applications for payment, insurance, bonding, contract termination, closing out procedures, warranties, and appeal procedures.

Contract Administration in the Owner's Behalf

The owner's legitimate interests in a construction contract must be assured through knowledgeable negotiation of the contract in the first place. Many owners are quite competent and able to care for their own interests. They are also fully capable of negotiating on the level of the most canny of general contractors. Others will undoubtedly need the skilled assistance of their legal advisors, architects, and other construction industry experts.

After contract provisions have been completely discussed, agreed upon, and reduced to mutually acceptable written terms, and the contract has been duly signed by both parties, then only upon the diligent follow-through of capable contract administration can the owner realistically expect to fully realize the benefits offered by the contract.

Architects who administer construction contracts in behalf of owners customarily apply the principles and procedures reflected in the various forms of standard agreements of the American Institute of Architects. The owner will usually engage the architect's services for this purpose by using the appropriate owner-architect agreement and by making certain that it contains harmonious conditions requiring the contractor's cooperation and participation. The premise of this book is that the architect's services will have been contracted for by the owner using the standard AIA form:

Standard Form of Agreement Between Owner and Architect, Fourteenth Edition, AIA Document B141, 1987 (Appendix A)

or a similar type of agreement. It is also based on the construction contract being formed substantially from one of the following two standard forms of construction agreement:

Standard Form of Agreement Between Owner and Contractor (where the basis of payment is a Stipulated Sum), Twelfth Edition, AIA Document A101, 1987 (Appendix B)

Standard Form of Agreement Between Owner and Contractor (where the basis of payment is the Cost of the Work Plus a Fee with or without a Guaranteed Maximum Price), Tenth Edition, AIA Document A111, 1987 (Appendix C)

The construction contract must also include the AIA General Conditions:

General Conditions of the Contract for Construction, Fourteenth Edition, AIA Document A201, 1987 (Appendix D)

The Contract Documents

The documents which comprise a complete construction contract in the United States are called the *contract documents* and consist of the agreement between the owner and contractor (usually called the agreement), conditions of the contract (general, supplementary, and other conditions), construction drawings, and specifications. (See Chapter 2, The Contract Documents)

Also included in the contract documents are addenda issued prior to execution of the contract, other documents which may be listed in the agreement, and modifications issued after execution of the contract. Examples of modifications are (1) a written amendment to the contract signed by both parties, (2) a change order, (3) a construction change directive, and (4) a written order for a minor change in the work issued by the architect.

Unless specifically enumerated in the agreement, the contract documents do not include other documents such as bidding requirements (advertisements or invitations to bid, instructions to bidders, sample forms, the contractor's bid, or portions of addenda relating to bidding requirements). Shop drawings, product data, samples, and similar submittals are not contract documents.

The foregoing understandings as to what constitutes the usual documents in a standard construction contract are to be found in Subparagraphs 1.1.1 and 3.12.4 of the 1987 edition of the AIA General Conditions. In addition, none of the following are contract documents: surety bonds, insurance certificates, requests for information, proposal requests, price quotations, clarification drawings, information bulletins, unexecuted change orders, and correspondence.

Contractual Relationships

The AIA standard documents are based on the contractual relationships in the conventional contracting systems prevalent in the United States and some other countries. These relationships are illustrated in Figure 1.1. When analyzing a given situation, it is helpful to review this diagram to determine and visualize the true contractual relationships and communication links between the entities.

Although communication should flow freely along contract lines, the single exception is that the architect, as administrator of the construction contract, controls communication between the owner and contractor. All parties involved should respect the proper communication procedures. This will facilitate systematic administration while avoiding confusion and unexpected liability consequences.

Introduction

Figure 1.1 Contractual Relationships

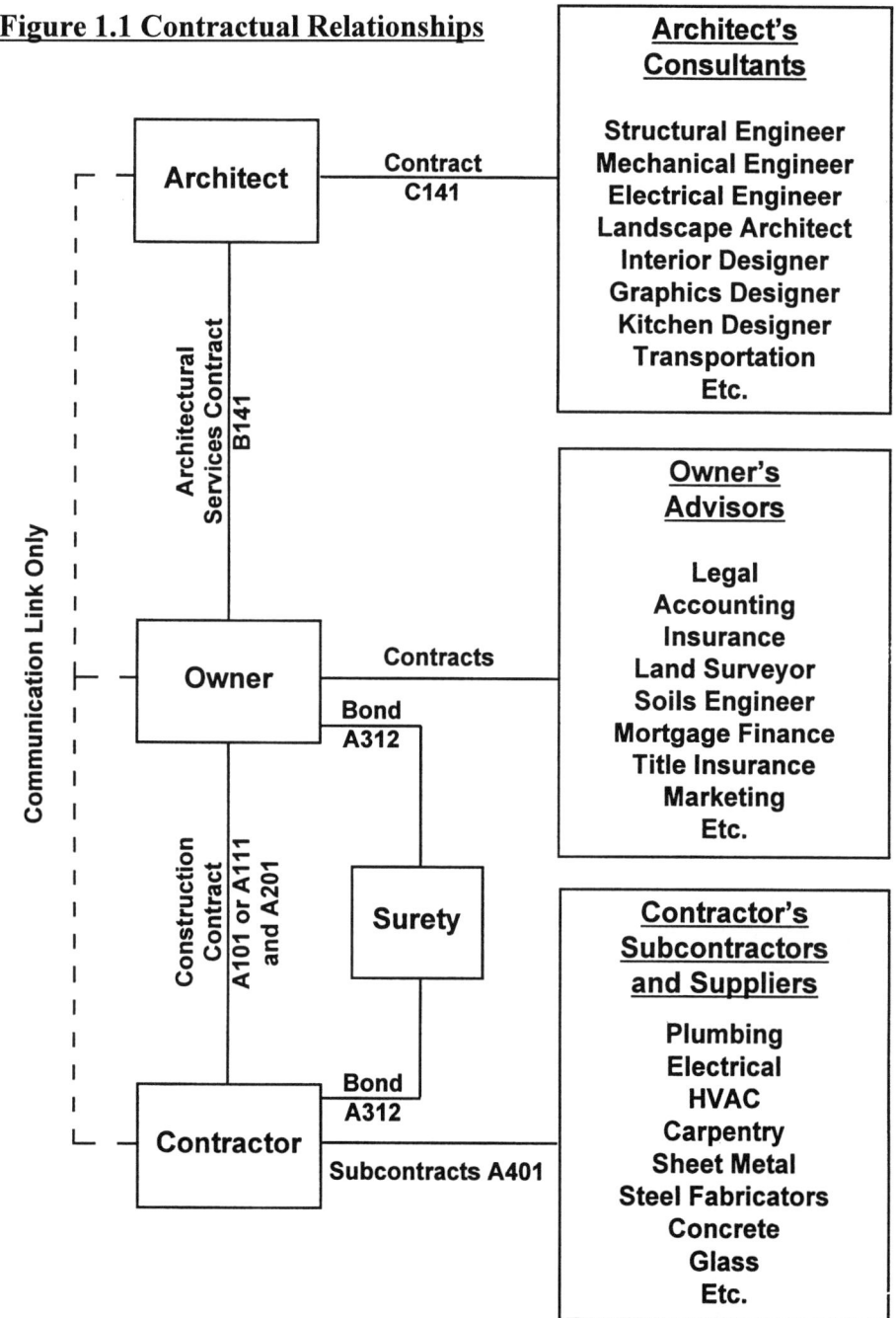

Fig. 1.1 Contractual relationships in the conventional construction contract using AIA form documents. The boxes represent the various entities involved in a typical building project. The solid lines connecting the boxes represent agreements between the entities, generally in the form of written contracts. (Note that the numbers of the appropriate AIA documents are shown.) The dashed line represents a communications link only. The solid line connecting Owner and Contractor is the *construction contract*, comprised of the contract documents. This is the contract which will be administered in the owner's behalf by the architect.

General Principles

All recommendations and procedures will be based on the use of the standard AIA documents with the objective of providing effective and efficient service to the client in full compliance with the offerings of the architectural services contract. Consideration is also given to the responsible limitation of professional liability where possible and to the elimination of unnecessary risk. Unavoidable risk should be controlled and minimized. This is in the interest of both the architect and the owner.

The AIA standard documents have evolved on a continual basis over decades of practical daily usage. They are periodically republished to reflect the developing trends in construction technology, techniques, and administration as well as legal and insurance implications and professional liability consequences. The latest issue of the core construction industry documents, all coordinated and consistent with each other is the 1987 edition. Some of the special purpose documents referred to in Chapter 2 were issued at later dates. All of the AIA Documents referred to are synopsized in Appendix G.

It is always advantageous to use standard form documents where possible, as the construction industry is accustomed to their use and most participants are acquainted with their provisions and the principles upon which they are founded. Customization to address the peculiarities of a specific project can be accomplished by means of deleted or added paragraphs amending the standard forms. In this way, only the deviations need be studied, rather than an entirely new and unfamiliar document. It is also recommended that the actual printed AIA forms be used, rather than retyping them, so that users can remain confident that there have been no basic document revisions. Although standard documents may be incorporated into contracts and specifications by citation, it is always preferable to include the actual documents for immediate reference by the user.

In this book, many of the standard AIA documents have been quoted in part or paraphrased for convenience. Professionals planning to take important administrative actions on their projects are not only advised but urged to refer to the actual documents in their entirety for the exact wording in proper context.

Although the opinions expressed in this book are based on many years of practical experience and on the orthodox and generally accepted interpretations of the AIA documents, they may not be directly applicable to all construction projects. Advice furnished by lawyers specializing in construction will be helpful in considering unique, complex, or unusual contractual situations which might require special analysis.

Introduction

Some of the chapters of this book have previously appeared in slightly abridged form in *Design Cost & Data for Management of Building Design* and in *L.A. Architect*, a publication of the Los Angeles chapter of the American Institute of Architects. Additional material used in expanding the book into this edition, now called *A Guide to Successful Construction: Effective Contract Administration*, previously appeared in *Punch List*, a publication of the American Arbitration Association, *Design Cost & Data*, and in *California Construction Law Reporter*, a publication of Shepard's McGraw-Hill.

2

The Contract Documents

Contract Documents Defined

There would be fewer disputes and disappointments in the construction industry if all parties to construction contracts clearly understood their obligations and were fully aware of their rights and privileges. These are to be found primarily in the contract documents.

The term *contract documents* is uniformly used in all AIA standard form agreements to designate the group of documents that comprise the construction contract between the owner and contractor. The term contract documents and the word contract may be used interchangeably.

Owners and contractors, when making claims against one another, will attempt to enforce the provisions of conversations, inferences of actions, and any scrap of paper developed in the course of their relationship. Similarly, architects frequently search the documents seeking authority to compel the owner or contractor to perform some duty. However, only those documents and promises that have been reduced to writing and incorporated into the contract can be relied upon as being legally enforceable.

Which Are the Contract Documents...

The AIA General Conditions (Document A201) contains a contractual definition of the contract documents (Subparagraph 1.1.1):

"The Contract Documents consist of the Agreement between Owner and Contractor (hereinafter the Agreement), Conditions of the Contract (General, supplementary and other conditions), Drawings, Specifications, addenda issued prior to execution of the Contract, other documents listed in the Agreement and Modifications issued after execution of the Contract. A Modification is (1) a written amendment to the Contract signed by both parties, (2) a Change Order, (3) a Construction Change Directive or (4) a written order for a minor change in the Work issued by the Architect."

...And Which Are Not

The same subparagraph (A201, 1.1.1) identifies some documents which should never be considered as contract documents:

"Unless specifically enumerated in the Agreement, the Contract Documents do not include other documents such as bidding requirements (advertisement or invitation to bid, Instructions to Bidders, sample forms, the Contractor's bid or portions of addenda relating to bidding requirements)."

Furthermore, Paragraph 3.12.4 clarifies the status of shop drawings and other submittals:

"Shop Drawings, Product Data, Samples and similar submittals are not Contract Documents."

More Non-Contract Documents

In addition, by inference, the contract documents do not include numerous other documents used in the construction process. For example, none of the following, unless specifically incorporated into the contract documents, are contract documents:

Correspondence, letters of transmittal, unsigned change orders, insurance policies, insurance certificates, performance bonds, labor and material payment bonds, title reports, topographic and boundary surveys, soil tests, material tests, laboratory reports, engineering calculations, environmental impact reports, materials certifications, inspection reports, subcontracts, purchase orders, payment requests, schedules of values, payment certificates, field observation reports, wording on shop drawing stamps, requests for information, requests for price quotations, cost breakdowns, price quotations, conference reports, certificates of compliance, manufacturers' literature, industry standards, guarantees, warranties, certificates of substantial completion, notices of completion, operating instructions, photographs,

The Contract Documents 13

certificates of occupancy, mechanics' liens, lien releases, building permits, building codes, or zoning regulations.

The architectural services agreement is not a part of the owner-contractor agreement and is therefore not a contract document.

Any written agreements between the owner and contractor which pre-date the construction agreement are not contract documents unless specifically incorporated in the construction agreement. (A201, 1.1.2)

If the owner, contractor, or architect wish to consider any of the normally non-contract documents to be included in the contract, they simply must so-provide in the supplementary conditions or in the agreement before it is signed, or afterwards as an amendment to the contract.

Two More Contract Documents

In order to rise to the stature of a contract document, a writing must be agreed to by the owner and contractor and it thus becomes a part of the contract or an amendment to it.

There are two notable exceptions to this general concept: construction change directives, which are signed only by the owner and architect, not the contractor (A201, 7.3), and an architect's written order for minor changes, which is signed neither by the owner or contractor, but solely by the architect. (A201, 7.4)

Both of these are contract documents by virtue of the definitions of Modification (3) and (4) in A201, 1.1.1, quoted above.

The Project Manual

The project manual, according to Subparagraph 1.1.7, is the volume usually assembled for the work which may include the bidding requirements, sample forms, conditions of the contract, and specifications. The project manual is not a contract document, although some of its inclusions will be and some will not.

Oral Agreements Are Contracts But Not Documents

Generally, the only obligations which can be enforced by the owner or contractor against the other are those which appear as a requirement in one of the contract documents. Likewise, no rights are acquired by either unless they are conferred by some provision of a contract document.

Some actions or oral statements, however, may serve to amend the contract, even though they are not contract documents. For example, an owner might ask a contractor to move a brick pier and agree to pay an additional sum. After the contractor has removed and reconstructed the pier, the owner cannot then repudiate the oral agreement even though there is no related contract document.

Enumeration of the Documents in a Contract

Confusion in a specific contract can be lessened or possibly avoided by careful enumeration of the documents which are to be considered part of that agreement. The AIA standard form owner-contractor agreements have blank spaces designated for the purpose of listing all of the contract documents:

> Article 9 in AIA Document A101, Standard Form of Agreement Between Owner and Contractor where the basis of payment is a Stipulated Sum, Twelfth Edition, 1987
>
> Article 16 in AIA Document A111, Standard Form of Agreement Between Owner and Contractor where the basis of payment is the Cost of the Work plus a Fee with or without a Guaranteed Maximum Price, Tenth Edition, 1987
>
> Article 6 in AIA Document A107, Abbreviated Form of Agreement Between Owner and Contractor for Construction Projects of Limited Scope where the basis of payment is a Stipulated Sum, Ninth Edition, 1987
>
> Article 1 in AIA Document A105, Standard Form of Agreement Between Owner and Contractor for a Small Project where the Basis of Payment is a Stipulated Sum, Small Projects Edition, 1993

Document Review

Properly managed architectural and engineering practices employ rigorous review procedures and the documentation is routinely checked at various stages of progress. All of the construction drawings, specifications, and all other contract and bidding documents should be included in a comprehensive review process. Extensive check lists are available and many have been produced by individual offices, reflecting past problems with the hope of eliminating them in the future.

Document review should always be under the close supervision of competent qualified senior personnel. The documents should be checked for compliance with the owner's program and instructions, applicable building codes, industry standards, and customary construction procedures.

Checking should also evaluate whether or not the documents, when executed in the field, will properly carry out the design intent. The checker should make sure that the documents reflect the proper use of materials and that the structural and other systems are practical, economical, and constructible. One of the most important aspects of document review is checking for physical coordination among the various engineering and design disciplines. It is also necessary to confirm that all engineering recommendations have been carried out on the drawings and in the specifications.

The final review before release of the documents for contract bidding should be comprehensive and thorough. Regardless of the scope and quality of document review, however, it is inevitable that some level of error or inconsistency will remain undetected. (See Chapter 24, Professional Standard of Care.)

Errors of Inconsistency

Of the significant errors, some will be in the form of inconsistencies between the various elements of the contract documents. For example, the specifications might not agree with the drawings or the agreement in some respect. However, some inconsistencies and conflicts occur within a single document. For example, one section of the specifications does not agree with another, or the floor plan is not in accordance with a section.

Those errors that are discovered by or brought to the attention of the architect during the bidding period, can be easily dealt with by prompt consideration of the technical problem and issuance of an appropriate addendum to the affected document. Most of the remaining errors will more than likely be discovered during the construction period. Some will remain undetected to surface years later or possibly never.

Those differences that are caused by conflicting requirements in two or more of the contract documents could sometimes be easily resolved if the documents could be ranked in a hierarchical order of authority.

This would be simple and effective for some types of inconsistency. For example, if the specifications stated that a certain group of light bulbs are to be 300 watts and the lighting fixture schedule on the electrical drawings required 500 watts, then the higher ranked document would govern. However, this settles only the contractual obligation of the contractor, leaving other unresolved problems. For example, if the drawings were designated as governing, then the 500 watt bulbs would have to be supplied by the contractor. However, if the 500 watt designation is incorrect, the contractor must be further directed to supply 300 watt bulbs and issue a credit for the difference in cost between 500 watt and 300 watt bulbs. Yet, if the error is not discovered until after the 500 watt bulbs have already been installed, the contractor must be reimbursed for the extra costs of relamping, restocking (if allowed by the supplier), supervision, overhead, and profit. The credit for changing to smaller bulbs will be consumed or exceeded by the associated costs.

Some will point out that this problem could have been prevented entirely if the same information had not been given in two places. The bulb size should be shown in the lighting fixture schedule on the drawings and eliminated from the specifications, or vice versa. This is undoubtedly true for this example and generally represents good advice. However, other situations often arise where the inconsistency is more difficult to perceive. For example, where an industry standard, code, or regulation is cited in the specifications and the size, gauge, thickness, or some other quality shown on the drawings is inconsistent with the cited standard, code, or regulation.

Precedence of the Contract Documents

Considering the extraordinary volume of data contained in the typical set of construction documents, literally hundreds of thousands of bits, it is surprising that we do not see more errors, inconsistencies, and anomalies. Fortunately, most errors encountered are trivial and cause no more than a momentary pause and possibly a minor degree of professional embarrassment to the architect with no economic consequence to owner or contractor.

Over the past several years, the concept of establishing a hierarchy of documents has never been embraced by the professionals serving on the AIA's document committees. On the contrary, the AIA recommends that a precedence of documents not be established, but rather that all documents are complementary. What is required by one document is as binding as if required by all. Subparagraph 1.2.3 of the AIA General Conditions provides:

> "The intent of the Contract Documents is to include all items necessary for the proper execution and completion of the Work by the Contractor. The Contract Documents are complementary, and what is required by one shall be as binding as if required by all; performance by the Contractor shall be required only to the extent consistent with the Contract Documents and reasonably inferable from them as being necessary to produce the intended results."

This does not directly resolve the problem of inconsistent or conflicting requirements but instead requires the architect to make a determination taking into account all relevant factors obtained from anywhere in the documents or reasonably inferred from them. The architect's decision must determine what the contractor is obligated to do and what adjustment, if any, is to be made in the contract price or time. This is the acid test of an architect's ability to be fair to both owner and contractor.

The Contract Documents

Establishing Priority of Documents

The AIA, recognizing that occasionally some situations require a precedence of documents to be established, recommends a priority ranking of the documents followed by a requirement that in case of inconsistencies in or between drawings and specifications, the higher cost condition should govern.

Guide for Supplementary Conditions, Fourth Edition, AIA Document A511, 1987, where an order of precedence is required, suggests that the following be added to A 201, 1.2.3:

> "In the event of conflicts or discrepancies among the Contract Documents, interpretations will be based on the following priorities:
>
> 1. The Agreement
> 2. Addenda, with those of later date having preference over those of earlier date
> 3. The Supplementary Conditions
> 4. The General Conditions of the Contract for Construction
> 5. Drawings and Specifications
>
> In the case of an inconsistency between Drawings and Specifications or within either Document not clarified by addendum, the better quality or greater quantity of Work shall be provided in accordance with the Architect's interpretation."

The recommended order of precedence considers the drawings and specifications to be of equal authority, and they are last after all other documents. The final paragraph, which relates only to inconsistencies between the drawings and specifications, settles only the matter of controlling the contractor's obligation for price and time but does not settle the issue of which requirement is correct or proper to be carried out. When the error is not discovered until after the questioned work is completed, the contractor must be paid all extra costs of changing from the more stringent or costly condition to the correct one. Apparently, the paragraph does not apply to inconsistencies between or among these two documents and the others listed. It also fails to answer the question of inconsistencies within the other documents.

Most architects who feel that it is preferable to have a precedence of documents provision in their contract documents tend to favor drawings over specifications, while construction attorneys seem partial to the agreement over all other documents. Apparently, each is more trusting of the documents with which they are most familiar or have the most control over. Some architects specify that large scale drawings will take precedence over small scale drawings and that figured dimensions take priority over scaled dimensions.

Obviously incorrect typographic or clerical errors must be interpreted reasonably. No contractor can be considered credible when claiming to be misled by a 40 inch thick concrete floor slab requirement which is obviously intended to be 4 inches thick, or by a number 40 steel reinforcing bar which is meant to be a number 4 bar. Neither is the owner to be taken seriously when demanding that the contractor furnish the 40 inch thick slab or the number 40 steel bars or requests a monetary credit in lieu thereof.

Architect's Interpretations

Architects who are required to render interpretations and make decisions resolving inconsistencies or conflicts between or among the documents must stay meticulously within the contractual guidelines of A201, 1.2.3. Regardless of what the architect intended to include in the documents, the decision and all inferences must be based on what can actually be found in the contract documents.

The architect-decision maker, being also the author responsible for the inconsistent or erroneous documents, must at all costs guard against self-serving rulings. If a fair ruling must favor the contractor and go against the owner, it is usually an indication that the documents are indeed imperfect. If this creates some problem between the architect and client, then this becomes a matter the two of them must resolve. It is of no concern to the contractor and should not enter into the architect's decision regarding the owner-contractor relationship. The contractor should not be responsible for any errors in the documents.

In determining what is reasonably inferable from inconsistent, erroneous, or incomplete documents, the design professional must exercise impartial judgment based on a credible and plausible analysis justified by the actual state of the documents. This decision will be final and binding upon the parties if not appealed to arbitration within 30 days. (See A201, 4.3 Claims and Disputes, 4.4 Resolution of Claims and Disputes, and 4.5 Arbitration.) If the architect's determination is not sustainable from evidence that can be found in the contract documents, fair, knowledgeable, and reasonable arbitrators will find no difficulty in overturning or amending the architect's decision in their award.

Identifying the Contract Documents in Controversies Involving the Correct Drawings

A problem that must be sorted out and conclusively proved in many construction industry disputes is exactly which documents comprise the contract. On the surface, the issue seems fairly straightforward, as most modern construction contracts include a listing of the contract documents.

The Contract Documents

In general, the contract documents include the agreement, the general, supplementary, and other conditions, the specifications, the drawings, the addenda if any were issued prior to the contract signing, and modifications if any were issued after the contract was signed. Gathering these documents together in one place should not be too difficult. The problem usually is in organizing, identifying, and interpreting the miscellaneous conglomeration of drawings that have accumulated after the contract was signed. There are also the questions of which sheets have been superseded and identifying which are missing.

Controversial Construction Drawings. When all of the construction drawings are collected and viewed together, there is often confusion as to the various editions or versions of some, many, or all of the sheets. The owner, contractor, subcontractors, and suppliers may have differing views on which drawings comprise the official set or the contract set.

The set that has been stamped approved by the building department may differ from some or all of the other editions and may or may not be the set on which the contract was predicated.

In addition to the bound sets, there will be numerous loose sheets. The final contract set usually will include some of each.

The problem can be solved easily if the various editions are dated, but care must be taken when comparing dates so that reproduction dates are not confused with edition dates. One cannot always assume that the contract drawings will all be dated at or about the contract agreement date or that they will all be the same date. Some architectural offices mark the relevant prints with a rubber stamp, "Contract Set", at the time they are issued.

This all is neatly resolved when the AIA General Conditions are a part of the construction contract. In the General Conditions, it is provided that the owner and contractor shall both sign the contract documents but that if one or both fail to sign them, the architect will identify the contract documents if requested. (A201, 1.2.1)

Causes of Confusion. It would seem simple enough to keep track of the contract drawings even if the owner and contractor had failed to sign them. However, identification problems also arise when changes in the drawings are made after the contract has been signed. Changes are made in the interest of reducing construction costs, to reflect unforeseen construction conditions, and accommodating changes in the owner's mind during construction. Changes are then made to changes. Partial changes are then made to partial changes. Some changes are later canceled. Some are wholly or partially reinstated or only partially canceled. Some of the changes, cancellations, or reinstatements are made hurriedly over night or over the weekend to accommodate construction progress in the field and sometimes with or without the formality of accompanying correspondence. Some changes are made over the telephone and followed later with a drawing. This tangled web gets more snarled when projects are under construction for protracted periods of time. Lack of adequate record keeping, changes in personnel, and passage of time render the problem exceedingly difficult to unravel.

Change Procedures

This all sounds like the architectural and engineering professions are not very disciplined. However, there are accepted procedures in the design community for designating changes in the documents and tracking the state of the contract. These procedures are generally understood by the construction industry. It is when the established procedures are circumvented or omitted that the confusion proliferates.

Dates on Drawings. The general date of a drawing, usually found in the title block, is the date when the drawing was started or finished. Some offices will conform all of the drawings in the original set to a single uniform date. The title block date is usually never changed after the drawing becomes a contract document.

New drawings originated after the contract is signed will usually be dated when started or finished or may possibly be conformed to a change order date. Printing dates are usually not on the originals, but are rubber stamped on the surface of the prints only. This is not the drawing date, but merely the date on which the reproduction was made. Some offices deviate from this procedure by recording printing dates on the original instead of on the surface of the print. Drawings of different printing dates may be identical if no interim changes have been made on the originals. Other dates may have been stamped on the surface of prints, such as the dates when the prints were received in various offices. These received dates may or may not establish the drawings as a part of the contract.

Drafting and Reproduction. To fully understand the dating and identification phenomenon, it is necessary to understand the relationship between drawings and their reproductions. All drawings are not made in the same way and they are not reproduced in the same way.

Original drawings are made using pencil or ink on tracing paper or sheet plastic. Whether the drawings are produced by hand drafting or by computer driven plotters, the principle is the same. The type of paper used is usually translucent so that reproductions can be made directly from the originals.

The drafted originals are always handled carefully in the architect's office. To lose or damage an original drawing would be a catastrophe, as all of its production time would be lost. Computer-produced original tracings, if lost or damaged, could be replaced quickly and inexpensively by merely replotting from the original computer program.

Reproductions. Years ago, the prints made from the originals were called blueprints as they had white lines on a blue background. It was a cumbersome wet process and the prints had to be dried. Although this process is no longer used, some still use the old term blueprint to represent any roll of printed construction drawings. The common procedure today is a dry ammonia vapor process that produces the familiar print with a white background and purple lines. The same process can also yield other colors of print called bluelines, blacklines, or brownlines. The background color can be other than white if different colors of paper are used.

If the original drawings are on opaque paper, such as with pasted up originals, the prints can be produced by xerography. Most architectural and engineering offices have dry electrostatic copiers that can produce same size prints up to 11 inches by 17 inches. Larger xerographic prints can be made at the local reprographic shop (that used to be called the blueprinter).

Opaque originals cannot be reproduced by the ammonia vapor process. To do so would require a reproducible tracing, made by xerographically copying the opaque original onto translucent paper.

If required, the reproducible tracing can be altered by eradicating unwanted information and by drafting in new information.

Half size prints are often used for convenience in carrying, for desk top use, and for economy. A half size print of a 30 inch by 42 inch drawing is only 15 by 21, thereby saving 75 percent of the paper volume. All scales on the drawings are halved. Readability is sacrificed, so the full size drawings should be used for most technical purposes, such as estimating and the actual construction. Full size and half size prints are identical, except for size, when made from the same originals.

Record of Drawing Revisions

A common practice in architects' and engineers' offices is to identify and record all changes made in the content of drawings. Deletions, additions, and alterations made before the incorporation of the drawings into a signed construction contract are not often recorded, as it would rarely serve any useful purpose. However, after the bidding drawings are released, all changes made in the original drawings should be faithfully recorded.

The prevalent system in use consists of a Revision Block on the face of each original drawing. This is usually in the form of a rectangular table where all changes are listed by date and content. As each change is made on the drawing, it is added to the list. Each change is assigned an identifying symbol, commonly a number in a small triangle, called a delta. Typically each change recorded in the revision block would include the delta number, date, and a brief description of the change.

The change location on the drawing is also marked with the delta symbol. In order to identify the changed portion of the drawing it is usually circumscribed by a billowing line resembling a cloud and generally called a cloud. To completely understand the nature and extent of any change, one must usually compare the changed area on the revised drawing with the same area on the previous edition of the same sheet. When a new revision is to be added to a previously revised sheet, the last added cloud is removed, and the new cloud is added to identify the new revision. The cloud is usually drawn on the back of translucent originals to facilitate its removal if necessary. Naturally, all the old deltas must remain. One must review all the old editions of a particular sheet to fully grasp its change history.

When an entire new sheet is added to describe a construction change, it should be properly dated and identified in its revision block.

Since the advent of CADD (Computer Aided Design and Drafting) some architectural offices are finding it more convenient to issue supplementary drawings to describe changes in the contract drawings. This is done by issuing computer generated copies of portions (8 1/2" x 11" or 11" x 17") of the original drawings, altered to illustrate the changes. The new drawing title or description and the date in the new title block, along with its related documentation should establish whether or not the drawing is a contract document.

Issuing Changed Drawings

As each new edition of a particular sheet is issued, the recipients should insert it in their complete set of contract drawings, ensuring that all editions of each sheet are available for comparison. Complete, up-to-date sets of the contract drawings should be maintained by the contractor's office, site office, the owner, and the architect. Major subcontractors and suppliers will also maintain complete sets.

When revised drawings are issued, they should always be accompanied by a letter of transmittal, explaining the significance of the new drawing. In some cases, the drawing is being issued provisionally, pending the contractor's quotation and the owner's acceptance of proposed cost and time changes. Other times the changed drawings are being issued for carrying out work in the field.

Alone, a changed drawing does not constitute a change in the contract. The change must have been authorized by an agreed change order or something equivalent. The drawing merely explains and illustrates the proposed or agreed change. Naturally, the change order and the changed drawing must be consistent. Neither will become a contract document until agreed upon by the contractor and owner.

There should be no problem in determining the status of the contract drawings at any moment in time, if the accepted procedures are followed.

3

Architectural Services Agreements

Architect as Construction Administrator

The scope of an architect's position as administrator of any specific construction contract is defined in the owner-architect agreement.

It does little good to specify the architect's duties and authority in the construction contract if the architect has not previously been retained by the owner to provide those same services. The agreement should carefully delineate the range of the architect's duties and scope of authority. The architect will be acting in the owner's behalf. Therefore only the owner can empower the architect. The contractor accepts the architect's authority by signing the construction agreement.

The architect's authority as administrator will be seriously compromised if the owner has not knowingly agreed to the principles underlying effective contract administration. This often requires the architect to embark on an owner-education process, as many owners would not necessarily be aware that there could be many problems in administrating the contract.

Standard Form of Agreement Between Owner and Architect, Fourteenth Edition, AIA Document B141, 1987 (Appendix A)

This standard agreement describes all of the architect's and owner's obligations and the extent to which the architect can act for the owner. There is a careful demarcation drawn between those services that are included as the standard basic services package and those that would be considered as contingent or optional additional work subject to additional charges. These should be thoroughly reviewed with the owner to make certain that all necessary services are included in the basic services category and that unneeded or unanticipated services are omitted. This joint review is a good way of explaining the administrative process and obtaining the owner's understanding and concurrence.

Phases of architectural services under B141:

1. Schematic Design Phase
2. Design Development Phase
3. Construction Documents Phase
4. Bidding or Negotiation Phase
5. Construction Phase -- Administration of the Construction Contract

Alternative Standard AIA Contract Forms

A Short Form of Agreement. For projects of limited scope and duration, one could use a shortened form of agreement such as:

Abbreviated Owner-Architect Agreement, AIA Document B151, April 1987

This agreement form reduces the standard five service phases of B141 to three by combining and abridging. Phases of architectural services under B151:

1. Design Phase (Combining phases 1 and 2 of B141)
2. Construction Documents Phase
3. Construction Phase -- Administration of the Construction Contract
    (Combining phases 4 and 5 of B141)

As with all abridgments, the shortening process omits some of the detail. B151, at 8 pages, is not much shorter than B141's 10 pages. In many cases, it would be more appropriate to use an edited down version of B141. B151 should not be used without a thorough review to assure that all required construction administration services are included.

Architectural Services Agreements

A Shorter Form of Agreement. An even further abridgment is available in this 4 page document:

> Standard Form of Agreement Between Owner and Architect for a Small Project, Small Projects Edition, AIA Document B155, 1993

The architectural services phases have been reduced to two by combining and condensing:

1. Design Phase (Combining phases 1, 2, 3, and 4 of B141)
2. Construction Phase

B155 is designed for extremely small projects where the architect's services are minimal. The form should only be used when these companion owner-contractor agreement forms and general conditions are used:

> Standard Form of Agreement Between Owner and Contractor for a Small Project where the Basis of Payment is a Stipulated Sum, Small Projects Edition, AIA Document A105, 1993, and

> General Conditions of the Contract for Construction of a Small Project, Small Projects Edition, AIA Document A205, 1993

In the shortening process only the absolutely bare essential elements of B141 have been carried forward. Many of the rights and safeguards protecting architect and owner alike have been eliminated. For example, the right to arbitration is not included. The architect's construction administration services are omitted from the architectural agreement and instead appear only in the general conditions. This has the advantage of avoiding inconsistencies between the architectural and construction contracts.

It is absolutely necessary to use these three forms together as a coordinated group as none will mesh exactly with the mainstream AIA documents.

The Small Project Editions, B155, A105, and A205, are handy for very small work assignments, but should be recognized as a limited substitute for the more familiar and complete AIA core documents, B141, A101 (or A111), and A201. When used, the Small Project Editions should be carefully reviewed making certain that all required provisions are included.

A Long Form. For projects where the architect or owner wish to break down the architect's services into finely defined individual tasks, a lengthier (32 pages) standard AIA form may be used instead of B141:

Standard Form of Agreement Between Owner and Architect with Descriptions of Designated Services and Terms and Conditions, AIA Document B163, 1993 (Appendix E)

This agreement form consists of three parts:

Part 1 - Form of Agreement
Part 2 - Descriptions of designated Services
Part 3 - Terms and Conditions

The phases of architectural services in B163 are broken down more than in B141 and consist of:

1. Pre-Design Phase (Phase 1 of B141 is subdivided into phases 1, 2 and 3 of B163)
2. Site Analysis Phase
3. Schematic Design Phase
4. Design Development Phase
5. Contract Documents Phase
6. Bidding or Negotiations Phase
7. Contract Administration Phase
8. Post-Contract Phase (An added phase not in B141)

B163 contains numerous subdivisions of each of the phases of service as well as detailed definitions of each of the finite tasks. This allows for the individual consideration by the owner and architect of each of the services that will be required for the project. Each task can be separately evaluated as to time and cost.

Specialized Agreement Forms

Some architectural service situations do not fit the standard agreement forms. The AIA has developed some special forms to be used in these cases. See Appendix G, Synopses of AIA Standard Form Documents, for detailed descriptions and recommended uses of these documents. When using the various agreement forms, it is important to make sure that the families of documents are matched up properly. That is, using the coordinated owner-architect, owner-contractor, general conditions, and construction manager agreements.

Standard Form of Agreement Between Owner and Architect, Construction Manager-Adviser Edition, AIA Document B141/CMa

Standard Form of Agreement for Interior Design Services, AIA Document B171

Architectural Services Agreements

> Abbreviated Interior Design Services Agreement, AIA Document B177
>
> Owner-Architect Agreement for Housing Services, AIA Document B181
>
> Standard Form of Agreement Between Owner and Architect for Limited Architectural Services for Housing Services, AIA Document B188
>
> Standard Form of Agreement Between Design/Builder and Architect, AIA Document B901

In preparing an agreement for architectural services, one must look ahead and anticipate what the architect should be doing during the bidding and construction periods, the times of greatest demand for contract administration.

Using Standard Forms. Although there is no requirement in competent architectural practice to use the AIA standard forms, there are sound reasons to do so. They utilize standard and consistent terminology. They have been developed over years of experience and attempt to answer the compelling needs of practical and realistic practice. They reflect legal principles that have been tested through generations of evolution in the documents, the architectural profession, and the construction industry. They have achieved a fair balance in the apportioning of risk and liability among owners, architects, and contractors. They realistically reflect construction industry practices and are in consonance with standard insurance coverage.

Whatever contract form is used, it should be carefully considered to assure inclusion of a comprehensive description of the architect's construction administration services. Only these services may be included as architect's duties in the construction contract documents.

Oral Agreements and Letter Agreements

Oral Agreements. Some architects are very reluctant to present lengthy agreements to their clients. Usually, their stated concern is that it would scare the client away, particularly in the case of small or casual assignments. So they proceed on the basis of an oral agreement and a handshake. If everything goes all right, there is no particular disadvantage to this procedure. But there could be serious disadvantageous repercussions if there is a disagreement over fees, scope of the assignment, professional liability, or any one of a dozen other topics of dispute. The two major practical problems with informal unwritten agreements are:

1. Many terms and conditions that should have been in a complete architectural services agreement were not discussed.

2. No one knows for sure, and cannot prove, what are the terms and conditions of the agreement.

One of the principal omitted terms is a dispute resolution mechanism, such as an arbitration clause.

Letter Agreements. One step above an oral agreement is the use of a letter agreement. This can be an informal memorandum on your letterhead wherein you set out the scope of the project and the fee plus anything else that was discussed and agreed upon. A refinement that will extend the value of this type of agreement is to incorporate by reference the terms and conditions of an AIA standard form agreement such as B141.

These comments about oral and letter agreements are not intended to legitimize such informality. Sound business principles require complete written agreements and that is my sincere recommendation.

Architectural Fees. How Architects Charge for their Services

Fee Schedules, Historically. Prior to 1972, various components of the American Institute of Architects (usually the state societies and local district chapters) printed schedules of standard fees to be charged by their members for various types of buildings or other design assignments. Usually they were based on a percentage of the construction cost. Complex building types would typically command a higher percentage fee than simpler or repetitive types. The charts reflected a declining percentage fee as the construction cost increased. Some schedules reflected a higher percentage for projects that required inordinately large amounts of professional time and effort such as building additions and alterations and for projects that were to be contracted on the basis of separate contracts or cost plus a fee.

The schedules were compiled from empirical data submitted by firms with experience in different types of projects of various size and scope.

Larger more established firms with a full complement of overhead would find that the prescribed fees would be inadequate and they would lose money when adhering to the schedules. It would be difficult for them to charge more than the scheduled fees as their clients would believe they were being overcharged.

At the same time, smaller newer firms with underdeveloped overhead structures would find the scheduled fees to be unattainable goals. It was hard, if not impossible, to find clients who would pay these idealistic fees, while their clients felt that the recommended fee schedule was far too generous for fledgling architects.

Architectural Services Agreements

Added to this general dissatisfaction with the published fee schedules was the AIA's attitude at that time that architects who did not follow the fee schedules were acting unethically, or at least unprofessionally. AIA members were also ethically proscribed from advertising their services and from openly competing with other architects on the basis of fees. All of these attitudes were similarly embraced by the medical, accounting, engineering, and legal professions.

Clearly, these perceptions of ethical and professional behavior were designed to benefit the members of the various professions. Furthermore, they were interpreted by the U.S. Department of Justice as conspiracies against the interests of the general public and a violation of the Sherman Antitrust Act. Accordingly, in 1972 and again in 1990, the AIA, under severe legal pressure, agreed to the removal of ethical restrictions on all forms of fee competition. This was formalized by consent decrees which prohibited the AIA from any further interference in the natural economic processes of free market competition and supply and demand. The decrees provided for government monitoring of the professional society on into the 21st century to assure compliance with the decrees. The AIA was required to pay substantial fines as well as carry the considerable burden of its own legal fees and of the monitoring expense.

Private Fee Schedules. The consent decrees do not prevent individual architects or firms from establishing and publishing their own fee schedules. Many firms do so adjusting the old AIA schedules to suit their individual production cost, overhead experience, and profit expectations.

Lack of Fee Study Committees. The decrees do not prohibit the AIA from having fee study committees nor sharing of information on production and overhead costs or operating profits of its members. However, national AIA and its components are very apprehensive and extremely reluctant to venture into these areas for fear of inadvertent or unintentional violation of the consent decrees. This is unfortunate, as architectural firms are severely handicapped in not having available the financial information and production cost data and norms needed for responsible financial planning and establishing fees.

Some of the national construction cost manuals list their versions of architectural fees for the benefit of construction estimators. However, they are seldom realistic as they are usually based on too small a sample and little or no professional input.

Practically no valid information on fees which architects charge has been available in the professional press for over 20 years. The AIA currently provides no guidance to its members as to what they should charge for architectural services.

Methods of Charging. So, how do architects know what to charge? And how do prospective clients know what to expect? And how can construction estimators get authentic information to use in their building cost estimates?

There are four basic approaches to the determination of architectural fees:

1. Percentage of construction cost
2. Multiple of direct personnel expense
3. Stipulated hourly rates
4. Agreed lump sum

All other techniques are variations or combinations of these basic systems. Each method has its advantages and disadvantages to the architect and to the client.

Percentage Method. The architectural fee is computed by applying the agreed percentage to the construction cost. Until such time as the actual construction cost is known, the updated cost estimate is used. The fee usually (but not always) includes the engineering consultants' fees but not reimbursable expenses. An advantage of the percentage method is that it automatically adjusts the fee to reflect construction complexity and volume as the construction cost rises. For the client, the fee is easy to understand, to visualize, and to calculate. It will not exceed the established economic relationship with total construction costs.

The main disadvantage falls on the architect as there is no reliable direct relationship between construction cost and the cost of producing architectural service. Thus, with the percentage fee, although the architect could have a windfall profit if the job goes smoothly and efficiently, it could easily be a disastrous loss when misfortune prevails. Either way, the client will not be aware of the architect's financial results and will not be shocked by fees of unexpected magnitude. However, when construction costs soar, architectural fees rise accordingly, producing a result not expected by the architect and usually unacceptable to the owner's sense of fair play. Some owners would be suspicious of an architect's choice of expensive systems and materials.

Percentage fee contracts must have a good definition of construction costs to which the percentage is to be applied. The percentage fee usually applies only to the architect's basic services. Additional services generally will be compensated for by an hourly rate method or an agreed sum as such services seldom involve a change in construction cost commensurate with the cost of services.

Multiple of Direct Personnel Expense. This is an hourly rate system where compensation is based on the architect's professional payroll multiplied by an agreed factor which varies in practice from about 2 to 3.2; most often the multiplier is around 2.5.

The hourly rates used in this computation must be carefully defined in the architectural contract and usually consist of the basic wage rate paid to each employee plus all customary and mandatory payroll taxes and employee benefits. The principle of the multiple is that it provides compensation for all direct labor costs plus indirect office operating costs and a profit. For example, if an employee's basic wage rate is $15.00 per hour and the wage-related costs are $6.00 per hour, this total of $21.00 is multiplied by the agreed multiplier, say 2.5, yielding a total hourly billing rate of $52.50 for that employee. The time of principals is usually expressed as fixed hourly rates not subject to the multiplier.

The advantage to the architect is that all costs are recovered plus a guaranteed profit. There will be no losses, but no excess profits.

Stipulated Hourly Rates. Some architects prefer to quote architectural fees on the basis of a specific charge per hour for each class of personnel or individual in the firm. This can simplify billing procedures and will provide an easy method of expressing architectural fees.

Architects who work on any kind of time and expenses basis must be prepared to reveal all relevant accounting and financial records to their client upon request. This is a standard and reasonable provision of architectural agreements. (B141, 10.6)

Fees based on monthly billings of time and materials are favored by many architects as they allow the architect to be paid for all direct costs, indirect overhead, and a profit.

Guaranteed Maximum Fee. Some clients would never agree to any fee basis lacking an upper limitation such as in the three previously described fee systems.

If the project goes smoothly and the architect has an efficient operation, the client will undoubtedly be pleased with the resulting moderate total outlay. But if the project is plagued by adversity or inefficiency, the total fee can become embarrassingly monstrous. This disadvantage to the client can be overcome by imposition of a ceiling or guaranteed maximum fee. The guaranteed maximum can be based on a percentage of construction cost or an agreed lump sum.

If inefficiency causes the fee to reach the guaranteed maximum, the architect may soon thereafter be reduced to a break-even or loss position as continuing new costs erode previous profits. Therefore, it is important to the architect that the level of guaranteed maximum be carefully established. Also, changes in the agreed scope of services must be diligently recorded as they occur so the guaranteed ceiling can be adjusted up or down accordingly.

Agreed Lump Sum. When the fee is set as a stipulated lump sum, the scope of services to be provided must be definitely known. The architect must then be able to estimate the cost of producing the work, including all labor, materials, consultants, overhead, and expected profit. The resulting fee must be reasonable in the market and may have to be negotiated before a lump sum agreeable to both sides can be established. The architect's estimate of production costs and overhead figures will be based on past experience and quotations from outside consultants. The amount to be included for profit and risk will be based on past experience, judgment, and evaluation of the economic climate.

Changes in the scope of architectural services must be recognized as they occur, so that supplementary agreements amending the previously agreed lump sum fee may be executed. The principal advantage of the lump sum fee is that both parties know at all times what the fee will be. The disadvantage to the architect is that inefficiency will reduce or eliminate the profit while efficiency (unless exceptional) will not usually increase the expected profit because lump sum fees are often heavily negotiated and thus are usually based on optimistic estimates of efficiency.

Engineering Consultants and Reimbursable Expenses. Architectural compensation should always be clarified as to whether or not engineering consultation fees are included. The fees for normal structural, mechanical, and electrical engineering are usually included in percentage fee and lump sum fee agreements but not in hourly rate agreements. In hourly rate agreements, all outside consultants' fees are considered to be reimbursable expenses.

Reimbursable expenses should be defined in the architectural agreement and usually include the costs of all reproductions, postage, couriers, transportation, telecommunications, building and planning permits, governmental fees, renderings, models, mock-ups, special insurance, and computer time.

Most architectural firms apply a billing multiplier to all reimbursable expenses, varying from about 1.1 to 1.25, most commonly 1.1. It is common practice to charge interest on all overdue accounts at a specified rate.

Termination of an Architectural Contract

Ending The Contractual Relationship. At the time architects and their clients enter into architectural agreements, there is usually great optimism and eager anticipation on both sides. The architect is not only excited about being selected but is savoring the design challenge, while the client is looking forward to eventual ownership of a successful building project. Both are usually expecting economic rewards as well as recognition from their peers. The interpersonal relationship is one of rapport and mutual pleasure. They regard each other with respect and esteem.

But later, conditions may sometimes change, causing one party or both to consider ending the arrangement.

There are various good and sufficient reasons why an owner or architect would want to terminate their agreement. There are also some reasons that would force them to discontinue the contractual relationship. In the unlikely event that this usually unpleasant experience must be faced, one must always look first to the architectural services agreement for guidance. Oral, informal, or brief letter contracts would not have anticipated this infrequent but possible contingency. Those firms that regularly use the standard AIA agreement forms will find that this subject is contractually covered in a practical and equitable way.

Article 8 of B141 is devoted to the subjects of termination, suspension, and abandonment.

Meaning of Termination, Suspension, and Abandonment. In order to accurately understand the meaning of the contract one must know how the language is used. The contract speaks of *termination* of the *contract*, but *suspension* or *abandonment* of the *project* and *suspension* of architectural *services*. Thus, the contract may be terminated if the project is abandoned, but a suspension may be of the project or of the architect's services. The contract remains in full force and effect during and following a suspension.

Reasons for Termination. The architectural services contract can be terminated only for certain limited specified reasons. If either party substantially fails to perform its obligations according to the contract, then the other may, if it wishes, terminate the agreement, after giving seven days' written notice. This is assuming, of course, that the breach was not the fault of the party wishing to terminate. (B141, 8.1)

Should the owner fail to make payments as required by the contract, this is considered a substantial breach and is a specified cause for termination. (B141, 8.4)

The architect could breach the contract in several ways, but none are specifically listed in the contract. Some which readily come to mind are:

1. Failure to meet the agreed schedule for completion of designs, drawings, or specifications
2. Failure to properly administer the construction contract and to issue certificates of contractor's payments when required
3. Failure to meet the agreed schedule for filing with public agencies
4. Failure to carry out the owner's legitimate instructions

However, the contract provides no criteria for determining what magnitude of non-performance or poor performance would be considered "substantial".

Breaches of the contract which might be committed by the owner could include:

1. Failure to provide program and site information in a reasonable time
2. Failure to provide legal, insurance, or accounting information when needed

These are not specifically mentioned in the AIA standard agreement and there is no specific guidance as to what would constitute a substantial breach of the contract.

One way of defining substantial would be if the breach of one party prevented the other party or the project from proceeding in an orderly fashion. Some have suggested that several or a substantial number of minor breaches constitute a substantial breach of the contract. However, these speculations being uncertain, the final determination would probably rest in the hands of arbitrators, if contested by either party.

As was previously pointed out, the contract may be terminated if the project is abandoned by the owner. The contract uses the term, permanently abandoned. If only temporarily abandoned, it would seem more like a suspension. If the project is permanently abandoned, the owner, after giving seven days' written notice to the architect, may terminate the contract. Should the owner neglect or fail to give the specified written notice, the architect may consider the project to be abandoned after ninety consecutive days of abandonment. Then the architect may declare the contract terminated by giving written notice to the owner. (B141, 8.3)

Termination Expenses. In all cases of termination which are not the fault of the architect, whether initiated by the owner or architect, the architect is entitled to receive termination expenses in addition to compensation for Basic Services and Additional Services rendered to date and reimbursable expenses. Termination expenses are those costs incurred by the architect which are directly attributed to termination. A method of computation of termination expenses is specified which is a percentage of architectural fees for Basic Services and Additional Services earned to the time of termination as follows:

> "8.7.1 Twenty percent of the total compensation for Basic and Additional Services earned to date if termination occurs before or during the pre-design, site analysis, or Schematic Design Phases; or
>
> 8.7.2 Ten percent of the total compensation for Basic and Additional Services earned to date if termination occurs during the Design Development Phase; or
>
> 8.7.3 Five percent of the total compensation for Basic and Additional Services earned to date if termination occurs during any subsequent phase."
> (B141, 8.7)

Termination expenses are not recoverable where the termination is the fault of the architect. In cases of suspension, termination expenses would not be applicable as suspension does not constitute termination of the project.

Architectural Services Agreements

Suspension of the Project by the Owner. The owner is entitled to suspend the project at any time without stating any reason. Should the owner do so for more than thirty consecutive days, the owner must compensate the architect for all architectural fees and reimbursable expenses incurred prior to notice of the suspension.

At the time the project is resumed, the architect's compensation will be equitably adjusted to provide for any expenses incurred by the architect for interrupting and resuming the work. (B141, 8.2)

Subparagraph 11.5.1 of B141 provides for revision in compensation if the time for performance of architectural services is extended beyond an agreed period through no fault of the architect. This would allow for appropriate adjustment to account for inflation, employee wage increases, or other relevant factors.

Should the architect's services be suspended for a long period causing the services, when resumed, to extend beyond the limitation in subparagraph 11.5.1, then the balance of the fee may be adjusted accordingly.

In the event an owner's suspension of the project is for less than thirty days, the contract does not provide any relief for the architect. This often imposes considerable hardship on the architect in assignment of personnel if the office does not have alternative projects to work on during periods of standing-by pending conclusion of the suspension. If the likelihood of this eventuality can be anticipated, an appropriate provision could be added to the contract at the time of negotiation. Possibly the thirty day allowable suspension period could be reduced to fifteen or twenty days or even eliminated. (B141, 8.2)

Suspension of Services by the Architect. Should the owner fail to pay the architect's invoice for services and reimbursable expenses when due, the architect may give the owner seven days' written notice to suspend services and may then suspend the services at the end of the seven day period with no further notice. The due dates of the architect's periodic payments are as inserted and agreed to in B141, 11.5.2.

If the architect suspends services under the contract in this manner, the contract provides that the architect shall have no liability to the owner for any delay or damage to the owner caused by the suspension of services. (B141, 8.5)

Additional Reasons for Termination. The owner may want or have to breach the contract, thereby causing termination for various reasons that are not the fault of the architect, and thus would incur termination expenses. Some examples:

  1. Change of mind about the suitability of the project
  2. Inability to obtain anticipated required governmental approvals

3. Economic reverses, including bankruptcy or insolvency
4. Inability to obtain expected equity financing or mortgage loans
5. Inability to make sales or obtain tenants
6. Inability to obtain full use of the property, rights of way, or easements in a timely manner
7. Unexpected appearance of environmental or neighborhood activists
8. Unexpected adverse legislation
9. Forces majeure, such as fire, earthquake, landslide, or storm which destroy the owner's property, or lengthy labor disruption, war or insurrection
10. Owner's death or disability

Similar reasons affecting the architect or the practice could cause the architect problems in proceeding with the contract. In such cases, the architect would not be entitled to receive termination expenses. The architect would probably be able to recover the value of any work done which is of any value to the owner. He or she would not be likely to collect for partially completed services which cannot be utilized by the owner. Some examples:

1. Discontinuance of the architectural practice due to bankruptcy or insolvency
2. Inability to proceed on account of the architect's death or disability

There are additional reasons either the owner or architect would want to conclude their relationship but none of these are legally excusable. Therefore, if the termination is not mutually agreeable, the party damaged by the other's unauthorized termination may be able to recover appropriate compensation through appeal to arbitration.

Owner examples:

1. Personality conflict with the architect
2. Uncooperative architect

Architect examples:

1. Personality conflict with the owner
2. Uncooperative owner
3. Office too busy and competent personnel unavailable
4. Inadequate fee

Considering the potential legal ramifications and economic repercussions of a contract termination, architects and owners should give serious consideration to conferring with legal counsel prior to making any final decisions about initiating a termination or if they have received a written notice from the other of an impending termination.

Architectural Services Agreements

Resolution of Disputes. Owners and architects who are able to negotiate conditions of their contractual terminations and suspensions in a civil, courteous, fair, and mutually agreeable manner are very fortunate. A friendly, objective, businesslike termination will limit losses, could save considerable extra expense and time, and leave the parties well disposed to future business relationships.

In those cases of termination in which there is hostility, animosity, and loss of respect, it is often impossible for the parties to arrive at mutually agreeable settlement terms. This is particularly true when economic losses are difficult to bear.

The contract provides that all irreconcilable differences be resolved by referral to arbitration in accordance with the Construction Industry Rules of the American Arbitration Association. (B141, Article 7)

Arbitration is final and binding and can be commenced or completed as rapidly as the parties wish. The arbitrator will be an experienced construction industry expert who knows the subject matter and, if properly informed by the parties during the hearing, will very likely arrive at a fair and equitable judgment, sensitive to all of the nuances. (See Chapter 22, Resolution of Construction Disputes.)

An arbitration is usually quite economical compared to litigating the same matter in the civil court system. However, it could still cost considerably more than a settlement negotiated between the parties. This might require some degree of compromise on each side to resolve the matter to mutual satisfaction. The economic value of early resolution and personnel time saved should also be considered when compromising to effect a settlement.

Construction Administration by the Architect's Consultants

Some of the architect's construction administration duties are actually performed by the consultants who designed that portion of the work. This would be true primarily in the work designed by civil, structural, electrical, and mechanical engineers, landscape architects, and other technical consultants. These specialized consultants, in most cases, would be better qualified than the architect to review the relevant shop drawings and to examine the work in the field. Their assistance will also be required in most decision making related to their areas of expertise.

It is important that the architect's agreement with technical consultants be consistent with the architect-owner agreement in respect to construction administration services. (See Chapter 8, Consultants and Advisors.)

4

Construction Contracts

Types of Standard AIA Contract Forms

The Two Main Forms. The American Institute of Architects has produced a range of standard form construction contracts that will accommodate practically any situation that is likely to arise in normal practice. The two mainstream forms of construction contract are:

> Standard Form of Agreement Between Owner and Contractor (where the Basis of Payment is a Stipulated Sum), Twelfth Edition, AIA Document A101, 1987 (Appendix B)

> Standard Form of Agreement Between Owner and Contractor (where the Basis of Payment is the Cost of the Work Plus a Fee with or without a Guaranteed Maximum Price), Tenth Edition, AIA Document A111, 1987 (Appendix C)

These two forms will accommodate any payment method. There are blank spaces to add in any desired additional agreed provisions. These forms should always be used in conjunction with the AIA general conditions (Appendix D), without which they would be incomplete.

The architect's contract administration duties and authority referred to in these contracts are in harmony with those set forth in the owner-architect agreement, B141 (Appendix A).

Alternative Standard AIA Contract Forms

Short Forms of Agreement. The AIA issues two abbreviated construction contracts to be used on small projects:

> Abbreviated Form of Agreement Between Owner and Contractor for Construction Projects of Limited Scope where the Basis of Payment is a Stipulated Sum, Ninth Edition, AIA Document A107, 1987

> Abbreviated Form of Agreement Between Owner and Contractor for Construction Projects of Limited Scope where the Basis of Payment is the Cost of the Work Plus a Fee with or without a Guaranteed Maximum Price, Second Edition, AIA Document A117, 1987

These two forms each contain their own abbreviated form of general conditions, so it would be inappropriate to append or incorporate A201. These forms are considerably shorter than the basic A101 or A111 used with A201. They should not be used without comparison and verification that they include all desired terms and conditions contained in the lengthier versions. Both of these abridged forms include arbitration clauses.

A Shorter Form of Agreement. For extremely small or simple projects of short duration, it may be appropriate to use AIA's small project forms:

> Standard Form of Agreement Between Owner and Contractor for a Small Project, where the Basis of Payment is a Stipulated Sum, Small Projects Edition, AIA Document A105, 1993, and

> General Conditions of the Contract for Construction of a Small Project, Small Projects Edition, AIA Document A205, 1993

These two forms must be used as a unit as the terminology is interlinked and consistent. In order for the architect's contract administration duties and authority to be consistent, the architect's services should have been retained by use of the companion owner-architect form:

> Standard Form of Agreement Between Owner and Architect for a Small Project, Small Projects Edition, AIA Document B155, 1993

Construction Contracts 41

The small project forms are considerably abridged and do not contain all of the procedures and safeguards of the longer versions. They do not provide for arbitration of disputes.

Construction Management (CM) Agreements. (See Chapter 5, Project Delivery Systems, for a discussion of CM.) The AIA issues standard construction contract forms for projects that will be built with a construction manager. Various forms are available to match the most prevalent configurations of Construction Management. See Appendix G, Synopses of AIA Standard Form Documents, for further descriptions and uses of these forms:

> Owner-Contractor Agreement Form, Stipulated Sum, Construction Manager-Advisor Edition, AIA Document A101/CMa. For use when there is a separate CM, not the contractor or architect. This construction agreement should be used in conjunction with General Conditions, AIA Document A201/CMa. The architect's and CM's services should be engaged by the owner using these coordinated agreement forms:
>
>> Architect: Standard Form of Agreement Between Owner and Architect, Construction Manager-Advisor Edition, AIA Document B141/CMa
>>
>> Construction Manager: Standard Form of Agreement Between Owner and Construction Manager where the Construction Manager Is Not a Constructor, AIA Document B801/CMa
>
> Standard Form of Agreement Between Owner and Construction Manager where the CM is also the Constructor, Stipulated Sum, AIA Document A121/CMc, AGC Form 565. General Conditions A201 should be used. The owner-architect agreement should be B141
>
> Standard Form of Agreement Between Owner and Construction Manager where the CM is also the Constructor, Cost plus Fee, AIA Document A131/CMc, AGC Form 566. General Conditions A201 should be used. The owner-architect agreement should be B141

Design/Build Agreements. (See Chapter 5, Project Delivery Systems.) The design/build entity should have an agreement with the owner and, in turn, must also contract with a contractor and an architect. These standard AIA form agreements are designed for design/build projects:

> Owner-Design/Builder Agreement, AIA Document A191. This form consists of two agreements. The first is for preliminary design and budgeting services, while the second is for final design and construction. The second may be canceled by the owner if so desired.

Design/Builder-Contractor Agreement, AIA Document A491. This form is also in two parts, allowing the second part to be canceled if the second part of A191 is canceled.

Design/Builder-Architect Agreement, AIA Document B901. This form is also in two parts, allowing the second part to be canceled if the second part of A191 is canceled.

Negotiated Contracts. What is Negotiable?

Everything is Negotiable. Anything can be discussed. Nothing is sacred. Every contract is unique. There is no practical limit to the number of permutations in modern construction contracts. The important objective is to achieve a mutually agreeable contract where both sides are satisfied with the price and can amicably accept all other terms and conditions. The owner and contractor each have aspirations to be achieved as well as rights to be preserved. Each must enter into the negotiation in a spirit of mutual cooperation and should realistically expect to yield some points in the effort to gain others.

If there is a genuine meeting of the minds, then the resulting written agreement should persevere all during the time of construction and through to the end of the warranty period. The true test of a satisfactorily negotiated construction agreement is that it anticipates all normal eventualities as well as the extraordinary situations and the occasional adversities.

No agreement will be so perfect that it provides for anything that can and will happen. The additional ingredients that will fill in the interstices between the expected and the unexpected in the construction process are the fair-mindedness and good intentions of the owner and contractor. To provide for the possibility that either party might not measure up to this ideal standard, then it is necessary to have a competent contract, a skilled contract administrator, and a practical dispute resolution procedure.

A complete and candid discussion of each component of the contract will greatly assist the parties in understanding each other's points of view and in achieving a true meeting of the minds. In the process, many of the usual subjects that trigger construction disputes will be identified and resolved or eliminated.

AIA Standard Contracts. The AIA standard form contracts are written in a manner that respects the interests of owners and contractors fairly and equitably. They were written by committees of practicing architects in consultation with contractors, construction lawyers, insurance advisors, and accountants. They have been developed over the past century to reflect the evolving practices of the construction industry and the interpretations of the judiciary. They are continuously reviewed, revised as necessary, and reissued periodically. AIA's construction agreements have been approved and endorsed by the Associated General Contractors of America.

Who Does the Negotiation? Realistically, architects would seldom be called upon to be the sole negotiator of a construction contract on behalf of a client. This is something most owners would not likely delegate. However, architects are frequently consulted on contracts and asked to sit in on contract negotiation sessions and to contribute to the discussion.

Terms and Conditions to be Negotiated. Any provision in the standard documents can be amended or removed by mutual consent. Any new provision can be added in if that is what both parties want. Any construction industry custom or practice may be altered or eliminated as long as it is legal, does not harm others, and will suit the needs of the owner and contractor. We will consider here some of the usual topics for discussion in the typical negotiation of a construction contract.

The Scope of Work. The scope and quality of work is logically the first item open for discussion. There is no point in talking about the contract price before the work has been defined. Presumably, the owner had previously decided what was to be included and the drawings and specifications were prepared on that basis. However, in the effort to control and reduce costs, it is advisable for the owner to examine the price/value ratios of the significant parts of the work. This can only be done effectively with the contractor's cooperation. Certain features would be omitted from the scope of the project if the owner knew the cost. Others might be added or the quality enhanced when the owner discovers how it affects the total contract price.

The contract price can be raised or lowered by selective upgrading or downgrading of the scope and quality of the work.

Both scope and price can be adjusted, upward or downward, by manipulation of the unit prices and allowances.

The architectural and engineering costs of preparing addenda to properly amend the contract drawings and specifications to reflect the agreed changes in scope and quality must be considered. In some cases, these costs will consume all or most of the expected savings.

The Contract Price. The total contract price is the most prominent and usually the most significant element of the contract to both owner and contractor. When there is no concurrence on price, there is very little else to discuss.

If the job had gone out to competitive bidding, all of the contract terms should have been stipulated in advance so that the tendered prices would have reflected them.

The price submitted by the low bidder would not then have to be adjusted later as additional contract terms became known.

Each term and condition of the agreement has a value, positive, neutral, or negative, to both parties. If an agreed term is withdrawn or a new term is imposed by either side after the price is agreed upon, then there may be a need to adjust the price accordingly. Conversely, when searching for ways of reducing the price, the value of some of the terms should be explored. The elimination or alteration of some terms could possibly be negotiated to arrive at a mutually acceptable revised price.

A Guaranteed Maximum Price (GMP). It has become quite prevalent for the owner and contractor to agree upon a price that will be treated as a guaranteed maximum. This is practical only when the precise scope of the project is known. The contractor will require completed contract drawings and specifications from which to calculate the guaranteed price. During construction, the contractor will then expect to be paid for all deviations in conditions depicted in the agreed contract documents. This is put into effect by raising (and occasionally lowering) the guaranteed maximum price.

The main reason so many GMP contracts have ended up in bitter controversy, owner-disillusionment, and expensive litigation is that the GMP was predicated on incomplete contract drawings and specifications. Simply stated, the owner and contractor both jumped the gun. The contractor was impatient to tie up the job and close the deal. The owner wanted to get the project under way and was lulled into complacency with the promise of a Guaranteed Maximum Price. The first two words of GMP are packed with reassurance and pleasurable meaning for the owner.

When the GMP is based on incomplete construction documents, the owner hopes that the contractor would have allowed funds for items not yet shown on the drawings nor expressed in the specifications. Also, there are the building department corrections not yet incorporated. The typical contractor's attitude is that there is no way that a construction estimator can accurately predict what will ultimately show up on the drawings or in the specifications. So consequently, their rational approach is to figure the job as close as possible to what is actually there on paper. The contractor also argues that there is no way to anticipate what the required quality will be. In reality, the contractor is concerned that making allowances for too much drawing and specification development will drive the GMP too high. Regrettably, the job will then go to a competitor who proposes a GMP based on the current state of the documents. In a competitive climate, the contractor has no other rational choice.

The contractor feels that it is fair to charge the owner for the actual cost of the building plus a reasonable fee. The fair-minded owner cannot seriously argue against this reasonable-sounding contractor argument but nevertheless feels cheated

Construction Contracts

after being led to believe that a guaranteed maximum price was being offered. A guarantee is a guarantee and a maximum is a maximum, the owner reasons.

Shared Savings. Theoretically the contractor's fee, added to the actual construction costs, could be less than the GMP. This difference rightly belongs to the owner. However, many owners (and most contractors) feel that conceding a portion of the "savings" to the contractor will serve as an incentive to the contractor to keep costs under control, through increased diligence and efficiency. The contractor's share of the savings would be subject to negotiation and in practice varies from 10% to 50%. The prospect of savings below the GMP is also an incentive to the owner to suppress unreasonable demands. After the GMP has been breached, however, the owner's incentive to be reasonable is gone and the contractor faces sure erosion of scheduled profits, and in extreme cases, actual out-of-pocket losses.

The Contract Time. The time of completion must be agreed upon. Basically, it should be a function of the time realistically needed to accomplish the work of the contract. If the owner requires completion by a certain date, it is up to the contractor to determine if this is physically possible and if it is willing to legally commit itself to meet that date. On the other hand, the contractor could state its opinion of the time required and the owner could decide if this is acceptable. When the completion date has been stipulated in the bidding instructions, then contractors who cannot accept the stated date should not submit a bid. Submission of a bid constitutes concurrence in and acceptance of the required completion date.

Liquidated Damages. After mutual agreement on a starting date, a construction time period, and a completion date, the parties may consider whether or not a liquidated damages provision will be included in the agreement. The theory here is that late completion of the project, once agreed upon, will cause the owner some additional expenses such as extra rent on the old premises, a double move if the old premises must be vacated, extra interest on construction funds, and lost rents. The costs of delayed completion, if caused solely by the contractor, should be paid for by the contractor. Liquidated damages must bear a reasonable relationship to estimated actual costs to be suffered by the owner. If set at an unrealistically high or punitive level, they will not be collectable by law.

A liquidated damages clause should provide for changes in the completion date for certain contingencies not in the control of the contractor. Change orders that affect the critical path of construction will often extend the completion date or, in some cases, could even shorten it. The number of days of change in contract time, if any, should be agreed concurrently with each change order at the time it is negotiated.

Inclement weather such as rain, wind, heat, or cold will also trigger contractor requests for extension of contract time for each construction day lost. Some clauses are written to allow extension for each and every day lost due to inclement weather. Others are written to allow only for days lost due to adverse weather that is above the normal expected in that locality at that time of the year. If the latter concept is agreed upon, inevitable disputes can be eliminated if the expected allowable days of adverse weather can be stipulated and stated in the contract.

Retainage. It is common practice for the owner to withhold an agreed percentage of all progress payments due to the contractor. The retained sums will then be released to the contractor at the end of the job and after the lien period has passed. The amount to be retained in each contract is subject to negotiation. Commonly, the rate is ten percent. This is not a magic number and could be higher or lower, depending on the owner's requirements or the contractor's views. Variations in application of the retention have been widely used.

Change Orders. When the first change order arrives shortly after construction is underway, the owner is often dismayed at the way a small item can grow when all the related charges are added on. A way to avoid this lack of understanding is to ask the contractor to prepare a pro forma change order billing for the owner's examination and approval during the contract negotiation.

The owner will then see that the typical bill is made up of a number of discrete elements. Contractors are justified in charging for all legitimate extra labor, labor burden, supervision, materials, sales tax, delivery, unloading, subcontracts, rentals, bond premium, overhead, and profit.

The amount of contractor's overhead and profit to be added onto change orders can be discussed and stipulated in the agreement. Most change orders include the work of subcontractors, so subcontractor mark-up also has to be addressed. It should be recognized that subcontractors usually charge a higher mark-up than do general contractors, so this should be discussed and evaluated.

A compromise sometimes agreed upon provides that the basic construction contract will include a certain total value of change orders without charging any additional overhead and profit. For example, that the first $10,000 of changes will not be marked up for contractor's overhead and profit.

The way credit change orders will be treated should always be discussed. It is common practice not to deduct related profit and overhead in credit change orders. If purchased items are to be returned, the credit will be reduced by any re-stocking charges incurred as well as costs of transportation, loading and unloading, and uncrating and re-crating.

If the owner and contractor have a complete understanding of how extra charges and credits are to be billed, a major source of dispute will be eliminated. (See Chapter 14, Change Orders.)

Surety Bond. The owner must decide whether or not a labor and material payment bond and performance bond will be required. If so, it should be specified in the bidding instructions and clarified as to whether the bond is to be paid for by the contractor and included in the contract price or kept out of the contract and paid for directly by the owner. (See Chapter 11, Construction Administration When the Contract is Bonded.)

Interest on Past Due Amounts. The AIA General Conditions (Paragraph 13.6) provides for interest to accrue on all sums not paid when due. The legal rate of interest prevailing at the place of building is to be used unless the parties agree on a different rate. To avoid disputes, it is better that the parties discuss this contingency and agree on a mutually acceptable stated rate. Some contractors are reluctant to discuss this as they would prefer to have all their invoices paid when due rather than to collect interest. If the rate is discussed, the contractor can make known its attitude on late payments and the owner can clarify its intentions in regard to the payment procedure and timing.

Progress Payments. The life blood of a contracting business is cash flow. This is produced by timely billing and prompt collection of all receivables. The typical contractual requirement is monthly billing for the value of all work in place on the last day of the month. Contractors strive to issue their bills on the first of the month with the aspiration of collection by the tenth of the same month.

Contractors are not likely to consider this a negotiable item. For a contractor to concede on prompt payment, the owner would have to yield something significant, perhaps a few extra points on late payment interest. Frequency and timing of billing and payments are important matters to be discussed and agreed upon.

On some small projects and residential work it is not uncommon to agree on a payment schedule based on certain points of progress achieved. For example, 20% of the contract price to be paid upon completion of the foundation, 20% upon completion of framing, 20% upon completion of roofing, 30% upon completion of the entire construction, and 10% upon expiration of the lien period. These percentages as well as the milestones are open to discussion and adjustment.

Contractors would like to be able to include in their monthly billing materials or equipment committed for but not yet incorporated into the building. They may be stored on the site or stored elsewhere or may still be in the process of manufacture in the fabricator's shop. Until such time as these items are incorporated into the building, they remain portable and could be lost by fire or theft and there is serious question as to the owner's title to the property. According to the AIA General Conditions (Subparagraph 9.3.2), it is a matter to be agreed upon between the owner and contractor, whether or not the value of these materials and equipment are to be included in progress billings. Understandings must be reached in respect to insurance, owner's title, place of storage, and transportation.

The Warranty Period. The warranty provision of A201, 3.5.1. does not carry with it any explicit time limitation.

The warranty period is established by statute, varying from state to state from 2 to 20 years. However, the length of the warranty period in any particular contract could be shortened or lengthened by agreement if that is what the parties want to do. The length of the warranty period has a monetary value and could be one of the terms to be discussed in arriving at an agreed contract price.

During the warranty period, the contractor is obligated to rectify all construction defects that are reported by the owner. Some specified components of the construction may have longer warranty periods. For example, the roofing, water heater, or fluorescent lighting fixtures might have two year or longer warranties.

The one year period mentioned in A201 (12.2.2), is a "correction period", during which time the owner can demand the contractor's return to the site to remedy any defects discovered in the work of the contract. This correction period is often confused with the warranty. Conceptually, the correction period is an *additional* remedy given to the owner when A201 is part of the contract. AIA documents provide the owner with a choice of three theories of remedy for defective work: warranty, correction period, and other remedies for breach of contract.

Types of Construction Contracts. In general, there are two basic types of construction contracts: Lump Sum and Cost Plus Fee. Cost Plus Fee contracts can be with or without a guaranteed maximum price. The contractor's fee can be determined in various ways. The type of contract and the agreement form to be used are both subject to negotiation.

Lump Sum Contracts. The stipulated price in lump sum contracts can be determined by competitive bidding or by negotiation. If by negotiation, the contractor should be willing to expose the complete cost breakdown and analysis as well as the overhead breakdown and expected profit. Complete and open candid discussion often yields a mutually beneficial agreement and an acceptable contract price.

Cost Plus Fee Contracts. In cost plus fee contracts, one of the most significant factors is the definition of what items are included in the Cost and what items are included in the contractor's Fee. In AIA's standard cost plus fee contract (Document A111, see Appendix C), there are comprehensive listings of the types of costs to be reimbursed (Article 7) and of the costs not to be reimbursed (Article 8). The contractor will have to pay from the fee all costs that are not to be reimbursed. Direct job overhead is included in costs to be reimbursed, while indirect overhead and profit are in costs not to be reimbursed.

A111 classifies the various costs in the most customary way but owners and contractors can disregard this and instead agree on any treatment they find suitable to their unique situation.

Trade discounts, rebates and refunds are usually credited to the owner (A111, Article 9), but this can be changed if desired.

The Contractor's Fee. The contractor's fee can be determined in any manner, in any amount, and at any rate agreeable to the parties. The most conventional ways are an agreed fixed amount or an agreed percentage of reimbursed costs. Various adjustments can be made, upward or downward, to create incentives or impose penalties for meeting or failing to meet stated economic or timing goals.

Unit Prices and Allowances. Unit prices and allowances should be reviewed, discussed and agreed upon. It is important for both parties to know what is included and how it will be administrated.

Many contractors, to save on minor cost accounting and invoicing, favor the procedure of setting an agreed lump sum to cover small tools and other maintenance costs. The same allowance procedure can be used to account for the costs of trucks and other equipment along with their operating and maintenance costs. This is also to the owner's advantage, as dozens of small bills will be eliminated and auditing costs will be reduced accordingly.

If there is a contingency allowance in the cost breakdown, there should be a complete discussion as to the character of expense that may be charged to it. The contractor usually views it as the account to charge for miscellaneous missed items and other estimating errors, while some owners erroneously consider it as an allowance for change of mind. A discussion and meeting of the minds on the purpose of the contingency reserve will eliminate later disagreement on this point.

Agreement Forms. Whatever the negotiation yields in terms of agreement on the various issues discussed, it all has to be reduced to writing. All of the AIA standard forms lend themselves conveniently for this purpose. In entering the agreed terms and conditions into the appropriate form, it is best to omit or amend the standard form provisions by reference to the numbered paragraphs. This is superior to striking out or obliterating parts of the form which would prevent one from knowing what has been omitted from a familiar document. The added, amended, or omitted provisions can be inserted in the blank places on the form or on attached addenda sheets.

Before finally signing the agreement, it would be advisable for the owner and contractor alike to have the final version reviewed by their respective legal, accounting, and insurance counsel.

Retainage: Protection for Owner and Surety?

Retainage Must be Agreed to. Most construction contracts provide for a system of retainage. This is a sum temporarily withheld by the owner from each progress payment due to the contractor. The purpose of the retainage is to provide the owner with a degree of financial protection in the event the contractor fails to faithfully complete all of the terms and conditions of the contract. The retainage also provides a financial incentive to some contractors to remain on the job.

A retainage cannot be exacted, unless provided for in the construction contract. The amount and terms of retainage are specified in A101, 5.6.1, 5.6.2, and 5.8. Similar retainage provisions will be found in A111. However, when A107 is used, retainage provisions, if required, must be added as they are not included in the basic form.

The total of sums withheld as retainage is then due and payable to the contractor at the end of construction of the project in the manner set out in the contract, provided that the contract has been faithfully performed.

Retainage is Contractor's Burden. On large projects the accumulated retainage becomes quite substantial and it could create a major financial burden on the contractor. This should be anticipated when the contractor is estimating the job and preparing the bid proposal. However, the contractor must consider that inclusion of realistic interest to carry the retainage could render the bid uncompetitive unless all bidders include such interest. To lessen the impact somewhat, contractors usually call upon their subcontractors and suppliers to carry their own share of the retainage. The sub-bidders will have the same problems with retainage as the contractor but on a proportionally smaller scale.

Subcontractors that complete their portion of the work early in the construction sequence are reluctant to wait months or even years until the project is complete to collect their retained sums, so they will strongly resist the contractor's lengthy retention of funds.

Some preferred subcontractors who have solid low bids may have sufficient bargaining power to negotiate subcontracts for themselves with retainage provisions either reduced or deleted.

Reduction, Elimination, or Release of Retainage. Some contracts provide for reduction of or elimination of any further retainage after the project reaches some specified degree of completion, such as fifty percent. For example, on a $1,000,000 project with a ten percent retainage, at the point of fifty percent completion, $50,000 would have been cumulatively withheld from payments to the contractor. The retainage clause could provide for reducing the retainage to five percent for the remainder of the project, so that the total sum retained by the end of construction would be $75,000 or 7.5 percent of the contract sum. Alternatively, the provision could have eliminated any further retention during the latter half of the project. By the end of construction, the percentage of retained sums would gradually diminish to five percent of the contract.

Often, any reduction or elimination of retainage depends upon the contractor's work being judged by the architect as progressing satisfactorily. This type of provision will place a great burden of pressure on the architect when the contractor is performing marginally and is economically dependent on the reduction, elimination, or release of the retainage.

When a construction contract is bonded, the surety's unconditional written consent should be required of the contractor for any reduction, elimination, or release of the retainage or any change in the contract's retainage formula. Neglecting to inform the surety and obtain its consent to a material contract amendment is an open invitation to their repudiation of the bond coverage.

An AIA standard form, Consent of Surety to Reduction in or Partial Release of Retainage, Document G 707A, is quite suitable for this purpose as it is worded in unconditional language.

Amount of Retainage. The amount to be retained may vary on various projects depending on the degree of protection required by the owner. Generally from 5 to 20 percent of the contract sum is retained with 10 percent being the most commonly used rate. On large and lengthy projects, it is common that further retention be discontinued after the fifty percent point.

Construction economy dictates reduction of the retainage as rapidly as possible provided that financial protection of all interested parties is not jeopardized. If contractors include the cost of carrying the retainage in their cost breakdowns, then obviously the cost will be transferred to the owner. Therefore, retainage should be specified at the least amount deemed sufficient to protect the owner's interest.

The rate and conditions of retainage should be stated in the bidding instructions to allow the bidders to assess the realistic economic consequences while they are estimating the project cost.

Retainage Conditions in Contract. Although the withholding of retainage is referred to in the AIA general conditions, that is not the source of the contractual requirement. The retainage rate and conditions should be included as a part of the construction contract. For example, when using A101 as the agreement, the retainage percentage should be inserted in Subparagraph 5.6.1 for work on the project site. The retainage percentage for work suitably stored on or off the site should be inserted in paragraph 5.6.2. Any provisions for reducing or limiting retainage should be inserted at paragraph 5.8.

The retainage is in addition to any sums withheld by the architect's refusal to issue a payment certificate in whole or in part to protect the owner from loss because of any of the reasons listed in Subparagraph 9.5.1 of A 201. The reasons listed include:

1. Defective work not remedied
2. Third party claims
3. Contractor's failure to pay subcontractors
4. Evidence that the work cannot be completed for the unpaid balance of the contract sum
5. Damage to the owner or another contractor
6. Evidence that the work will not be completed within the contract time and the unpaid balance is insufficient to pay actual or liquidated damages,
7. Persistent failure to carry out the work of the contract.

Defeating the Retainage.

Some inventive contractors, using creative accounting methods, have succeeded in neutralizing the effect of retainage. By subtly manipulating the figures in the schedule of values, the owner's financial protection can be lost. Basically, the system involves overstating the value of work which will be completed early in the construction process and undervaluing work which will be completed toward the end of the job. This process is known in the construction industry as "unbalancing the schedule of values" or "front end loading". Although it is fairly common, it is still an inappropriate practice, involving intentional misrepresentation of financial information. It is done for the sole purpose of inducing the architect to approve and the owner to pay sums of money they would not have if they knew the truth. This practice is very difficult for the architect or owner to detect.

Some argue that this process is necessary for the purpose of recovering the immediate cash outlay for job mobilization and heavy general conditions expenses incurred at the start-up of the project. This is a valid concern and would be more properly resolved by including mobilization costs as a separate item in the schedule of values. The schedule of values should be an accurate document reflecting actual financial conditions. Obviously, this procedure would enhance the contractor's cash flow and thereby enable undercapitalized contractors to operate.

Table 4.1

Schedule of Values and Payment Request #3

For a 3800 square foot veterinary office building in a suburban community.

Schedule of Values & Payment Request #3

Description of Work	True Costs (Column 1)	Front End Loaded (Column 2)	Percent Completed To Date (Column 3)	True Cost (Column 4)	Front End Loaded (Column 5)
General Conditions	21,400	24,610	40	8,560	9,844
Concrete	30,400	34,960	90	27,360	31,464
Masonry	26,200	30,130	95	24,890	28,624
Carpentry, rough	24,400	28,060	100	24,400	28,060
Carpentry, finish	8,600	7,310	0	0	0
Roofing	8,200	9,430	100	8,200	9,430
Insulation	400	460	100	400	460
Doors & Frames	8,400	7,140	0	0	0
Windows & Glazing	4,950	4,210	0	0	0
Flooring	1,825	1,550	0	0	0
Painting	9,620	8,175	0	0	0
Specialties	2,150	1,825	0	0	0
Equipment	23,700	20,145	0	0	0
Plumbing	11,600	12,180	70	8,120	8,526
Heating	17,340	18,200	60	10,404	10,920
Electrical	15,040	15,800	70	9,828	11,060
Site Work	63,520	54,775	15	9,528	8,216
Landscaping	8,100	6,885	0	0	0
Total Cost	**$285,845**	**$285,845**			
Total Completed to Date				$131,690	$146,604
less 10% retainage				(13,169)	(14,660)
Total earned to date				$118,521	$131,944

Manipulating the Schedule of Values. The accompanying illustration (Table 4.1) shows, for a fictitious project, a schedule of true values in column 1 which has been adjusted in column 2 to produce a front end loaded schedule of values. Front end loading means raising the values of work performed early in the job and lowering the values of work done later. Both schedules of value add up to the same total $285,845, the contract sum.

The contractor's overhead and profit are properly distributed in columns 1 and 2 to each of the work items in the schedule of values. Alternatively, overhead and profit could have been listed as a separate work item and proportionally billed each month to progress as the job proceeds. Both methods are commonly used and both are proper and acceptable. When the contractor submits the front end loaded schedule of values to the architect for review, it would be extremely difficult to discover that it had been improperly adjusted and no longer represented the correct figures. After the architect's approval, it becomes the basis for all future payment requests. (A 201, 9.2.1)

Columns 3 and 5 show the figures that would be included in the contractor's Payment Request no. 3, after three months of construction.

Column 3 shows the percentage of completion of each item of work.

Column 4 shows the correct amount earned by the contractor based on the correct schedule of values in Column 1.

Column 5 shows the amount of money applied for by the contractor based on the front end loaded schedule of values in column 2.

In this example, the falsified current payment due the contractor of $131,944, even though retainage of $14,660 has been deducted, is greater than the correct total value of the work completed to date of $131,690. In effect, the retainage protection for the owner and surety has been completely eliminated.

As the job progresses to completion, the values of completed work and the retainage gradually approach the correct figures, and upon completion the totals of payments due and retainage will be correct and will equal the contract price.

Should the contractor become insolvent or otherwise incapable during the first months of the project, the owner and surety will discover to their consternation and economic detriment that there is no retainage protection. There will be insufficient funds remaining in the undisbursed account to pay replacement contractors to complete the work of the contract.

Practical Recommendations. The key to avoiding this difficulty lies in a correct analysis of the schedule of values submitted by the contractor pursuant to the requirements of A201, 9.2 which requires:

> "9.2 SCHEDULE OF VALUES
>
> 9.2.1 Before the first Application for Payment, the Contractor shall submit to the Architect a schedule of values allocated to various portions of the Work, prepared in such form and *supported by such data to substantiate its accuracy as the Architect may require. (Italics Added)*
>
> This schedule, unless objected to by the Architect, shall be used as a basis for reviewing the Contractor's Applications for Payment."

The architect should require the contractor to submit substantiating data to support the schedule of values. If there remains any suspicion that the contractor is submitting false, misleading, or erroneous data upon which to predicate contract payments, the architect should seriously consider recommending that the owner obtain the opinion of a building cost consultant.

Construction lenders, the surety, and the owner are all relying on the architect to make a reasoned and skilled judgment of the contractor's schedule of values. The retainage will be ineffective when the schedule of values is based on falsified cost data.

Should a defaulted contractor's surety lose its expected financial protection, the actions of the architect who has approved the faulty schedule of values will certainly be scrutinized.

Liquidated Damages and Bonus Clauses

Many owners and their legal advisors feel that contractors require some financial incentive to complete their projects on time. Even in the absence of contractual incentives, an extended construction period will cost the contractor the daily cash value of continuing superintendence, security, temporary facilities such as site office, telephone, power, toilets, transportation, fencing, and possibly hoisting and scaffolding. Home office support services must also be continued. In addition, there is the interest loss on delayed payments and retainage.

Some feel that these actual cash losses do not provide a sufficient economic stimulus for the contractor to complete on time. When the completion date is of economic importance to the owner, additional pressure can be applied by charging the contractor for overrunning the established date or by rewarding early completion.

These ends may be accomplished by inclusion of a penalty or bonus clause in the agreement.

When a construction contract is not completed on time it is recognized that the owner could suffer certain monetary damages such as loss of rental value, interest on invested funds, and rentals on duplicate premises, to name a few. If the owner incurs such losses they can be charged to the contractor in the form of liquidated damages.

Establishing the Construction Time. To analyze this problem it is important to know how the contractual completion date is usually set. There are two common possibilities: either the owner (or architect) sets a time period (or date) which is required or convenient, or it is established by the contractor at the time of bid submission.

In the first case the owner-established date might provide for a reasonably workable construction period, or it might be impractically short. Architects, in trying to establish a realistic time period, usually refer to the experience of previous contracts or will confer with contractors. Sometimes contractors will agree to unrealistic completion dates just to get the contract. The self-delusory expectation is that good fortune will somehow resolve the practical problems, so they take a chance.

In the second case, the contractor sets the date or time period. Sometimes in competitive bidding, each bidder is asked to submit a proposed completion time along with the bid. The contractor with the shortest construction period might be favored over the lowest monetary bid, or the low bidder could be asked to meet the lowest quoted construction time. In the haste and pressure of bidding, insufficient thought is usually given to a realistic construction time schedule. Some bidders get carried away in the spirit of optimism and competitiveness and take an unnecessary chance on a tight or impossible time schedule.

It is seldom in the owner's best interest to create an artificial demand for unrealistic early completion. If, on the other hand, it is an economic or practical necessity to have occupancy at a certain time, then the contractor should be contractually involved in the effort to achieve the required deadline.

Contract Clauses. A liquidated damages clause providing for the contractor to forfeit a stipulated sum to the owner for each day of delayed completion will no doubt serve as an effective financial motivation to the contractor to finish on time. There is no question that it focuses the contractor's attention on the time schedule.

Some contractors feel that if it is of value to the owner to gain early occupancy, then the contractor should be paid an early completion bonus. It could be a single stipulated sum for timely completion and/or an agreed sum per day of early completion. If this is what the owner and contractor agree to, then a bonus provision could be added to the liquidated damages clause.

Legal formbooks are loaded with appropriate clauses. The following are suggested by the American Institute of Architects in its Guide for Supplementary Conditions (for use with General Conditions of the Contract for Construction, AIA document A201, 1987 Edition, and Instruction to Bidders, AIA Document A701, 1987 Edition.), AIA document A511, 1987 Edition:

> "9.11 Liquidated Damages: The Contractor and the Contractor's surety, if any, shall be liable for and shall pay the Owner the sums hereinafter stipulated as liquidated damages for each calendar day of delay until the Work is substantially complete:
> _____Dollars ($_____)
>
> 9.12 Bonus: The Owner shall pay as a bonus to the Contractor a sum of _____ Dollars ($_____) for each calendar day preceding the established date of Substantial Completion that the Work is determined to be substantially completed.
>
> These paragraphs are to be added to Article 9 of the supplementary conditions for the information of all bidders. They should also be added to the general contract agreement at Article 3 of A101 when the basis of payment is a stipulated sum and at Article 4 of A111 when the basis of payment is the cost plus a fee."

Some owners cannot use the building if delivered early so they would not care to pay a bonus. Their only interest would be in compelling timely completion and being compensated for monetary losses when the completion is late.

If such clauses are going to be included in the agreement, they must be available for the contractor's reference and evaluation during the bidding period before the price is set. If they are imposed as new requirements after bid prices are revealed, an appropriate adjustment in the contract price would have to be considered.

Deciding the Amount. The amount of liquidated damages should approximate the magnitude of actual monetary loss the owner would suffer if the project is not completed and ready for use at the designated time. The amount might be expressed as a lump sum for the first day of lateness with a smaller amount for each additional day thereafter, or it could be the same amount for each day the project remains uncompleted.

Amounts considerably greater than actual monetary losses would be considered punitive and should not be specified as courts and arbitrators are equally reluctant to enforce penalties. The actual damages likely to be incurred should be realistically estimated by the owner and can be rounded off for convenience.

Administering the Contract Provisions. When these types of contract clauses are included, it is vitally important to make a record of the day that construction actually started, the number of calendar days of construction time contemplated, and the target completion day. The date of commencement of construction is usually stipulated in the agreement and sometimes is stated in a notice to proceed issued by the owner or the architect on the owner's behalf. (A201, 8.1.2, Date of Commencement and 8.2.2, Owner's Notice to Proceed.)

The term completion must be defined in a coherent and practical manner. (See Chapter 19, Closing Out the Job.) The usual definition will be the date of substantial completion. This is defined in the AIA General Conditions as the date certified by the architect to be the date on which the work is suitable for its intended use, although there may still be miscellaneous punch list items left to be done. (A201, 8.1.3, Date of Substantial Completion, and 9.8.1, Substantial Completion.)

Construction time should be expressed in calendar days or specific calendar dates. When expressed simply in days or in working days, disputes will inevitably develop as to exactly what is meant and how the time is to be calculated.

Most contracts anticipate that the completion time will be extended by construction changes that impact the critical path. All changes should be in writing, always with the time component noted. The time element should be agreed upon at the same time that the physical change and the price adjustment are discussed and not be allowed to slide to be taken up at some later time. Often the time factor if immediately known would have been unacceptable to the owner and the change order would have been canceled. The owner's pressure to maintain the completion date, if excessive, often leads to unfair decisions on the time element of change orders.

If the proposed completion date is important to the owner, then the owner must actively cooperate by suppressing all changes that will affect the critical path and by timely performance of all contractual requirements. This would entail primarily making payments when due and making all decisions when needed. The owner could also interfere with the contractor's timely completion by inexpert coordination of separate contractors, such as kitchen equipment installers, carpet layers, or landscape contractors.

Determining Completion. The most contentious issue of all occurs when the contractor suddenly declares (with an air of finality) the project to be completed or announces that it will be completed on a particular day, say, 3 days hence. The owner, responding with equal vigor, declares that the work is nowhere near complete and certainly will not be done in 3 days. The contractor's response will always be

Construction Contracts

that the remaining work is inconsequential, nit-picking, and consists solely of routine punch list items. The contractor insists that the owner can move in, minor pick up work will be continued, and the job will be 100 percent complete in no time at all. The owner's attitude is generally that after moving in they do not want messy painters, noisy carpenters, untidy plumbers and electricians, and other disruptive workers interfering with the quiet, orderly, serene, operation of their building. The owner also is sure that the contractor's normally excellent judgment is being clouded by the desire to avoid paying liquidated damages. Of course, the contractor is utterly convinced that the owner has not overlooked the prospect of reducing the final payment by the exaction of several days of liquidated damages.

The best solution to this problem is to invest the final determination of completion to an impartial professional such as the architect. According to the AIA General Conditions (A201, 4.2.9, 8.1.3, and 9.8.2), the architect will determine the date of substantial completion. The architect would employ objective standards that will be fair to the interests of both the owner and contractor. Whenever the completion determination would result in an exchange of money to or from the contractor or owner, there is always the potential for distrust, dissatisfaction, and a legal claim. Therefore, the determination must be fairly made, based on prespecified criteria, and properly documented.

If the contract clause provides for a different degree of completion, such as 100 percent completion, or phased completions, the architect will use this as a measure and will make an appropriate and fair determination.

Cost Plus Contracts. Liquidated damages provisions in cost plus contracts should stipulate that the contractor shall pay the liquidated damages to the owner as a direct reduction of the contract price and not be treated as a reimbursable cost.

Effectiveness of Financial Incentives. Contractors, unsurprisingly, do not favor financial threats like liquidated damages and will try to avoid the eventuality of paying them. Thus, if they have the opportunity of including an allowance for possible liquidated damages in their breakdown of estimated costs, they will do so. If all competitive bidders make such an allowance, the owner, in effect, will pay the liquidated damages if the project runs late and the contractor will gain a windfall if completion is on time or early. This would be equivalent to a bonus and would serve as a positive incentive for timely completion. A bidder who does not allow for liquidated damages could end up as the low bidder. Timely completion in this case would be motivated primarily by the fear of having to pay liquidated damages.

Without a liquidated damages provision, the contractor's main incentives for timely completion would be in the avoidance of extended overhead costs and protection of its reputation in the community. However, when a contractor is working on several jobs at once, some with liquidated damages provisions and some without, the prudent course would be to apply limited resources to the avoidance of paying liquidated damages.

A contractor attempting to earn the early completion bonus could overspend on overtime, supervision, and multi-shift work and not be ahead at all, particularly if the job is completed only on time or a few days late, earning no early bonus and possibly incurring a few days' liquidated damages.

Contractors under the pressure of overrunning the completion date will tend to be more aggressive in asserting extra time claims arising from change orders and inclement weather. Some will seek to shift responsibility to the owner and architect for every conceivable change in conditions.

Ideally, owners considering liquidated damage clauses will be better off with a competent contractor who has a reasonable time period in which to construct the project. Time extensions fairly granted and a realistic liquidated damages provision will not yield timely completion when the allotted construction time is insufficient or the contractor is incapable.

The problems inherent in various combinations of unrealistic contractual clauses, incompetent contractors, and insufficient construction time will not be resolved by harsh, unyielding contract administration. Moreover, timely completion cannot be achieved by imposition of a liquidated damages or bonus clause when the construction period is insufficient.

Allowances in Construction Contracts

Purpose of Allowances. The establishment of cash allowances in construction contracts is a convenient method of allocating construction funds to portions of the work which cannot be specified with sufficient particularity for competitive bidding at the time of contracting. Primarily, this includes items which have not yet been selected pending the availability of new models or the arrival of updated catalogs. In some cases the owner has not as yet established definite criteria for certain equipment or furnishings, but this should not preclude proceeding with general construction. It also includes items of superficial or decorative nature which will be selected at a later time when colors, textures, furniture, and interior designs are more definitely established.

The types of purchases most frequently encountered as cash allowances are those such as finish hardware, lighting fixtures, special equipment, graphics, building signage, floor coverings, window treatment, and wall coverings. It is a flexible way of including in the contract items that are not yet designed, chosen, or specified. Allowances are practical for work of indefinite scope or where the quality, configuration, and other specific characteristics have not as yet been determined.

Contract Provisions for Allowances. Disputes between owners and contractors can easily develop over the issues of accounting and billing for cash allowances and related expenses. Therefore, it is important that the contract covers all aspects which could be incorrectly administered or misunderstood. Those contracts which include A201 will have the decided advantage of a practical set of conditions. Paragraph 3.8, Allowances, is based on the principle that the designated cash allowance is to be used only for the net purchase price of the denominated item. Any adjunct or related costs to be incurred by the contractor should be anticipated and included in the contract sum. Paragraph 3.8, Allowances, of A201:

> "3.8.1 The Contractor shall include in the Contract Sum all allowances stated in the Contract Documents. Items covered by allowances shall be supplied for such amounts and by such persons or entities as the Owner may direct, but the Contractor shall not be required to employ persons or entities against which the Contractor makes reasonable objection.
>
> 3.8.2 Unless otherwise provided in the Contract Documents:
>
> .1 Materials and equipment under an allowance shall be selected promptly by the owner to avoid delay in the Work;
> .2 Allowances shall cover the cost to the Contractor of materials and equipment delivered at the site and all required taxes, less applicable trade discounts;
> .3 Contractor's costs for unloading and handling at the site, labor, installation costs, overhead, profit and other expenses contemplated for stated allowance amounts shall be included in the Contract Sum and not in the allowances;
> .4 Whenever costs are more than or less than allowances, the Contract Sum shall be adjusted accordingly by Change Order. The amount of the Change Order shall reflect (1) the difference between actual costs and the allowances under Clause 3.8.2.2 and (2) changes in Contractor's costs under Clause 3.8.2.3."

Owner's Selections. During construction, the owner must further instruct the contractor how each cash allowance item is to be expended. The contractor will need to know, with specificity, what is to be purchased and from which sources it is available. This may require additional drawings, specifications, or description lists to be prepared by the architect. These additional instructions should be forwarded in writing to the contractor to avoid misunderstanding and possible error.

It is the owner's duty, without prodding by the contractor, to make all allowance selections in time to meet the contractor's previously issued progress schedule. This means that purchasing lead time, delivery time, and all other related time factors must be taken into consideration. If the owner's untimely or inappropriate selection causes any construction schedule slippage, the delay is chargeable, on a day for day basis, to the owner.

The owner's further direction to the contractor, usually prepared by the architect, should include all necessary information which will be needed for purchasing, including manufacturers' names, model numbers, colors, textures, sizes, capacities, and electrical and mechanical characteristics. If the owner desires the purchase to be made from a particular source, that should be stated. However, the contractor is not required to deal with any persons or entities to which reasonable objection has been lodged. (A201, 3.8.1)

Logically, the burden of locating purchasing sources of exotic or hard to locate items remains with the owner, although this is not mentioned in Paragraph 3.8 of A201.

Contractor's Purchase Cost and Related Costs. In its fiduciary capacity to the owner, the contractor is obligated to use its best efforts, skill, and purchasing power to obtain the most favorable price, terms, and other conditions of purchase. Obviously, it is unethical for a contractor to solicit or accept any unreported secret commissions or kickbacks from vendors. All price reductions due to rebates, refunds, and discounts should be applied for and obtained by the contractor and credited to the owner. Allowances are to be priced at the net cost to the contractor including all sales taxes and delivery to the site. (A201, 3.8.2.2)

All of the contractor's costs relating to cash allowance items, other than the actual purchase, should be included in the contract sum, not in the allowance. (A201, 8.2.3) These adjunct costs include handling, unloading, uncrating, cleaning, and secure storage. Also included are the contractor's installation labor, administration, supervision, interest, insurance, bonds, overhead, and profit. Related costs also include all required permits, inspections, certifications, and testing.

Cash allowances that are to be carried by subcontractors should be accounted for in similar manner.

The Contractor is responsible for losses due to damage to allowance items while under its care, such as while they are stored or during installation.

Should re-stocking charges be incurred, they are chargeable to the owner or contractor depending on who caused the expense.

Specifying Cash Allowances. In addition to the requirements for administration of allowances in Paragraph 3.8 of the A201, a schedule of cash allowances should be included in Division 1 of the project specifications. A reference to each cash allowance item should be made in the applicable trade sections to clarify specifications and installation standards.

Construction Contracts

The cash allowance procedures described here and in A201, 3.8 are the orthodox, commonly accepted practices in the administration of construction contracts by architects using the AIA standard contract documents. Should the owner or contractor wish to stray from these customary practices for any reason, it is a simple matter to add appropriate amendments to the contract. Any desired amendments to the AIA General Conditions should be included in the supplementary general conditions.

5

Project Delivery Systems

The Traditional System

By far, the most common system of design and construction of buildings is the traditional system, in which the architect designs the building and prepares the contract documentation and a contractor builds it. The architect is the owner's impartial professional advisor, the owner's agent during the construction period, and administrator of the contract. The mainstream AIA documents and most of our legal precedents are based on the traditional construction system.

It is only natural that we would try to devise ways of eliminating the shortcomings encountered in our old ways of doing things.

The Search for an Ideal Contracting System

All through the years, we keep hearing about creative new systems that will miraculously resolve all the imperfections inherent in our old worn out methods of buying construction.

Each new system that appears on the horizon promises faster construction at less cost and with more efficient management and accountability. Everyone involved wants to believe this, so the usual skeptical analysis is often bypassed.

It usually develops that the new contracting system is merely a copy of one of the old ones but with a new twist or two and a newly coined name.

Owners are always desperately searching for ways to save money and time and to fix responsibility in their construction projects. The construction industry, on the other hand, is constantly seeking new ways to market its services, lessen competition, and limit liability. These two opposing groups of objectives fit perfectly and are the basic motivation for most newly arrived contracting systems. Anyone who does not immediately embrace the newly invented system is branded obsolete and out of touch.

Whenever economic prospects of the construction industry need a fresh stimulus, the hue and cry is raised for the invention of new project delivery methods.

Indispensable Entities

Quite often, newly invented construction delivery systems are devised by manipulating the entities involved and by redistributing their conventional duties and responsibilities.

In addition to the owner, the essential entities required for every project are a designer and a contractor. From there on, as additional entities are introduced into the equation, the cost generally goes up. Some new systems are created by involving the financing entities with the construction entities.

Innovative systems that do not add entities are frequently based on reducing or combining the number involved. Either way is flawed because the result will be either higher costs or lack of control and accountability. The combining of entities generally creates conflicts of interest, real or potential.

Changing the traditional distribution of duties often results in confusion and any expected cost savings are many times dissipated in overlapping or losses in responsibility.

New unfamiliar alignments of the participants' inter-relationships and responsibilities trigger two significant, usually unexpected, problems. First, the parties will have to learn, understand, and accept their newly defined and unfamiliar roles. And second, new documentation must be devised, perfected, and explained to the parties. Even then the learning curve will take its toll of the first few projects.

Some Alternative Construction Delivery Systems

New construction delivery systems are usually invented by the contracting industry in the continuing effort to expand construction markets. It is also a good way to carve out a niche to corner some of the market, at least temporarily.

Package Deal. A *Package Deal* usually combines construction and architectural services with some or all of the needed elements of real estate acquisition and temporary or permanent financing. *Turnkey* and *Design-Build* are other words for types of package deal. Each of these systems means different things to different people as none are standard. Whatever new system is agreed upon must be defined and each participant's role must be determined. Whichever is the lead entity must assume the duties of coordinator.

Design/Build. In this system the design and construction entities are merged in their offering of services to the owner. Although a number of permutations are possible, three different organizational configurations are prevalent:

1. Contractor leads. The architect is subordinate, as a subcontractor or employee of the contractor.
2. Architect leads, in one of two ways:
    a. The general contractor is the architect's subcontractor or employee.
    b. The architect deals directly with the trade contractors and suppliers, eliminating the general contractor.
3. Contractor and architect join forces as partners in a joint venture.

The owner might feel safe in the hands of the design/build architect but may not realize that in some permutations the architect's loyalty and financial interests would then lie with the design/build team, not with the owner. The owner may be unaware that the architect would not serve in the usual role of impartial professional advisor. There will be no one to judge the contractor's performance of the contract on behalf of the owner. The architect's conflict of interest should be resolved by full disclosure to the owner.

Sometimes the architect starts under employment by the owner, to be later assigned to the design/build contractor. In other situations, the owner's architect (referred to as a bridging architect) does not continue after the design development phase but is replaced by the design/build contractor. The owner's interests have a greater chance for realization when the architect is working directly for the owner until after completion of the design development documents.

The design/build entity in most states must be properly licensed as architects and as contractors, as they will be offering to provide or furnish architectural and contracting services. There is no specific licensing of design/builders.

The architect involved in a design/build entity will find that the normal professional liability insurance coverage will be inadequate. Usually, architects are judged on the basis of the professional standard of care. This means that there will be liability for errors and omissions only when there is negligence. There is generally no liability for non-negligent error. The normal architect's professional liability insurance applies to the situation where the architect is not closely involved in the actual construction and contracting.

Contractors, on the other hand, are subject to the legal doctrine of strict liability, which does not require a showing of negligence. Architects, along with their construction associates in a design/build entity, will share in the responsibility for jobsite safety and the means and methods of construction as well as for the design. When the architect is involved in such matters as construction warranties, guarantees, and workmanship, professional liability insurance will not cover. Contractor's general liability insurance is required.

Architect-led design/build operations may have difficulty in obtaining performance and payment bonds unless they are in a strong financial position.

Architects who decide to enter into business relationships with contractors should be extremely careful to clarify their legal, ethical, and insurance positions.

Owners are usually attracted to design/build by its concentration of responsibility and contact in a single entity. The price to be paid is the loss of an independent architect. Some owners are enticed by the lure of lower costs or faster construction time, but these are not guaranteed and do not always materialize.

A Closer look at Construction Management

What Does it Really Mean? And Who Does It? Loosely speaking, anyone who manages construction is a construction manager. However, this discussion is not about the usual day-to-day management activities of the regulars in the construction industry, it is about construction management and construction managers.

The term construction management, more familiarly CM, has become a permanent entry in the lexicon of the U.S. construction industry. Unfortunately, there is no widespread understanding or agreement on exactly what it means and what its ramifications are. The dictionary does not give us much help, as there is no entry for *construction management* and combining the definitions of *construction* and *management* does not yield a coherent answer.

Although the term has been in frequent use for at least a quarter of a century, there is still considerable confusion as to the basic concept of CM and how it fits within the framework of conventional relationships among contractors, owners, and architects. There is still no universally accepted understanding of the scope and compass of the services and responsibilities of a construction manager.

A definition of CM was advanced in 1982 by the Construction Management Association of America:

> Activities over and above normal architectural and engineering services --conducted during the pre-design, design, and construction phases -- that contribute to the control of time and cost.

This is a simple and understandable working definition, but it lacks any definite suggestion as to who the construction manager may be and who would, could, or should render these services.

Qualification Standards for CMs. There is no specific governmental regulation of CMs nor is there any system of registration or licensure. It is only when a CM infringes upon the architectural, engineering, or contracting disciplines that licensing is required. (This may be gradually changing, however, as a "consultant to an owner/builder" has now been brought within the definition of the term "contractor" for purposes of the contractors license law in California.) Moreover, there is no accepted qualification standard of education or experience as a prerequisite for any person or organization to call itself a construction manager.

Diverse Backgrounds of CMs. Construction management is still a developing profession. The services offered by a particular CM will depend largely on the CM's previous educational background, training, and experience. They will generally be heavily weighted to include the services usually performed by the discipline most familiar to the CM. CMs have originated from various segments of the building industry as well as from general business and includes contractors, engineers, architects, construction estimators, accountants, lawyers, general business persons, and jacks-of-all-trades. The quality of CM services offered in the marketplace ranges from the competent and comprehensive to the incompetent and meager, from the valuable to the worthless.

Function of the CM -- What CMs Do. When the CM's background has been in building contracting, the services offered will be similar to those normally offered by contractors, sometimes including the option of a guaranteed maximum construction price and a guaranteed completion date.

There is no standard understanding of the scope or quality of the CM's functions. In any specific project the CM's services and responsibilities will be as quantified and qualified in the CM's employment contract for that particular project.

The services likely to be offered by the CM will also depend on the stage at which the CM is engaged by the owner.

Early CM Appointment. When the CM is selected and engaged before the owner has been exposed to any architects or contractors, the CM will become heavily involved in the pre-design activities including selecting and employing of architects, engineers, and other design professionals. In this case, the CM will also be influential in programming the project, instructing the architect, and negotiating the architectural contract. The CM will be the main communication link between architect and client.

In many similar situations, the CM will completely isolate the architect from the client. The architect's professional position would be considerably subservient to that of the CM. Architects accustomed to the conventional close owner-architect relationship will not be pleased with this new unfamiliar situation and the diminished role.

Later CM Appointment. When the CM is added to the team after employment of the architect, the CM's duties will be focused more on cost containment and contractor selection, followed by monitoring of construction operations, control of project cost and construction time, and managing and accounting for all funds.

CM's Expanded Services. In many cases, the CM's duties will be expanded to encompass some parts of the architect's traditional services as well as a portion of the contractor's accustomed duties. Accordingly, some owners would expect a suitable portion of the architectural and contracting fees to be conceded to the CM fee. The CM fee will not often be totally recoverable from these concessions however as the addition of a CM on the job is to lessen the owner's normal duties and thus will result in additional fees. The added services must be paid for and the additional entity must be compensated.

There is no practical limit to the number of permutations of CM duties and techniques being offered in the construction market place.

Why Owners Find CM Attractive. Owners often find that their need for a project which must be designed and constructed will generate enormous demands on their own time and services. They will be required to deal not only with architects and contractors but with lawyers, insurance people, financial experts, land surveyors, engineers and other specialized designers, real property experts, construction cost consultants, and numerous other technical advisors and consultants, each with their own unique vocabulary. Many owners do not have the time nor expertise to make all of the necessary technical decisions and to issue responsible and appropriate instructions to their advisors and consultants. Their inexperienced or untimely responses can result in substantially increased building costs and elapsed time.

Other owners who have ample expertise personally or available in-house are looking for a reliable single-contact system to help simplify their involvement in the purchase of design and construction services.

Frequently, owners have found that the conventional system of selecting and engaging an architect to prepare design and bidding documents and later selecting and hiring a contractor subsequent to a bidding procedure not only requires a monumental amount of time, effort, and expertise but does not always produce a satisfactory result. Often, the desired budgetary and time constraints have been exceeded, adversely affecting the project's economics for years into the future.

There is always the hope that a new system such as CM will solve all the old problems. For these reasons, the services of a construction manager can be very attractive to many owners.

Some Problems with CM. The major risk to the owner in selecting an organization or individual to provide CM services is the possibility of making a poor selection. The owner must exercise caution and common sense when choosing and engaging a company or individual to furnish CM services. An incorrect choice of CM could be disastrous. Since the CM has the owner's ear, and considerable owner-granted authority, an incompetent CM can do immeasurable damage before the owner becomes aware of an unsatisfactory or deteriorating situation. An abrasive CM can easily lose the respect and cooperation of architects and contractors and may cause more harm than good in advancing the owner's interests.

Whenever a CM is added to the usual construction team, a period of reeducation must be endured, so all members will be aware of their duties and responsibilities. They must also learn how all members interrelate under this new unfamiliar alignment. The team leader, whether it is the owner, CM, or architect, must sense this void and immediately fill it. The team leader must seize the initiative before the project starts to insure that the construction is successful from start to finish.

Standard CM Contracts. There is a wide range of standard construction management contracts available from which to choose. Standard form contracts are published by and can be obtained from the American Institute of Architects (AIA), Associated General Contractors (AGC), Construction Managers Association of America (CMAA), and National Society of Professional Engineers (NSPE). These standard agreements vary considerably, as they are each based on the differing viewpoints of their authoring organizations. Review of these documents will confirm that there is little uniformity in the duties and responsibilities of CMs nor in the terms and conditions of their employment. Some CMs write their own contracts, usually based on one of the standard forms of contract.

Two Types of CM. The AIA and the AGC, recognizing that CMs who are involved with actual construction are different from those who furnish unbiased technical counsel, have jointly issued two new documents:

> Standard Form of Agreement Between Owner and Construction Manager where the Construction Manager is also the Constructor, Stipulated Sum, AIA Document A121/CMc and AGC Document 565, 1991.

> Standard Form of Agreement Between Owner and Construction Manager where the Construction Manager is also the Constructor, Cost Plus Fee, AIA Document A131/CMc and AGC Document 566, 1994.

All AIA CM documents are now differentiated as to whether the CM is an *advisor* (owner's agent) or a *constructor* (contractor). This attempts to resolve the issue of conflicts of interest which are likely to arise when the CM is under the pressure of a guaranteed maximum price and a guaranteed completion date backed by liquidated damages.

The old AIA CM form is suitable for use when the CM is of the advisor type, not a constructor:

> Standard Form of Agreement Between Owner and Construction Manager, AIA Document B801, June 1980 Edition.

This form contemplates that the CM is a separate entity from the architect or contractor.

When the CM is the General Contractor. When the CM's services include a guaranteed maximum construction price or a guaranteed completion time, the CM is in reality serving as a general contractor and, as such, must be licensed as a general contractor in most states. This would also be the case if the CM signs subcontracts in its own name or provides direct labor on its own payroll even if there is no guaranteed maximum construction price or completion time. When a general contractor calls itself a construction manager, the owner should realize that any additional services not normally provided by contractors are not without some possibility of bias and could include an element of conflict of interest. This is particularly applicable when the CM is attempting to render impartial technical advice or opinion regarding the quality of the contractor's own work.

The CM as the Owner's Impartial Agent. The CM (advisor type) will provide the owner with impartial technical advice, will competently carry out the owner's objectives, and will act as the owner's representative to the construction industry. The CM will be the owner's agent with all duties, powers, responsibilities, and limitations bestowed by the employment contract and will function in a fiduciary capacity. The CM will act as intermediary, facilitator, and coordinator of all of the owner's advisors, consultants, contractors, and suppliers without interfering with any of their functions.

The CM should enhance, not impede or inhibit the free flow of administration and communication among all those involved in the project. At all times of the construction process, the CM should keep the owner informed of the financial status of the project and the advantages and disadvantages of all proposed courses of action.

The CM can be an employee on the owner's staff or can be a separate contractor. When the CM is an employee of the owner, the AIA terms it a Project Manager. The CM could also be the design architect.

Fast Track Construction

The construction industry is a fertile field for colorful and racy jargon. New words and terms spring up most regularly, usually for the purpose of promoting something to gain economic advantage. A stylish new term of the past 20 years is *fast track construction*. It is a no-nonsense term that rolls easily off the tongue and conjures up images of expedition, economy, efficiency, and directness of purpose. The word track connotes a direct, unswerving route to a predetermined destination. The term as a whole promises accelerated construction and we can all accept that time is money. But, is it reasonable to expect that fast track construction will always live up to these prospects?

The answer is a qualified yes. It could work well and often does. But only under ideal conditions. Only if the owner, architect, and contractor could all conduct themselves as a team. Otherwise, the results can be disappointing and distressful to all concerned. According to one realistic construction industry expert,

> "Experience indicates Fast Track construction *will be delayed* and *cost more* than any other method."
> H. Murray Hohns, Preventing and Resolving Construction Contract Disputes, Van Nostrand Reinhold Company, New York, 1979.

What Is Fast Track Construction? The normal process of construction scheduling involves the performance of a series of discrete functions, one after the other in a predetermined sequential order. The customary logical order is: programming, design, governmental approvals, bidding and negotiation, contract award, construction, and finally, completion. This is shown on the first line of the accompanying diagram, Figure 5.1. Each activity is virtually completed before the next may be commenced. To perform all of these functions will take a certain amount of time. The usual way of shortening the time scale is by increasing productivity, that is, to complete each function as efficiently as possible and to start each new phase immediately upon completion of the preceding phase. Everything is done in proper order and no time is wasted.

Figure 5.1 Fast Track Schedules

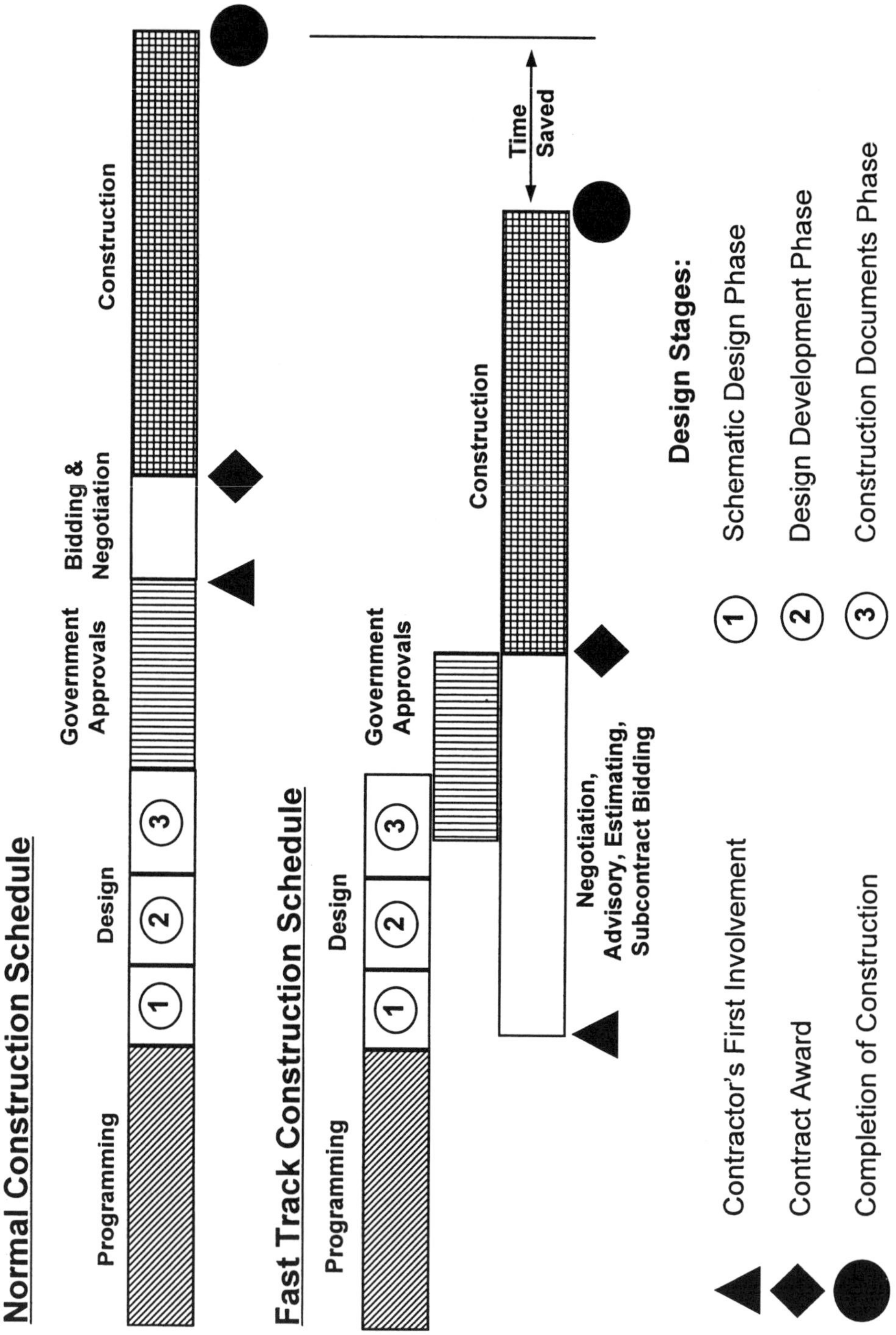

Normal Construction Schedule

Programming | Design ① ② ③ | Government Approvals | Bidding & Negotiation | Construction

Fast Track Construction Schedule

Programming | Design ① ② ③ | Government Approvals | Construction

Negotiation, Advisory, Estimating, Subcontract Bidding

Time Saved

Design Stages:
① Schematic Design Phase
② Design Development Phase
③ Construction Documents Phase

▲ Contractor's First Involvement
◆ Contract Award
● Completion of Construction

Project Delivery Systems

Another way of saving total elapsed time would be by *compression* of the time schedule. More specifically, by overlapping some of the functions, doing two things at the same time. This would be accomplished by starting a new phase of work where possible before the preceding phase is completed. Time saved by concurrent work will accumulate and appear at the end of the construction period in the form of early overall completion. Organizing the project to produce early completion by the technique of concurrent or overlapping time scheduling is the essence of fast track construction.

Understanding Fast Track. The usual procedure for overlapping functions entails the contractor's earlier involvement in the project. However, there is no accepted standard system or approach. Every fast track project can be different.

Some owners will select and engage a contractor to confer with the architect during the early stages of design. The contractor's advisory input during the early stages will assist the architect in making practical and economical choices from readily available materials and systems. Architects and engineers should profit from the experienced contractor's review and pragmatic judgment.

The contractor then starts producing provisional construction budgets and time schedules that are then continuously refined and updated as more precise design information is simultaneously developed by the architect.

Upon the owner's approval of the completed design development phase documents, the contractor can start in earnest seeking subcontractor interest and more definitive prices. At this juncture the size and character of the project will have been fairly well determined and described. The architectural, structural, mechanical, and electrical systems, materials, and other elements will have been resolved in principle, but the drawings and specifications will not yet be sufficiently advanced for actual construction or for submittal to governmental agencies for their review. Any suggestions to be advanced by the contractor or required by the owner should be made at this time, as it will be simple and economical to incorporate them into the next stage of the architect's work.

The third and final design stage, the construction documents phase, comprises an enormous amount of coordination, synthesis, and sheer production of documents, both graphic and verbal. It involves the intimate cooperation of structural, civil, mechanical, and electrical engineering and architectural personnel. All of the drawings and specifications must be properly interrelated and consistent. Precise details are being finalized and everything must be technically correct. Usually, changes in the program during this high volume production phase are very complicated and therefore costly. They create confusion and delay and significantly increase opportunities for error.

As the construction documents progress, the contractor will be able to monitor the architect's use of materials and selection of systems and to gradually refine the cost breakdown and time schedule.

By the time the construction documents are about 60 to 70 percent complete, they may be submitted to governmental authority for review, approval, and issuance of building permits. At the same time, the contractor can stabilize the breakdown of estimated costs and determine a guaranteed maximum price (GMP). As the contract documents are still incomplete, the GMP is necessarily predicated upon the contractor's assumptions and predictions of what will be included in the final state of the documents. In the GMP, the contractor must have included all work that is shown on the drawings and in the specifications at that time, but also items that are reasonably inferable or expected.

If the GMP and time schedule are acceptable to the owner, the contract may be executed and the contractor may start construction as soon as the building permit is issued.

The time compression described in this scenario, shown in the second line of the diagram, consists of:

1. The governmental approval period commences before the design period ends
2. The bidding and negotiation period is completely concurrent with the design and governmental approval periods
3. The construction period commences immediately after governmental approval

Faster Track. Even more schedule compression is possible by other variations in technique. For example, by obtaining limited governmental approval of the foundations so that they may be constructed concurrently with completion of the governmental review process. This is illustrated in the third line of the diagram. This technique carries with it an additional formidable risk: If the building officials, as they continue their review, later require any adjustments in the foundation design, it may be costly to correct the affected members in the field if they have already been constructed. This is a normal condition imposed by governmental authority as a quid pro quo for premature or piecemeal approval. In this case, all of the time gained could be lost and then some. In addition, the cost of reconstructing the footings or grade beams would be wasted. Some jurisdictions will not allow "foundation only" permits.

There are many other possible variations in arrangement of the work schedule to compress the time scale. The maximum theoretical amount of time which can be saved will vary from one project to another depending upon their individual characteristics.

Selecting the Contractor. Inasmuch as the contractor must be brought in at a very early stage, it is difficult to obtain genuine general contractor price competition. If bidding competition is required, the contractor cannot be selected until after the drawings and specifications are available, at least to the level of 60 or 70 percent completion. This would effectively rule out any contractor input during the early design stages when the most significant cost-affecting decisions are being made.

Contractors bidding from incomplete documents will be at a distinct disadvantage. If they price the work to include all that is inferable from the unfinished documents, they will not be competitive with the contractors who infer something less. The bidder who infers the least could end up with the low bid and will almost certainly be later involved in a dispute with the owner as to what was reasonably inferable.

For these reasons, it would make better sense to select a contractor on the basis of comparative experience, reputation, recommendation, and interviews. The price competition at the level of subcontractors and suppliers will be quite sufficient to protect the owner's interests. As a part of the selection process, the contractor remuneration system or formula can be discussed, negotiated, and adopted.

An experienced owner may, usually with a construction professional's assistance, choose a suitable contractor who has the required financial and practical resources to accomplish the assignment. If the owner has reservations concerning the contractor's qualifications or character, the contract cannot possibly succeed. Mistrust will breed dissatisfaction, and ultimately, unresolvable disputes. Any expectation of saving time by the fast track process will be long gone. There will be only contention, delay, and higher costs, as could have been reasonably predicted.

An owner who requires general contractor bidding competition and who cannot abide the uncertainties of bidding from incomplete documents should not get involved with fast track construction.

The Fast Track Construction Contract Form. The contract payment system in widespread use in fast track is based upon reimbursement of the actual cost of the work plus a fixed or percentage fee for the contractor. This can be with or without a guaranteed maximum price. A suitable contract form for this application is:

> Standard Form of Agreement Between Owner and Contractor where the Basis of Payment is the Cost of the Work Plus a Fee with or without a Guaranteed Maximum Price, Tenth Edition, AIA Document A111, 1987 (Appendix C)

> This should always be used in conjunction with General Conditions of the Contract for Construction, Fourteenth Edition, AIA Document A201, 1987 (Appendix D)

Possibly, the contractor will be paid an additional consulting fee for advisory and cost estimating services during the design period. In this system, the contract could be executed either before any services are performed or, alternatively, after the GMP is agreed upon, prior to commencement of construction. This could be accomplished by two contracts, the first for advisory and costing phase services and the second for the construction phase, or, if desired, by a single contract cancelable at the end of the advisory service if the construction is canceled or the owner decides to part with the contractor.

The Guaranteed Maximum Price (GMP). Although owners may regard the GMP as inviolate, contractors will rightly expect it to be appropriately adjusted to reflect any deviations from the drawings and specifications and their reasonable inferences. They will also expect that the contract time will be fairly adjusted to reflect unforeseen occurrences and unbudgeted changes in conditions. This is an area rich in the probability of misunderstanding. The owner and architect will generally assume that all inferences in the incomplete documents are obvious and should have been understood by the contractor, therefore no change in price or time is expected or justified when the completed documents are gradually, or finally, issued.

Unsurprisingly, the contractor will take an opposing view. When the contractor's position is found correct by a fair contract administrator or an arbitrator, then the GMP must be raised and the all-important completion date will have been breached. The only practical solution for resolving cost and time overruns at this point is to immediately reduce the scope or quality of the project. This will be a serious disillusionment to an owner who thought the maximum price and time were absolutely guaranteed.

The Architect's Problems. Under normal time scheduling, architects and their engineering consultants, take great pains to check their work product, to cull out errors, and to make sure that the work of all disciplines is coordinated. Even with the most rigorous reviewing and checking procedures some errors and anomalies will inevitably slip through. In the fast track process, incomplete drawings and specifications are released for bidding, governmental review, and construction. Although architects and engineers have checked their work at this point, some errors and omissions will undoubtedly be later uncovered during the completion of the documents and in the field as construction progresses. Also, as the drawings and specifications are being further developed, situations will arise where it would seem advisable to change previously released designs and details.

To correct errors or to change to more advantageous designs will require more change orders than would be common under normal construction scheduling. This will be an inconvenience to the contractor, sometimes an embarrassment to the architect, and usually added cost to the owner. There is also the added risk of jobsite confusion and construction delay.

Project Delivery Systems

Will it Save Time or Money? Although the fast track construction method has the potential of saving time, there can be no reasonable guarantee that it will save money. The monetary saving is primarily in the value of the time saved, if it materializes. The final construction cost will be about the same as any other cost plus contract with or without a GMP. However, in fast track contracts the construction contingency allowance must, prudently and realistically, be higher to cover the always expected but unknown costs of faulty predictions. So it is distinctly possible that a normal construction schedule with a smaller contingency reserve will produce a lower final total cost. It is also possible that the misunderstandings and erroneous inferences due to incomplete documents inherent in the fast track method will cause a loss of some or all of the overlapping advantages and thus yield a completion date on or after that of a normal time schedule.

Fast track construction should not be attempted unless the owner is under pressure to achieve a specific mandatory completion date. The owner, architect, and contractor must all be completely acquainted with the process and must enter into it fully aware of the inherent problems. They must be willing to engage in the give and take of compromise, and to assume reasonable attitudes. Even with all of the extraordinary efforts expended to achieve the time savings available in fast track construction, it is possible to lose them all in the event of labor disruption, material unavailability, unpredicted inclement weather, fire, accident, earthquake, or any of a number of other unfortunate occurrences.

Will the Conventional System Work?

In all cases, conventional contracting systems should not be abandoned without exploring their inherent advantages. A simple direct way of attaining the owners' objectives would be to start with the three basic entities, owner, architect, and contractor, and to try to get each of them to perform their traditional roles properly.

All of the documentation is already in existence. All of the technology is already in place. Nothing has to be devised, created, invented, rearranged, or explained.

The main reason that the traditional system occasionally fails is simply that one of the three principal participants, did not perform its role properly. Sometimes it is the architect or contractor, but more often it is the owner. Most of the alternative project delivery systems are designed to appeal to owners as a way of limiting risk, lowering costs, or lessening duties. These understandable objectives are often hard to realize.

Owners could probably achieve more of their aspirations by channeling their efforts into making the conventional system work better. The solution is in the hands of the owner. Ultimately, it is the owner who decides what system will be used. Here are some brash but serious suggestions for raising the quality and perception of the owner:

> Enter the real world. There is no magic contracting system out there waiting to be discovered. Realize that there is no free lunch. You cannot make something out of nothing. Everything has to be paid for. It is not realistically possible to find a system that functions automatically without anyone having to think, do their job properly, or assume responsibility. The project belongs to the owner. The owner is the real coordinator.
>
> Educate yourself properly to be the owner. The controller of the purse strings must be responsible and realistic. There is no higher authority, such as a governmental agency or insurance company, to protect your interests if you make a mistake. If you do not have the ability to conduct yourself properly in a construction contract, then you must educate yourself or hire someone to represent your interests.
>
> Engage a good architect. You cannot force an incompetent architect or contractor to perform properly just by devising a tougher contract. You must do the proper research to find a suitable architect. The architect must be relevantly educated, qualified, and experienced to undertake the project at hand. The architect must have a properly equipped and staffed office and must be financially stable.
>
> Select a good contractor. This may be accomplished by competitive bidding or by direct selection. If by competitive bidding, then appropriate research must be applied so that all bidders are qualified for the type and scope of work. The low bidder or the directly selected contractor should be properly experienced, and should have sufficient financial resources.

6

Selecting the Contractor

Negotiated Contract or Competitive Bidding

Before the contract drawings and specifications are completed, the owner and architect should start considering how to select a suitable contractor for the project. Often the owner will already be acquainted with a good contractor, perhaps from previously completed construction. The owner might feel very confident and secure in the hands of a familiar, tested, and trusted professional. When the owner wishes to proceed with a known contractor, it is simply a matter of negotiation with the chosen firm to determine the contracting method, the formula for the contractor's remuneration, and the cost of the project.

A negotiated contract with a selected contractor is favored by many experienced owners, as it provides opportunity for the architect and owner to consult with the contractor during the design and construction document phases. This enables the owner to confirm in advance of actual construction that the architect's design is realistic and that the scope and quality of the project are within budgetary limitations.

Other owners are reluctant to proceed with consideration of only one contractor because of the apparent lack of competition and the fear of overpaying. Actually, competition would not be completely lacking, as the selected contractor would obtain competitive bids from the numerous subcontractors and suppliers. A negotiated contract is advisable only when the owner has complete confidence in the competence and integrity of the selected contractor.

Establishing the Bidding Conditions

For those owners who require competitive proposals from several contractors, the architect and owner should discuss the entire matter prior to establishing the bidding procedures. The architect will need to know the owner's requirements and decisions so they can be properly reflected in the supplementary conditions and bidding instructions. The owner must decide the number of bidders and who will be invited to bid.

It is the owner's prerogative to establish the insurance and bonding requirements. The owner must also decide on the contracting method and form of contract to be used. Decisions on these topics should be formed by the owner in consultation with the architect and the specialized insurance, legal, and financial consultants.

Written instructions for the owner's insurance, bonding, contracting, and bidding requirements can be conveniently transmitted to the architect by use of AIA Document G612, Owner's Instructions Regarding the Construction Contract, Insurance and Bonds, and Bidding Procedures, 1987 Edition. (See Appendix F.) Upon the architect's receipt of the owner's decisions on all of these matters, the contract documents and bidding documents can be completed.

Bidding Practices

Contractors do not charge a fee for preparing a competitive proposal for a construction contract, although the expense, time, and trouble of preparing a careful and responsible bid are substantial. Most contractors keep a running account of the number of unsuccessful proposals submitted for every contract successfully signed up. The unsuccessful proposals must be charged to the contractor's general overhead, which eventually must be recovered proportionately from the active contracts undertaken.

Owners must reserve the right to reject any and all proposals submitted, as this is necessary in the event that the lowest bid exceeds the budget or if there is any suggestion or suspicion of collusion among the bidders. There is, however, an implied promise that the bidding procedure will be fair, just, and equitable and that the owner will enter into negotiations for a contract only with the lowest responsible bidder. If it were not for this tacit understanding, contractors would not be willing to commit the considerable resources necessary to prepare a competitive proposal. The owner's right to reject bids should not be used as a deception for favoring a contractor who was not the low bidder or did not enter the bidding competition at all.

Recommended Bidding Procedures

In the interest of establishing uniform bidding conditions respecting the legitimate concerns and objectives of owners and contractors alike, a written standard was devised by committees of the American Institute of Architects and the Associated General Contractors in 1948. It continues to develop, has been revised and updated several times, and now exists as Recommended Guide for Competitive Bidding Procedures and Contract Awards for Building Construction, June 1982 Edition, AIA Document A501, AGC Document 325(23). Architects who wish to quickly inform their clients of accepted ethical bidding procedures should present a copy of this nominally priced booklet to each of their clients for consideration.

Architectural offices that conduct bidding procedures perceived by the construction community to be unfair or unethical have increasing difficulty finding contractors to bid their clients' work. The procedures outlined in the AIA/AGC Guide are reasonable and provide fair and equal opportunities for all bidders without compromising the interests of the owner.

The architect must obtain the owner's approval of the list of bidders, contracting method, and bidding procedures to be used. The architect and owner should review the guide together, discuss its recommendations, and resolve any owner reservations. The owner will undoubtedly recognize that sticking to the procedures will be beneficial to all concerned.

An owner might ask why negotiations should be confined solely to the low bidder when the owner may have a preference for one of the other bidders. No contracting firm should be asked to expend its resources preparing a proposal if it would not be acceptable should it become low bidder. If a particular contractor on the list is so attractive to the owner that it would be given favoritism over the low bidder, the owner should forego the bidding procedure and enter into a negotiated contract directly with the favored contractor.

Selection of Bidders

One of the first decisions to be made is the number of contractors to be placed on the bidding list. In public work, bidding procedures are governed by law and little discretion can be exercised by the owner or architect. Generally, there is no limit to the number of contractors who may bid the job. In private contracts, the owner may control the number and identity of contractors who will be allowed to submit proposals. Too few contractors on the bidding list will not provide sufficient price competition while too many diminishes the incentive for serious competition. Judgment as to the length of the bid list will depend on such factors as size, nature, and complexity of the project and current competitive conditions in the community.

Usually a list of from three to five prequalified bidders will provide adequate price competition.

The contractors invited to bid should be carefully selected so that the owner is willing to enter into a contract with the lowest bidder. It is best to choose contractors who are appropriately experienced for the type of project and who are not too small or large for the size of project. The bidders should be approximately the same size, as overhead costs will be similar. Although contractors will not always agree, it is advantageous to select bidders local to the job site, as it stands to reason that the service will be better than that provided by contractors who travel long distances. A carefully screened list of prequalified contractors will save the time, inconvenience, and mutual disappointment of rejecting a low bidder for reasons that could have been discovered prior to bidding. Each prospective bidder should be required to submit evidence of relevant experience, qualifications of its personnel, equipment, and other resources, and financial capacity and stability prior to being invited to bid. A standard AIA form questionnaire, Contractor's Qualification Statement, Document A305, can be used by contractors as a uniform method of transmitting pertinent information regarding their organizations. References from financial institutions and prior owners should be verified by the owner.

The pre-bidding data received from prospective bidders should be organized and submitted to the owner for final screening and compilation of a bid list. The owner will have access to its bank and other financial advisors for credit checking and final determination of each contractor's financial stability.

Invitations to Bid

The contractors selected for inclusion in the bid list should each be sent a written invitation to bid, outlining the general scope of work and informing them that the bidding documents will be available to all of the bidders at the same hour and day. Each of the bidders who accept the invitation should be given identical information on which to base their proposals.

Instructions to Bidders

In the interest of fairness, it is important that all bidders receive identical information at the same time and that sealed bids are due at the same time and place. The American Institute of Architects has developed a standard Instructions to Bidders, 1987 Edition, Document A701, which contains a set of uniform conditions and directions. The standard instructions are coordinated and consistent with the general conditions and construction contracts of the AIA. The instructions are not intended to be a contract document; therefore any provisions of the document intended to remain in effect after execution of the contract must be included in the supplementary conditions of the contract.

The bidders are required by the instructions to bidders to examine and compare the bidding documents and the contract documents of other concurrent work on the site and to report to the architect at once any errors, inconsistencies, or ambiguities discovered. This is an idealistic requirement and impractical to enforce, particularly in respect to those who do not become low bidder.

Bidders and sub-bidders requiring clarification, interpretation, or additional information must make their requests in writing to the architect at least 7 days before the bids are due. Bidders are instructed not to telephone or visit the architect's office during the bidding period. The architect will collect all requests for additional information and disseminate an addendum simultaneously to all bidders at least 4 days before bid time giving requested interpretations, clarifications, deletions, additions, and corrections of the bidding documents. The addenda should include all determinations of acceptable substitutions of specified materials, products, and equipment.

A practice favored by many architects and contractors is to have a pre-bid conference at the onset of the bidding period for general orientation, to introduce the bidders to the site, to answer questions, and to explain and identify the important features of the project. At this conference, the architect or an assistant should take notes so that any questions answered on the day or resolved later can be included in an addendum to be sent to all bidders.

The Bid Package

The miscellany of information to be distributed to each of the bidders is termed the bidding documents, also called the bid package. The Bidding Documents are defined in the AIA Instructions to Bidders, Paragraph 1.1, and "...include the Bidding Requirements and the proposed Contract Documents. The Bidding Requirements consist of the Advertisement or Invitation to Bid, Instructions to Bidders, Supplementary Instructions to Bidders, the bid form, and other sample bidding and contract forms. The proposed Contract Documents consists of the form of Agreement between the Owner and Contractor, Conditions of the Contract (General, Supplementary and other Conditions), Drawings, Specifications and all Addenda issued prior to execution of the Contract."

Each bidder should receive identical bidding documents, free of charge, usually by posting a deposit which is refundable when the documents are returned in good condition. The bidders should be allowed to retain the documents until after the contract is awarded or until the bidder is eliminated from the competition. The architect should determine how many sets of documents are reasonably needed by the bidders considering the type and size of work and the length of the bidding period. The bidders need sufficient sets of the bidding documents to obtain several sub-bids on each subdivision of the work.

It is a faulty practice for bidders to separate the drawings and specifications of certain trades and issue only partial sets to subcontractors as they would then not be aware of the interrelationships between their work and that of other related or adjoining trades. The bidders should also be issued the contract drawings and specifications of work to be performed concurrently on the same site by separate contractors or by the owner.

In some cities, there are public plan services where the bidding documents of projects being bid are on display for subcontractors and suppliers to examine them, prepare estimates, and submit proposals to the bidders. Where available, this is very beneficial to owners since it provides greater exposure to subcontractors and a resulting diversity of sub bids.

It is false economy to tightly limit the number of bidding documents issued to each contractor. After bidding, all of the documents will be returned and can then be reissued to the awarded contractor who will need them for construction. Any bidder who requests additional bidding documents is usually allowed to purchase them, but they still must be returned after the bidding period.

The architect should promptly return the bidders' deposit upon receipt of the returned documents, deducting only the amounts necessary to replace damaged or missing sheets of drawings or pages of specifications.

Bidding Period

The bidders require enough time to review and assimilate the great volume of documentary material, visit and examine the site, make all of their material and labor takeoffs, contact their sub bidders, coordinate their financial information, and prepare their proposals. They will also need time to work out the rationale of their construction sequences, procedures, and techniques and prepare a tentative proposed construction schedule. The architect should determine the amount of time reasonably needed by the bidders and set a date and time upon which bids will be due. Traditionally, and for practical reasons, bids should not be made due on a Monday or the day before or after any holiday. Afternoons are better than mornings. Bidding activities must compete with other procedures in the firm involving the same personnel. It is best to avoid having bids due on the same day as other important bid openings in the same community, as they will involve many of the same contractors, subcontractors, and suppliers.

Selecting the Contractor 87

If it becomes necessary to extend the bidding period, all bidders should be notified in writing at least 3 days before the original due date. In addition, an immediate telephone call to each bidder would be appreciated, as it provides additional notice.

Base Bid, Alternate Bids, and Unit Prices

Some owners need the flexibility of adding to or eliminating portions of the work from the base bid as a practical technique for meeting the project budget. The base bid will be the amount tendered to provide the basic specified project. Alternate bids are the sums to be added to or deducted from the base bid if certain specified work is added to or eliminated from the basic project.

Unit prices are quotations for adding or omitting specified units of certain materials, such as square feet of terrazzo flooring, cubic yards of mass excavating, or lineal feet of 4-inch diameter vitreous clay sewer pipe. Unit prices are needed to facilitate pricing of change orders, should they occur.

Alternate and unit pricing create a certain amount of confusion in the contractor's office, particularly when sub bids are being evaluated and bid proposals compiled at the last minute before bids are due. It is best not to require any more alternate bids or unit prices than are absolutely necessary to serve the owner's legitimate purposes.

Bid Bond and Proposal Form

Part of the process of keeping conditions fair and uniform for all bidders is the usual requirement that bids be submitted in a standard format. A blank form of proposal should be included in the bid package and bidders should be instructed to submit bids, filling in all blanks with typewriter or manually in ink. All bids should be submitted in an opaque sealed envelope marked on the outside with the name of the project and the bidder's name and address.

Also included in the sealed envelope, if required by the bidding documents, should be the bid security, which can be in the form of a cashier's check or a bid bond. The purpose of the security is to reimburse the owner for any losses suffered by the low bidder's refusal to enter into a contract on the terms stated in the bid and to post performance and payment bonds, if specified. Should the contractor default for any of these reasons, the bid security will be forfeited to the owner as liquidated damages, not as a penalty. The amount forfeited is limited to the difference between the low bid and the next responsible bid up to the face amount of the bid security.

The amount of the bid bond is usually a specified sum equal to approximately 10 percent of the contract budget. Sureties usually charge their principal, the contractor, nothing or a nominal annual service charge for bid bonds. However, inasmuch as they will have to furnish performance and payment bonds if their principal is low bidder, the credit and status checking of the contractor will be as thorough as if the construction bonds were being applied for. The bonding company will want to ascertain that their contractor would not become overextended or undercapitalized by obtaining new work. If a contractor is unable to obtain a bid bond, this is a good indication of that bidder's lack of financial capacity.

In the event that the bid bond becomes forfeited, the contractor will have to reimburse the surety for any funds paid out, as the contractor must indemnify the surety as a condition of obtaining the bid bond. Bid bonds should be written on AIA Document A310, Bid Bond.

A recent technological development is the submission of bids by telephonic facsimile transmission. If this is acceptable, it should be so stated in the bidding instructions and provision made for prior submission of bid bonds by mail or hand delivery.

Opening of Bids and Determining the Low Bidder

On private work, it is up to the owner to determine whether or not the bids are to be opened in the presence of the bidders. As a courtesy to the bidders and as a demonstration of the fairness and openness of the process, the bidders should be invited to witness the opening of the sealed bids. Prior to the opening of bids, the architect or owner should prepare a blank schedule on which to record and summarize the bids as opened. A copy of the filled in schedule should be provided to each of the bidders whether present at the opening or not. It is not customary to make any determinations or announcements at the opening. The architect should merely thank the bidders and inform them that the owner will take the bids under advisement. It is understood that the architect or owner will contact the successful contractor after the bids have been analyzed to determine which is the lowest acceptable bid.

The low bid is the combination of base bid and alternate bids which will be actually contracted for. Thus it is possible for the contractor with the lowest base bid to lose the job if the accepted alternates are not also sufficiently low to comprise the low combination.

Negotiations leading to an executed contract should be held with the low bidder only. The low bidder may be asked to submit additional prices for adding to or deducting from the work of the project to make the contract sum fit the budget. But if the scope of the project is radically or materially changed, all bids should be rejected and the project re-bid. The owner has the right to reject all bids and to waive irregularities or informalities in any bid. It would be unethical for the owner to use these rights as a pretext to overcome the fairness of the bidding procedure, to favor, or to discriminate against any particular contractor.

Should the low bidder discover that an arithmetical or clerical error has resulted in an excessively low bid, the contractor should be allowed to withdraw the bid. In this case, the next lowest bid becomes the low bid. If the error is not arithmetical nor clerical but one of judgment or misevaluation, the owner is not obligated to allow withdrawal of the bid. However, it is questionable judgment to have a contractor on the job who is facing a certain loss by proceeding with the contract. The contractor's incentive to recover the losses is in direct conflict with the owner's objective of obtaining a project of a specified quality. Many owners would find it contrary to good business principles to force a contractor into an unprofitable contract even when it has the legal right to do so.

Award of Contract

The form of construction contract which the low bidder will be required to enter into is stated in the bidding instructions as AIA Document A101, Standard Form of Agreement Between Owner and Contractor (where the basis of payment is a stipulated sum). (See Appendix B.) If some other payment method or form of agreement is contemplated, it should be stated in the supplementary instructions to bidders.

In *stipulated sum contract,* the contractor guarantees the total cost of the project. If the contractor is efficient and the project runs smoothly, the profit will be as anticipated in estimating or even better. Conversely, when inefficiency and obstacles are encountered, there will be less profit and possibly out-of-pocket losses. In either case, the owner's position remains unchanged.

In *cost-plus-fee contracts*, the contractor will be allowed to recover all direct costs along with a fee for services. The most important features of a cost-plus contract are competent definitions of "costs to be reimbursed" and "costs not to be reimbursed." The second category is presumably included in the contractor's fee. The contractor's total remuneration for profit, overhead, and services is in the sum stated as the contractor's fee. It would be unethical for the contractor to secretly derive any fees or profit from any of the subcontractors or suppliers.

The contractor's fee in a cost-plus contract can be based on an agreed percentage of the reimbursed costs or can be an agreed fixed fee. Most cost-plus contracts are limited in amount by a stipulated guaranteed maximum price stated in the contract. Any change orders issued during the construction will serve to raise or lower the guaranteed maximum price accordingly. An excellent standard form of cost-plus-fee contract is AIA Document A111. (See Appendix C.)

Considerations in the General Contractor's Decision To Bid or Not To Bid

Some do not realize how costly it is for a general contractor to prepare a responsible competitive bid for a construction contract. To submit a properly thought-out proposal will involve the time and efforts of a considerable portion of the typical contractor's staff. Bidding activities must be interwoven among the normal high priority operations of keeping the previously contracted work serviced and operating in the field.

Contractors will not normally elect to risk and possibly waste their funds on a bidding competition unless the conditions appear advantageous. Ordinarily, they will not participate in all available bidding opportunities but will selectively decide which to enter on the basis of numerous practical and subjective considerations.

Probability of Getting the Job. This is probably the most significant overall factor, all other considerations being secondary. If any one of the known conditions indicates that there is little likelihood of actually being awarded a contract even after committing considerable time, funds, and effort, the firm will extend polite excuses and let the competition squander their resources.

In evaluating a bidding opportunity and assessing its chances of obtaining a favorable contract, each contracting firm applies its own criteria. The factors of most interest to general contractors are those that have to do with fair bidding conditions, the owner's ability to obtain construction funds, and the owner's apparent willingness to proceed with the work.

Availability of the Firm. A seriously managed contracting firm will not normally be interested in bidding for more work if they are already excessively committed to bidding other projects. They could spread their facilities too thin with the result that insufficient consideration would be given to each project being bid. Errors in judgment and mathematics could easily creep in. Mistakes on the high side would merely lose the project and waste all the efforts already expended, while those on the low side could win an unwanted job at an excessively low price guaranteed to produce a loss.

Construction firms would be reluctant to take on new bidding assignments if the bid due date is coincidental with or too close to other bids they are already preparing.

During periods when the firm's capacity is fully committed in field work under construction, thought must be given to finding new work to begin as old work is gradually being completed. At these times, bidding will be limited to those projects that can be started as their work load starts tapering off. Work would not be bid if new construction must be commenced before the firm's field capacity is available.

Size of the Project. The physical size and dollar volume of the work will have to fit in with the firm's organizational structure, physical capacity, and financial limitations. Very large firms cannot make a profit on small jobs as their usually ponderous and highly developed infrastructure will be inefficient, whereas smaller personal service contractors may find them profitable. Conversely, large contractors are very efficient on extensive projects that are beyond the capacity of smaller competitors. Contractors who are allowed to bid work out of their size class are usually wasting their time or risking the possibility of obtaining work that will prove to be unprofitable.

Contractors of all sizes must fit their projected work load into their bonding capacity. Each firm will know the largest size of single project their bonding company will allow as well as the total value of all bonded projects they may undertake at one time. Bonding capacity is based on the firm's construction experience, character, personnel and equipment, office procedures, financial history, and financial resources. There is no point in taking on new bidding assignments if the firm's bonding limitations would preclude their accepting the work.

Location of the Project. Normally, work that is not near the contractor's main office will not be as easy as closer projects. Each firm has its own criteria as to its preferred range of operations. Some large firms can operate efficiently anywhere in their state or even the entire country if it is their size and type of work. Other equally large firms and most small firms may specialize in certain geographical areas. Contractors will many times restrict their bidding activities to those projects that are reasonably within their normal operating area. However, in difficult times contractors may have to expand their working range to obtain enough work.

Type of Project. The nature of the proposed project must fit into the contractor's experience and capabilities. Contractors who are highly experienced in hospitals and colleges are unlikely to attempt bidding competition with others who specialize in regional shopping centers, warehouses, or tract housing. They will tend to limit their bidding activities to the building types most familiar to their personnel and customary subcontractors.

Available Personnel. A serious limiting factor for contractors is availability of competent supervisory personnel. In particular, it is of extreme importance that the superintendent assigned to the job be capable of managing the work in the field. The superintendent must also be intimately acquainted with the firm's administrative procedures. It would be risky and time consuming to break in a new superintendent on an important new job. Consequently, contractors are reluctant to bid new work when a dependable, experienced superintendent is not readily available. The risk is that upon being awarded the contract, the firm would not be able to proceed promptly with field work in the most responsible and economical way.

Contractor's Capital Requirements. The amount of operating capital required to finance the proposed project between contract payments must be considered. For some projects, new equipment will have to be acquired. It would not be wise to stretch the firm's financial resources out too thinly, as it would risk collapse of the whole business and would seriously compromise the firm's bonding capacity. If new capital requirements, including interest charges, are beyond the firm's comfortable capacity, then the job should not be bid.

The Competitive Environment. Contractors are usually very interested in knowing the number of bidders involved in a competitive situation. Many would simply decline to bid if they deem the number of competitors to be excessive.

Prospective bidders would also be quite sensitive to the identity of the other bidders on the bid list. They would not willingly compete with certain other firms that appear to have any undue competitive advantage, such as a smaller overhead. There is also the perception, real or imagined, that some competing firms figure their jobs too close, do not allow for a proper profit and overhead markup, or do not provide appropriate contingency allowances. Some firms have the reputation among other contractors of bid shopping their subcontracts or other practices that would impair fair bidding competition. Inclusion of such contracting firms on a bid list can sometimes motivate reliable firms to decline the bidding invitation.

Reputation of the Architect. Contractors, when evaluating the advisability of bidding a project will often consider the perceived characteristics of the architectural firm and their consulting engineers. Some architects and engineers have built a fine reputation for excellent quality of contract documents and reasonable, fair contract administration. At the other extreme are design firms well known among contractors for ambiguous, incomplete, and impractical documentation.

A few architects and engineers are known for their one-sided, arbitrary, unreasonable, or heavy handed contract administration. The contractor has to decide whether or not to bid work emanating from these offices. If a contingency reserve is added into the bid to cover harsh architectural decisions, it might cause the bid to be overpriced and all of the bidding costs will be lost. Some contractors would just not submit a bid.

Construction Funding. Practical contractors reason that it is useless to expend resources to bid a job that cannot proceed by reason of the owner's inability to obtain the necessary funds. It is important in the contractor's decision to bid that there is a sincere belief that the owner is financially capable and that the owner actually intends to award a contract and proceed with the work.

A Fair Bidding Procedure. Considering the high cost of submitting a bid, most contractors want to be taken seriously and treated with consideration. They would

Selecting the Contractor

not knowingly participate in a bidding process that was unfairly administered. A fair process involves standard conditions for all bidders.

The normal expectations would be that all bidders are treated equally and provided the same information. The complete list of bidders should be made known to all bidders. The bidding documents should be made available to all bidders at substantially the same time and all bids should be due at the same time. All addenda and answers to all questions should be released to all bidders at the same time. The bidding period should be of sufficient duration to enable bidders to properly evaluate the job, depending on its type and size.

After bids have been submitted and opened, a summary of all proposals should be provided to each bidder. This is small recompense to the unsuccessful bidders. After return of bidding documents, complete and in good condition, the bidders' deposits should be immediately refunded.

Award to the Low Bidder. The implicit understanding is that the lowest responsible bidder will be awarded the contract at the price submitted. Without this understanding, contractors would be reluctant to participate in the bidding process at all. Why should they? They only bid to obtain the work, not for bidding practice. If contractors do not sincerely believe that the low bid will be seriously considered, they will not bid the job.

If a contract cannot be awarded at the bid price, then any further price negotiations should be restricted to the low bidder.

The Most Important Factors. Of all of the factors covered in the above headings, the final three (available construction funding, a fair bidding procedure, and award to the low bidder) would be most crucial and virtually non-negotiable with most experienced contractors. The other matters discussed could be subject to some measure of flexibility under specific circumstances.

Construction Industry Folklore - Unlikely Bidding Systems

From time to time, we keep hearing about innovative, creative, and unusual construction bidding systems. We hear that someone has devised an imaginative system that cures the supposed evil of the low bidder. Most of the stories of this type are doubtful at best.

One method that has made the rounds and resurfaces periodically is supposedly from a far-away place where the people are far more ingenious than we are. The system involves inviting a number of general contractors to submit proposals and then awarding the contract to the second lowest bidder. In all fairness and practicality, this intention would have to be announced in advance to the bidders. Can you imagine the confusion this would cause among the bidders? Contractors submit bids for the express purpose of obtaining work and not for the educational benefit of bidding experience. It costs a great deal of money, proportionate to the size of work, to prepare a competent and reasoned proposal. Normally, a bidder reasons that the way to get the job is simply to submit the lowest bid. Therefore, all efforts are directed toward assembling the lowest combination of subcontracts, elimination of all unnecessary costs, promotion of efficiency, and shortening of the construction period. By what rational logic could a contractor bid to be second? It is easy to know how to aim for the lowest position but how does one aim for second place?

Another variation of the same system is even more fascinating. In this version, the supposedly sophisticated owner or architect has decided to eliminate the lowest and highest bids on the basis that they must be considered out of range. Then the remaining bids are averaged and the bidder closest to the average is awarded the contract. This imaginative scheme would severely tax the resourcefulness of the most clever estimator in the quest for a successful bid price. This unusual bidding method keeps coming to our attention, but it is difficult to visualize American contractors participating in it.

Bizarre bidding systems will continue to be discussed and seem firmly ensconced in construction industry folklore.

The Low Bidder

Whenever severe contractor shortcomings come to light, some cynics will conclude that it is on account of the low bidder always getting the contract and that there is something inherently suspicious about a low bidder. This is dubious logic because substantially all contracts are awarded to low bidders, so obviously all contractors in distress would have been a low bidder. Outstanding projects are also built by the low bidder.

Often, the reasons contractors are in some sort of trouble could have been discovered by the reasonable review of qualifications prior to the onset of the bidding procedure. If inexperienced and undercapitalized potential low bidders are eliminated before bidding, then the eventual low bidder will be of a higher quality. There is no valid cause to eliminate a low bidder for any ground that could have been unearthed before contractors have incurred the significant expense of preparing their proposals.

After weighing these considerations, it seems safe to assume that the low bidder system is here to stay.

7

The Preconstruction Jobsite Conference

Commencement of the Construction Phase

Effective and amicable communications are essential to the smooth running construction process which culminates in a successful building project. Assume for the purpose of illustration that the contractor prequalification process is completed, the bidding procedures have been concluded, the low bidder has been determined, and the contract price has been established and is within the owner's budget. The owner and contractor have signed the agreement and have identified all of the other contract documents by signing them. So the physical execution of the project may begin.

What Is a Successful Building Project?

A construction project is a very complex organizational mechanism involving the disciplined acquisition, application, and coordination of such diverse resources as materials, equipment, trained personnel, and capital to precisely carry out the terms of a technical contract within the constraints of a rigorous time limit and a fixed budget. So, to be considered successful, a construction project must result in faithful execution of the architect's design as expressed in the contract documents, within the owner's cost and time expectations, and must yield a reasonable profit to each of the contractors and suppliers. In most cases, this is performed on a new building site utilizing an unfamiliar combination of subcontractors and suppliers that have never before worked together to produce a unique project from an untested set of contract documents.

This sounds like a nearly impossible goal to achieve with an acceptable degree of precision and efficiency. However, the construction industry under these conditions routinely produces thousands of successful projects each year.

Preconstruction Jobsite Conference: A Communication Tool

In the interests of improved communications, mutual cooperation with contractors, and a heightened awareness of the contract requirements, many architects and engineers regularly schedule a conference at the jobsite before commencement of construction. The best time for this meeting is after the contract is signed, all required insurance is in effect, and the surety bond, if any, is issued, but before the owner has authorized the construction to proceed.

It would be best to announce in the specifications the architect's intention to have this meeting, as most contractors will welcome it as an indication of impending cooperation and good will. The owner and contractor should both be invited to attend this initial site conference, and the contractor should be instructed to have the superintendent and representatives of major subcontractors and suppliers present.

Since this meeting will be the first construction phase visit for the architect, notes should be taken so that a written observation report can promptly be prepared and mailed to all participants and interested absent parties. It should cover all subjects discussed and decided. All questions or issues raised by attendees should be answered in the report.

The following items should be addressed at the preconstruction jobsite conference.

1. Errors, inconsistencies, or omissions in the contract documents which have been discovered by the contractor
2. Contractor's use of the site
3. Security provisions
4. Noise and dust control
5. Contractor's use of owner's water, power, telephone, and toilet facilities
6. Hours of operation
7. Contractor to receive property data from owner
8. Architect's explanation of dimensioning system
9. Contract documents
10. Architect's explanation of design intent
11. Specification substitutions
12. Ordering long lead items
13. Progress schedule

14. Submittal schedule
15. Shop drawings, samples, and product data
16. Weather delays
17. Jobsite record keeping
18. Communications
19. Architect's job visits
20. Construction methods
21. Safety procedures
22. Contractor's payment requests
23. Testing and inspections
24. Notice to proceed.

Errors in the Contract Documents

The contractor should be asked if any errors, inconsistencies, or omissions in the contract documents have been noted by the contractor and reminded that as any such irregularities are discovered, the architect should be promptly notified. (AIA Document A201, Subparagraph 3.2.1)

Matters Pertaining to Use of the Site

Discussions are for the purpose of clarifying the information and requirements of the contract documents. If anything is decided which changes or extends the contract requirements, the architect should be quick to point this out to the owner and contractor so an appropriate change order can be prepared. It is poor practice to allow changes in the contract requirements to go undiscussed, as it will usually result in misunderstanding and controversy. The owner will usually interpret silence as an indication that there will be no extra cost, while the contractor will assume it means that the cost will be negotiated later. Naturally, the roles and attitudes will be reversed when contract conditions are eased, which should result in a credit to the owner.

Some of the matters which should be discussed relate to the contractor's use of the site. During construction, the contractor will be in responsible charge of the site. Before construction starts the contractor should inform the owner how the site will be organized as to general location of (1) materials storage, (2) debris storage, (3) chemical toilets, (4) jobsite office, (5) workers' parking, (6) truck parking, (7) signs, and (8) temporary fences. The architect and owner are free to offer alternative suggestions for the contractor's consideration and to voice reasonable objections.

Security provisions and noise and dust control should be discussed, particularly in alteration and addition projects where the owner will continue to occupy some portion of the building or site. Consideration should be given to the effect of noise and dust on neighboring properties.

The conditions of the contractor's use of the owner's water, power, telephone, and toilet facilities should be discussed.

Hours of operation on the site should be discussed and reasonable limitations decided upon. If the owner was aware of the necessity of unusually restrictive working hours, this should have been specified in the supplementary conditions to enable the contractor to price the job accordingly.

The owner should provide the contractor with copies of the boundary survey, topographic survey, legal description of the property, and reports of the foundation investigation. The datum points of the land survey should be identified to the contractor. The horizontal and vertical dimensioning rationale of the contract drawings should be discussed and explained by the architect if deemed necessary by the architect or contractor.

Contract Documents

The architect should make certain that the contractor has sufficient copies of all of the contract documents so that all of the subcontractors can know the full extent of their contracted obligations and their relationships with related or adjoining trades. The owner is obligated by the AIA General Conditions (Subparagraph 2.2.5) to furnish, without cost to the contractor, such copies of the drawings and project manuals as are reasonably necessary for the execution of the work. The contractor should be reminded that a complete set of the initial drawings, specifications, and addenda should always be kept in good condition at the jobsite as well as a complete set of all change orders, construction change directives, and other modifications. The documents should be marked as changes happen to serve as a record of changes and selections made during construction. In addition, the contractor should maintain at the site a set of all approved shop drawings, product data, and samples. (A201, 3.11.1)

Design Intent

The architect should take this opportunity to briefly explain the design concept and objectives to the builders present. This is the time to point out and discuss the special details and features of significance and identify what is important. Models, perspective renderings, and other design presentation drawings, if available will help illustrate the design intent.

Specification Substitutions

Most specifications provide that all contractor-initiated substitutions be proposed during the bidding period so that all bidders can be on an equal footing. However, sometimes that proves to be somewhat idealistic if some normally available specified item is just not obtainable. The architect should suggest to the contractor that purchase orders be placed with all subcontractors and suppliers at once, so that unavailable specified items become quickly known. Also, all long lead orders should be placed immediately to assure that the job will not be delayed. The specification provision on substitutions and application of the phrases "or equal" or "equivalent" should be explained. Usually, they mean that the proposed alternate item should be comparable in function, capacity, quality, and appearance and approved by the architect prior to its utilization.

Progress Schedules

The construction progress schedule required by Paragraph 3.10 of the AIA General Conditions should already have been submitted by the contractor and should now be available for discussion. If any of the subcontractors present have any doubt about the practicality of the time schedule and their own ability to comply, they should now voice such reservations.

The submittal schedule should also be discussed and the contractors advised to commence production of shop drawings in time to meet the schedule. All product data and samples should be immediately prepared or accumulated, organized, and marked for submittal on time.

In discussing time scheduling, it would be appropriate to review the contract documents in respect to allowable extensions of time for a justifiable delay such as that caused by inclement weather and other events that cannot be anticipated or controlled. The AIA General Conditions, Subparagraphs 8.3.1 and 4.3.8.2 provide that the only justifiable adverse weather delay is when "weather conditions were abnormal for the period of time ...." The architect's proposed administrative application of the contract for contractor's time extension requests should be discussed to avoid later arguments.

Record Keeping on the Jobsite

Most contractors require their job superintendent to maintain a daily log on the jobsite and to record all relevant information on a current basis. The superintendent's log should be preserved as a permanent chronological record of all significant events occurring on the jobsite. In the event that the contractor on this project does not usually keep a daily log, the practice should be suggested and urged, although it cannot be absolutely required unless previously specified. (See Chapter 13, Role of the Construction Superintendent.)

Communications

Although the communication system among the parties during the construction period is defined in the AIA General Conditions, it is advisable to review the requirements so all may understand and abide by it in practice.

>All communications between the owner and contractor should be channeled through the architect.

>All communications by and with the architect's consultants should be through the architect.

>All communications to or from the subcontractors and suppliers should be through the contractor.

>All communications with separate contractors should be through the owner. (A201, 4.2.4)

>The superintendent on the jobsite is a representative of the contractor, and communications given to the superintendent are as binding as if given to the contractor. (A201, 3.9.1)

>All important communications should be in writing or if given orally should be confirmed in writing.

Architect's Job Visits

The architect's intentions should be expressed for the benefit of the contractor as to the scheduling of periodic jobsite visits. Some architects make it a practice to visit the site on a regular weekly, bi-weekly, or monthly basis. Others will plan to visit only once for each contractor's payment request or upon the completion of certain agreed stages of the work. The frequency will be based on the type and magnitude of the project, the architect's professional judgment, and the agreement between owner and architect. The contractor or superintendent should be present for all site visitations to explain conditions and operations to the architect and to receive pertinent instructions. The subcontractors who have questions or problems might plan to be present for relevant discussion and decisions.

Promptly after each site visit, the architect should prepare and issue a written observation report to the owner with copies to the contractor and all other interested parties. The report should include the date, time, duration, weather conditions, persons present, work now being accomplished, percentage of work completed by trade, work progress compared to schedule, work scheduled before next visit, questions raised by contractor or owner, determinations made by the architect, and any questions or actions which remain pending for appropriate future attention. (See Chapter 10, Site Observation and Administration of Construction.)

Construction Methods and Safety Procedures

The architect must be careful not to exceed the authority bestowed by the owner-architect agreement or the contract documents. It is the contractor's sole prerogative to determine, control, and be responsible for the construction means, methods, techniques, sequences, and procedures and for coordinating the work. (A201, 3.3.1) The architect is specifically excluded from authority in these matters. (A201, 4.2.3 and 4.2.7) The architect's viewing of the work during its execution is for the purpose of determining if the work that is being performed will be in accordance with the contract documents. (A201, 4.2.2) The architect's opinion on these matters should be expressed in the periodic written observation reports.

Safety precautions and accident prevention programs on the site are the responsibility of the contractor. (A201, Article 10) The superintendent is the contractor's representative with the specific duty of carrying out all programs for personal safety and the protection of property unless the contractor designates some other person in writing to the owner and architect. (A201, 10.2.6)

Contractor's Requests for Payment

The contractor should be reminded that periodic applications for payments should be only for the percentage of work which is in place prior to or on the date of the architect's site verification. The architect cannot authorize payment of funds for work not actually observed to be in existence. Payments for the value of work in progress in fabricators' shops or stored off site cannot be authorized by the architect unless the owner and contractor have previously agreed otherwise in writing. The architect can allow the value of materials or equipment suitably stored on the site even though it has not yet been incorporated into the work. (A201, 9.3.2)

The contractor and architect should agree on a procedure so the architect does not unexpectedly receive payment requests with the contractor's expectation of an immediate site visit and approval. The payment requests and architect's site visits could be entered in the contractor's construction schedule or submittal schedule.

The subcontractors present could be reminded that the architect, if requested, is authorized to inform them of the status of payments to the contractor in respect to the work of the subcontractors. (9.6.3) (See Chapter 18, The Architect's Relationship With Subcontractors)

Testing and Inspections

The contractor is responsible for paying for and making all necessary arrangements for specified materials testing and inspections, utilizing independent testing laboratories or entities acceptable to the owner. Any tests or inspection requirements imposed after bidding will be paid for by the owner. (A201, 13.5.1) The contractor should reveal the identity of the testing agencies for the owner's consideration.

A very high percentage of construction defects are associated with roofing installation. Accordingly, many owners deem it advisable to employ a separate independent roofing inspector to observe the roofing work continuously during its installation. If this is the owner's intention, the contractor should be informed so that appropriate scheduling can be provided.

Notice to Proceed

Construction on the jobsite should not "jump the gun" by commencing before all required insurance is activated, surety bond issued, and mortgage liens recorded. Should this happen, bonds and guarantees will have to be arranged for by the owner to satisfy the mortgage lenders, title guarantee companies, and surety. The events and operations expected to be covered by insurance or bonds would not have been covered. Therefore, the AIA General Conditions contains a procedure that, if followed, will prevent these serious problems. The contractor is not to start any work on the site whatsoever, prior to the receipt of a written notice to proceed given by the owner. The architect, if authorized by the owner, could issue the notice to proceed. In the event that, for whatever reason, the contractor has not received a written notice to proceed, the contractor is required to give a written notice to the owner no less than five days before commencing work. (8.2.2)

Private Meeting

An additional meeting is needed among the architect, owner, and contractor to discuss other contractual matters which are of no direct concern to the subcontractors and suppliers. This meeting could be held at the time of signing of the contract documents or at the beginning or end of the preconstruction jobsite conference. Various matters need to be discussed in the interest of avoiding future misunderstandings and mutual distrust.

The following items should be addressed at the private meeting:

1. Owner's financial arrangements
2. Unit prices and allowances
3. Contractor's overhead and profit

4. Owner's separate contractors
5. Preconstruction submittals
6. Owner's insurance
7. Contractor's insurance
8. Liquidated damages
9. Architect's decisions

Owner's Financial Capability

The contractor has the right prior to signing the agreement and from time to time during the construction to request the owner to promptly furnish reasonable evidence that adequate financial arrangements have been made to fulfill the owner's obligations under the contract. (A201, 2.2.1) Should the owner fail to promptly furnish the requested reasonable evidence, the contractor may exercise its option to refrain from executing the agreement or commencing the work or to stop the work and terminate the contract. (A201, 14.1.1.5) (See Chapter 20, Termination of the Construction Contract.)

Unit Prices and Allowances

The contractor and owner should be reminded that any unit prices quoted in the bid proposal or later negotiated before the contract was signed are subject to change under certain circumstances. If the quantities originally contemplated are so changed by a change order or construction change directive that the original unit prices become substantially inequitable to owner or contractor, the unit prices are equitably adjusted. (A201, 7.1.4) Such would be the case if too few units were actually required, in which case the contractor would be undercompensated. If there were an exceptionally large number of units it would be to the disadvantage of the owner. If the owner and contractor cannot agree on new unit prices, the architect will have to decide the matter, subject to appeal in arbitration.

Contractor's invoices for allowance items should be prepared using the general principles of Paragraph 3.8 of AIA Document A201. Arguments and misunderstandings will be alleviated if both parties are aware of these provisions for the billing of allowances. The owner should be reminded that selections of materials, equipment, and vendors for allowance items should be presented promptly to the contractor to avoid being the cause of a construction delay. All items for owner's selection should be entered in the contractor's construction progress schedule or submittal schedule. If the architect is to make these selections for the owner, then the architect should conform to the schedule.

Contractor's Overhead and Profit

According to Subparagraph 7.3.6 of the AIA General Conditions, the contractor is entitled to add a reasonable amount for overhead and profit when submitting prices for extra work. Each of the subcontractors will include their overhead and profit in quotations to the contractor for changes in their contracts. The architect is charged with determining what is reasonable. Future misunderstandings could be minimized if the contractor's expectations and the architect's opinions in this matter were discussed with the owner. The owner should realize that subcontractors normally use a different markup than general contractors. This discussion would not be necessary if the contractor's and subcontractors' overhead and profit percentages were specified by the architect or required to be submitted by the contractor as part of the bidding procedure.

Separate Contractors

It is the owner's prerogative under the contract to engage separate contractors to perform other parts of the work. The contractor is obligated to cooperate with the owner and separate contractors. It is the owner's responsibility to coordinate the separate contractors with the work of the contractor. (A201, Article 6) The owner should disclose to the contractor the likelihood and extent of separate contractors being on the premises during the work of the contractor, so that related problems and costs may be anticipated.

Preconstruction Submittals

After the contract has been awarded and before any work is commenced, the contractor should promptly submit to the owner, through the architect, all specified submittals including the following documents required by the AIA General Conditions:

> **List of subcontractors and suppliers proposed for each principal portion of the work.** The architect should respond in writing if the owner or architect, after due investigation, has reasonable objection to any proposed person or entity. Failure to promptly respond constitutes notice of no reasonable objection. (5.2.1) Should a subcontractor or supplier be eliminated, the contractor is required to submit an alternate not objectionable to the owner and architect. The contract sum must be adjusted, up or down, to account for any difference in subcontract bids. (5.2.3)
>
> **Certificates of all required insurance.** (11.1.3) The architect should refrain from offering any opinions as to the adequacy, terms, or suitability of insurance certificates or the underlying insurance policies.

These are matters which the owner should take up with insurance advisors and legal counsel.

Surety Bond. Bonds covering faithful performance of the contract and payment for labor and materials, each in the full amount of the contract, if required by the owner. (11.4) If the owner required the bond before bidding, the contract bid price will include the bond premium. If the bond requirement was imposed by the owner after bidding, the premium can be billed as an extra to the contract. The owner should submit the bond to its insurance advisor or legal counsel for review.

Schedule of Values. This is a detailed cost breakdown allocating values to each trade and division of the work to be used as a basis for contractor's payment requests. The contractor's overhead and profit may appear as a segregated line item or may be proportionally distributed among each of the work items. This should be submitted by the contractor to the architect for review before the first application for payment. (9.2 and 9.3)

Construction Progress Schedule, showing all construction operations and completion within the time limits of the contract. It should be revised as necessary during construction to reflect changes in conditions and should be kept current. The schedule and updated versions are to be submitted to the owner and architect for information only, not for their approval. (3.10.1)

Submittal Schedule. This is submitted for the architect's approval and should be in synchronization with the construction progress schedule. It will show when all submittals are to be forthcoming from the contractor and when the architect is expected to return them after reviewing. (3.10.2)

Owner's Insurance

The owner, if it wishes the coverage, must carry its own liability insurance. (A201, 11.2.1) The owner is required to carry property insurance to cover the perils of fire and extended coverage and physical loss or damage, boiler and machinery insurance, and loss of use insurance, including the interests of the owner, contractor, subcontractors, and sub-subcontractors in the work. The owner is required by the contract to furnish a copy of all required insurance policies to the contractor before any losses occur. Therefore, it must be done prior to start of construction. (A201, 11.3) The owner should be advised to review the entirety of Article 11, Insurance and Bonds, of AIA Document A201 with its insurance advisor or legal counsel. (See Chapter 9, Construction Insurance.)

Liquidated Damages

If the contract contains a liquidated damages provision, the contractor and owner will benefit from an open discussion of the matter. The length of time allocated for construction will in most cases have been established or accepted by the contractor and could be a slight bit optimistic. A prudent contractor will apply for all justifiable time extensions as the construction progresses to avoid the possibility of having to pay any liquidated damages to the owner. Conversely, the owner will often take a hard line on the granting of time extensions. The architect should make it clear to the parties that all decisions on this and other matters will be fair to both parties.

Architect's Decisions

In the event of even a minor dispute between owner and contractor during these initial conferences, the architect must make the final decision as provided in Paragraph 4.3 of the AIA General Conditions. The architect can give the parties a feeling of security in the architect's hands if all such decisions and contract interpretations are handled firmly, fairly, and skillfully. To start with, both sides must always be given the opportunity to fully express their positions. The architect must set aside all perceptions of pressure from either party and make all decisions in accordance with the written provisions of the contract. All interpretations and decisions must be fair, just, and equitable, siding with neither party. If the disagreement is based on ambiguity or errors in the contract documents produced by the architect, the decision is that much more difficult. Usually, the faulty documents must be construed against the owner in favor of the contractor. The economic consequences of the architectural or engineering error must be taken up separately between the architect and client.

If improved communications among architect, owner and contractor can be promoted by use of the preconstruction jobsite conference then it will be well worth the time and trouble.

8

Consultants and Advisors

Liability for Mistakes of Outside Consultants

There is no question that architects, as prime design contractors, bear the legal responsibility for professional malpractices committed by their consulting engineers and other technical consultants. Although they cannot completely eliminate this burden, architects can try to conduct their practices in a manner which will at least minimize exposure to this contingent liability.

Individually and as a profession, architects have spent decades trying to convince prospective clients that they should look to the architect as "captain of the ship" in determining the design and administering the construction on the site. Indeed, they have persuaded them that the architectural profession should decide and fashion all of the built environment. It logically follows then, that the legal consequence of centralization of the design prerogative is that clients need look no further than their architects for affixing responsibility for design errors and omissions.

It has been made increasingly and painfully aware that architects can be held financially responsible for their our own professional shortcomings. One way to control, but probably not eliminate, the exposure to risk of malpractice claims is to conduct their practices as carefully, diligently, and skillfully as they are able, thereby limiting the circumstances for error.

Limiting Exposure to Liability

The use of well-worded professional service contracts such as those provided by the American Institute of Architects is the first step in limiting liability exposure. In addition, it is logical to avoid situations which are conducive to professional negligence. The objective is to create a positive climate in which a higher quality of professional output is likely, and accordingly there should be less possibility for error. The most elementary possibilities for consideration are:

> Adequate compensation should be required for all professional assignments. If the fee is not sufficient to cover the time necessary to perform all of the contracted services competently and completely, the chances of making mistakes are greatly increased.
>
> All assignments should be properly analyzed, designed, and produced. All procedures and work product should be carefully monitored and checked. Slighting any of the required services on account of limitation or depletion of available fee is not a legally or ethically acceptable excuse for incomplete or hastily done work.
>
> Assignments beyond the technological expertise of the available personnel should not be accepted. Various firms have specialized in particular building types, thereby raising the required standards for anyone performing those types of work. Avoid misleading portrayals of your firm's experience or abilities when making representations to prospective clients.
>
> Employ only competent personnel. Utilize only skilled and qualified persons on each assignment. They should be appropriately supervised, commensurate with their experience and ability. All assignments carried out by interns and recent graduates should be closely supervised and checked by fully qualified persons.
>
> All assignments should be approached skillfully and competently, exercising judgment and taste to the standard of care of architects practicing in the same community. In order to know what the standard is, it is necessary to pursue continuing education both as teacher and student, read the professional press, associate with other architects, and observe other architects' work product as well as their work. The standard of care requires that all professional activity must be pursued with due diligence and reasonable skill.
>
> Novel and adventurous design solutions should be undertaken only after frank and open disclosure to the client of all the attendant risks

and possible disadvantages as well as the advantages. The client's informed approval should be obtained in writing. Disclosures should include the intended specification of materials and equipment which have undergone only limited field experience.

All outside consultants should be carefully selected to assure that they possess the requisite technical competence and appropriate experience.

Limiting Liability for Consultants' Shortcomings

No matter how perfectly one is able to raise one's own professional level of performance, being only human, one can still make mistakes. Added to this there still remains the risk of liability for errors made by consultants. This underscores the necessity for cautious and discriminating choice of outside consultants. The search for competent engineers and other consultants who are suitable and acceptable as design collaborators and technicians becomes a very personal problem. Mutual trust and respect must ever be present in the relationship. Working with the same expert advisors over years and numerous projects will gradually result in the development of mutually accepted understandings and safe procedures.

Although most contracts between architects and their consultants consist merely of a brief memo or an oral discussion, there appears to be sufficient mutual understanding between the parties to get them through most of the demands of the relationship. A friendly give and take attitude and professional courtesy will help maintain a climate of fairness and equity between the parties.

When a minor misunderstanding or lapse in communication occurs, there is little or no argument over the resultant redesign or redrawing required or fixing of blame. Even when larger sums of money are involved, polite negotiation will usually settle the matter based on past understandings, mutually accepted operating principles, and the expectation of a continuing relationship.

However, in the event of sizable or catastrophic errors or omissions which involve large-scale economic liability, the controversy between the architect and consultant will no longer be a quiet matter to be decided between friendly and respectful colleagues. It will instead be decided by insurance carriers and by judges or arbitrators after lengthy and often heated disputes involving third parties. In this unfortunate situation, the consultant relationship would have been better defined if it had been exactly spelled out, as it would be in a complete written contract.

AIA Agreement Forms. The American Institute of Architects has over the years developed a series of standard form contracts which can be used to formalize the agreements between architects and their consultants. The AIA agreements cover the appropriate subjects peculiar to the architect-consultant relationship and establish the mutual rights and responsibilities of the parties. AIA architect-consultant agreements are in harmony with the AIA owner-architect agreements and the AIA General Conditions which is part of the usual construction contract. All of the AIA standard form agreements have uniform terminology and procedures and meshing arbitration clauses.

The most widely used of the AIA architect-consultant contracts is AIA Document C141, Standard Form of Agreement Between Architect and Consultant, 1987 Edition. This newly-revised agreement is suitable for use when the prime owner-architect agreement is AIA Document B141, 1987 Edition. The AIA publishes three other variations of the architect-consultant agreement to accommodate situations where other versions of the prime owner-architect agreement are used. (See descriptions of the C-Series of documents in Appendix G.)

Fair Liability Apportionment. The use of a written contract with consultants will not necessarily minimize or eliminate liability but will at least apportion unavoidable liability fairly, so that the parties will generally be liable only for their own errors and omissions. Although architects must answer to their clients for consultants' shortcomings, they have legal recourse to their consultants for indemnity. This recourse is only as effective as the indemnitor's financial ability to fund it. This can be overcome if consultants carry professional liability insurance.

Owners' Versus Architects' Consultants and Advisors

The principal area for minimization and possibly virtual elimination of contingent liability for certain consultants' malpractice is in prevailing upon the owner to engage certain consultants and advisors. Those who provide information about the owner's property or advice to the owner, as contrasted to those who assist the architect in design, should be employed directly by the owner. This includes land surveyors, programmers, geotechnical consultants, cost estimators, financial advisors, real estate feasibility advisors, marketing experts, insurance advisors, accountants, economists, lawyers, and construction schedulers.

Those consultants who jointly participate in the design process with the architect should be engaged by the architect. They should be under the direct control and direction of the architect. Primarily, this includes the normal engineering disciplines such as civil, structural, electrical, and mechanical engineers. In addition, landscape architects, graphic and interior designers, kitchen consultants, construction specifiers, and all others who participate in the architect's design and production of the contract documents should be hired by the architect. It is the architect's responsibility to coordinate the work of the consultants who participate in the design. All of the consultants' construction drawings and specifications must be

Consultants and Advisors

coordination-checked to make certain that the work of each discipline is properly interfaced, mutually compatible, and not in physical conflict.

Consultants such as vertical transportation, commercial kitchen designers, acoustical engineers, environmental impact consultants, hazardous substances consultants, inspectors, and materials testing laboratories, if engaged primarily in furnishing information solely on existing conditions, design criteria, or construction testing, should be employed by the owner. If these specialized consultants are to participate in the design function, they could be hired by the architect.

Consultants' work should be properly identified as to authorship by use of the usual title block credit. In the event of an error in the consultant's work, it is important that it be easily distinguishable from the architect's work or that of other consultants. If this system of assigning consultants is adopted, the architect can at least minimize, if not avoid, responsibility for mistakes made by the owner's advisors and information providers.

Owners often do not wish to get involved in the arcane particulars of selecting, negotiating, and hiring of technical consultants and will request the architect to administer these details. In such cases, the architect should hire these consultants in the owner's name or have the owner sign the agreements. The architect should make certain that the owner receives copies of all agreements and correspondence to and from these consultants along with copies of all reports and other works. All owner's consultants should be instructed to direct all invoices to their client, the owner.

If any portion of the work product of the owner's consultants is to be included in the contract documents produced by the architect, it should be presented with full authorship credit. No misleading impression should be given that it is part of the architect's work product.

9

Construction Insurance

Financial Responsibility for Accidental Losses

Unfortunate accidental occurrences causing severe property damage, personal injury, or death on a construction site could easily bring economic ruin to those who are financially responsible. It is customary to fund this major liability by means of insurance. Owners and contractors who enter into construction contracts owe each other the obligation of providing insurance to cover certain specified risks.

Insurance is an extremely important element of the construction contract, so it is not surprising that over 10 percent of the text of the AIA General Conditions is devoted to this complex subject.

AIA General Conditions

The AIA General Conditions of the Contract for Construction, Document A201, was extensively revised in 1987 (Fourteenth Edition). The insurance portions were carefully considered, written, and edited by construction industry legal and insurance experts in consultation with professionals outstanding in the field of construction administration and documentation. Construction professionals are usually not expected to be insurance experts but need to know how to cope with the insurance aspects of construction contracts and their administration.

The AIA General Conditions specifies the insurance requirements in broad general terms, while the Supplementary Conditions written by the architect must specify the specific insurance coverage required for the project, the interests to be insured, the policy limits, the perils to be insured, the insurance contract term and the deductible amount.

It is a very risky practice for architects to provide insurance advice to their clients. Many professional liability insurance policies carried by architects specifically exclude coverage for the furnishing of insurance advice. The architect should firmly but respectfully decline to answer any insurance questions. Should an owner rely on an architect's erroneous insurance or surety advice to its detriment, the professional would likely be held liable for the resultant damages.

Owner's Instructions Regarding Insurance and Bonds

The AIA Owner-Architect Agreement, Document B141, also revised in 1987, provides in Paragraph 4.8 that the owner will furnish all legal, accounting, and insurance services as may be necessary at any time for the project. Therefore, the architect should obtain the insurance requirements directly from the owner. Owners are not normally insurance experts either, so they will have to confer with their own legal and insurance advisors. There is a convenient form devised by the AIA entitled Owner's Instructions Regarding the Construction Contract, Insurance and Bonds, and Bidding Procedures, Document G612, 1987 Edition, which architects can present to their clients. (See Appendix F) It is in the format of a nine page questionnaire in three parts, A, B, and C. Part B, Owner's Instructions for Insurance and Bonds, pages 4 through 7, can be used by the owner's insurance advisor to instruct the architect in respect to the owner's insurance and bond requirements. The specifier can then rely on these instructions when writing the insurance specification in the Supplementary Conditions. (See Chapter 27, Obtaining the Owner's Instructions.)

Certificates of Insurance

Evidence of insurance specified to be carried by the owner and contractor must be provided as proof of its existence and its terms and conditions. Certificates of insurance are commonly issued free of charge by insurance carriers, agents, and brokers when requested by their insured. Certificates of the contractor's insurance should be addressed to the owner, while certificates of the owner's insurance should be addressed to the contractor. In the interest of promoting uniformity in construction industry documentation and administration, the AIA has issued a standard form of Certificate of Insurance, Document G705, which many construction insurers have adopted. This document may soon drop out of common use, as a new insurance industry form is taking its place and is now recommended by the AIA. The Agency Company Organization for Research and Development issues

Construction Insurance

the new form, designated as ACORD 25-S (3/88). One of the most important aspects of an insurance certificate is the statement that the policies will not be canceled or allowed to expire unless 30 days' written notice has been given to the addressee of the certificate.

According to the AIA General Conditions, Subparagraph 4.2.4, all communications between owner and contractor shall be through the architect. Consequently, the insurance certificates from each party will flow through the architect's administration to the opposite party. When the architect transmits insurance certificates or surety bonds, the covering letter should merely inventory the enclosures, but should not comment on the sufficiency of the carrier or the surety or the adequacy or conditions of the coverage. The architect should advise the client to seek legal or insurance advice when appraising the insurance coverage underlying the certificates.

All specified insurance and bonds must be in force before any construction is undertaken at the jobsite. Once construction operations are under way, insurance companies and sureties are reluctant to issue insurance coverage or bonds.

10

Site Observation and Administration of Construction

Construction Observation: How Much is Enough?

The owners of a project under construction always seem to find time, even if it's not convenient, to make frequent visits to their construction site. Seeing their visions materialize is currently the most exhilarating and fascinating event in their lives. They do not always understand why their architect appears to be less absorbed in the physical evolvement of their undertaking than they are. The usual reason for this common misunderstanding is that their architect hasn't made it clear what they should expect during the construction period. Perhaps they would be more understanding if they were fully aware of all the behind-the-scenes activities of the architect while the construction progresses in the field.

Residential homeowners as clients are generally more subjective and emotional than clients for commercial, industrial, or institutional projects. However, all clients are similar in at least one respect: their construction undertakings are very important to them. Their projects represent major commitments of funds, proportional to each client's economic situation. All clients need to be cultivated, encouraged, respected, and kept informed.

Architectural Service Agreements

The standard architectural services contract describes the basic services to be performed by the architect throughout each of the phases of professional service. The most recently revised contract is AIA Document B141, Owner-Architect Agreement, Fourteenth Edition, 1987. (See Appendix A.) This agreement is harmonious with the AIA General Conditions, which is a part of the construction contract.

Also available from the AIA are numerous other owner-architect agreement forms. They each have their own descriptions of the architect's construction period services.

Unless otherwise indicated, paragraph numbers given parenthetically in this chapter refer specifically to AIA Document B141, although the commentary applies equally to architectural services provided under any of the other contract variations.

Assume that the construction contract is based on A101 or A111 and includes the General Conditions of the Contract for Construction, Fourteenth Edition, AIA Document A201, 1987. (See Chapter 3, Architectural Services Agreements.)

Construction Phase

According to the agreement (2.6.1), the construction phase commences with the award of the construction contract. Whether the architect is being paid on the basis of a lump sum fee or a percentage of the cost of construction, approximately 80 to 85 percent of the total architectural fee will have been expended for all of the services performed prior to start of construction. This leaves approximately 15 to 20 percent of the fee available to cover construction phase services.

Even with hourly rate contracts, knowledgeable owners are likely to become disgruntled if fee billings during the construction phase appreciably exceed 15 to 20 percent of the total architectural fee. Architectural practices are usually conducted in a rational, businesslike manner and with anticipation of a reasonable profit. The normal expectation is that preconstruction phase services will not consume more than their allotted proportion of the fee and that the 15 to 20 percent of gross fee reserved for construction phase services will be available. All too often, the early phases of schematic design and design development consume more time than is allocated. The construction documents phase is not only complicated but voluminous and not highly conducive to shortcuts or other timesaving methods. The bidding or negotiation phase will sometimes consume considerably more time than was expected, particularly when the lowest bid is in excess of the owner's budget. In this case, if a fixed limit of construction cost has been agreed upon, the architect at its own expense will have to modify the construction documents as required to comply with the fixed limit. (5.2.5) This will also entail additional dealings with

Site Observation and Administration of Construction

the owner to determine acceptable economic alternatives and with contractors to obtain the cost differences and to finalize the construction agreement.

The honorable architect will do all that is required by the contract during the final phase even if it means only breaking even or, worse, suffering an economic loss. The construction phase ends when the architect issues the final certificate for payment or 60 days after the date of substantial completion, whichever occurs first. (2.6.1)

The construction phase therefore could last up to two months longer than expected. The normal process of constructing an uncomplicated residence, if it goes smoothly, would take a minimum of 5 to 7 months and with some complications could require 10 to 12 months. Difficult or complex projects or those in distress will take longer. A hospital estimated to require 18 months for construction might unexpectedly take 24 to 30 months or more. A simple 5 month warehouse project might run an unanticipated 7 or 10 months.

The difficulty in estimating the actual length of the construction time complicates the process of realistically budgeting the architectural services fee during the administrative period.

Construction Phase Services

Advice and Consultation. The services which the architect must provide during the construction phase are varied and numerous. The architect must always be available for advice and consultation with the owner. (2.6.4) The contract does not impose any limitation on the amount of time which might be required for this sensitive and significant duty. Some clients require considerable personal attention, explanation, and reassurance during the construction period. This time-consuming activity can rarely be delegated to less experienced members of the staff.

Site Visitation. Visiting the construction site, as one of the architect's most conspicuous activities, has a great potential for causing misunderstanding with clients. According to the architectural contract, the architect must visit the site "at intervals appropriate to the stage of construction...." (2.6.5) This means that the architect must exercise professional judgment in determining frequency and timing of site visits. The architect should be present to observe work or events of major impact on the structural or design integrity of the final result. Some parts of the work need to be examined during their execution and before they are covered by subsequent operations, while other components need only to be observed upon their completion. The contract further explains that the site visit is for the architect "to become generally familiar with the progress and quality of the Work completed and to determine in general if the Work is being performed in a manner indicating that the Work when completed will be in accordance with the Contract Documents." The architect is required to keep the owner informed of the progress and quality of the work. (2.6.5)

Reporting to the Owner. Keeping the owner informed of construction progress and quality is best accomplished by a system of regularly issued written reports. This also serves to keep the client aware of the extent of services being performed on its behalf by the architect. It will also impart to the owner the sense that the project is being properly monitored and is not out of control.

The agreement also says that the architect will "endeavor to guard the owner against defects and deficiencies in the Work." This is not a guarantee or assurance by the architect but is a promise to exercise skillful and informed observation of the contractor's work with the expectation of detecting and preventing noncompliance with requirements of the contract documents.

Providing Communication Channel. The owner and contractor are required by their construction agreement to channel all of their communications to each other through the architect. (General Conditions of the Contract for Construction, Fourteenth Edition, AIA Document A201, 1987, Subparagraph 4.2.4) The administrative burden generated by this provision consumes time throughout the construction period.

Certifying Contractor's Payments. All payment applications made by the contractor must be reviewed, correlated with construction progress, and certified by the architect. (2.6.9) No payment certification should be issued unless based upon a concurrent site observation visit.

The architect's certificate for payment is a representation to the owner that the architect has personally conducted on-site observations, that the work has actually progressed to the point indicated, and that the work is in accordance with requirements of the contract documents, to the best of the architect's knowledge, information, and belief. (2.6.10)

Reviewing Submittals. During construction, the architect is also obligated to receive, review, and act upon the contractor's submittals of specified shop drawings, samples, and product data. (2.6.12) This necessitates the operation of a methodical record-keeping system to keep track of the status of all submittals received from the contractor, indicating the dates on which they were received, referred to consultants, or returned to the contractor for correction, rejected, or approved. A standard form published for this purpose is AIA Document G712, Shop Drawing and Sample Record. The contract documents should be carefully reviewed to make certain that all specified submittals have been received from the contractor.

The architect must also receive and act upon construction and submittal schedules, certificates of required insurance, and warranties tendered by the contractor. When

Site Observation and Administration of Construction   121

sending along the contractor's insurance certificates and warranties to the owner, the architect should avoid giving any opinions as to their adequacy. These would be considered technical insurance and legal matters beyond the purview of normal architectural competence. Should the owner rely upon erroneous opinions to its detriment, the architect would have little if any legal defense available. If advice on these subjects is needed, the owner should be advised to consult with its own insurance or legal counsel.

Processing Change Orders. Construction changes required by the owner or contractor or caused by unexpected conditions will result in change orders or construction change directives which must be prepared by and further administered by the architect. (2.6.13) (See Chapter 14, Change Orders.)

Resolving Disputes. All matters in contention between the owner and contractor are to be submitted to the architect for consideration, interpretation, and decision. (2.6.15, 2.6.18, and 2.6.19) All interpretations should be based on careful study of the contract documents and consideration of all relevant factors, and the decision should be in writing. (2.6.16) (See Chapter 22, Resolution of Construction Disputes.)

Considering the wide range of time-consuming duties contractually assumed by the architect to be performed during the construction period, it is usually difficult to keep the time expenditure within the fee allocated to this phase. None of these activities can be safely slighted, least of all the time which must be spent at the construction site. The architect's exhaustion of the fee reserve is not a legally or ethically acceptable excuse for failure to perform the contracted duties completely and competently.

Budgeting the Fee for Construction Phase Services

The fee available for construction phase services should be budgeted as realistically as possible, by predicting the time needed each week to perform all of the known duties and activities which have been contracted.

When the fee is determined as a percentage of construction cost, projects of lesser cost will not usually generate sufficient fee to cover all of the services promised in the AIA Owner-Architect agreement. Architects in such cases have only two options: increase the fee or eliminate some of the services. Either option requires concurrence of the client and must be discussed and agreed upon when the contract is originally negotiated. It is too late to address these matters months later during the construction period.

Budgeting the fee is primarily a function of the length of construction time. If the construction becomes unpredictably protracted, the architectural time budget will be irretrievably compromised. Partial economic relief is offered by the agreement in subparagraph 11.5.1 in which a time period (in months) for performance of Basic Services is to be inserted. Any services provided after expiration of the agreed time span would then be compensated on the same basis as for additional services. (10.3.3 and 11.3.2) The value of this provision will be realized only when the time period inserted proves to be realistic, not so short as to cause client resistance nor so long as to be ineffective. Even with this possibility of billing for services provided during prolongation of construction time, some clients will not take kindly to additional architectural cost. The architect must then make a business decision which could risk the continuing good will of the client.

Construction Observation

The architect's site visits should be regularly scheduled or at least announced in advance so the contractor is represented at the site as well as the owner, if desired. The architect should keep notes of visual observations, relevant comments offered, and oral directives given by any of the parties present.

Architect's Field Report. The architectural agreement indicates that the purpose of the architect's site visit is to become "generally familiar" with the progress and quality of the work and to determine if the work is being done in a manner which will yield results consistent with the contract documents. The agreement further explains that the architect is not required to make exhaustive or continuous on-site inspections. The architect is required, however, to keep the owner informed of the progress and quality of the work. (2.6.5) Most architects discharge this continuing obligation by sending the owner a written report of each site visit with a copy sent to the contractor. The report should be a complete record of the proceedings at the architect's jobsite visit and should include:

1. Date
2. Time
3. Duration
4. Weather conditions
5. Persons present
6. Percentage of work completed by trade
7. Work progress compared to schedule
8. Work now being accomplished
9. Work scheduled before next visit
10. Questions raised by contractor or owner
11. Determinations made by the architect
12. Any questions or actions which remain pending for appropriate later attention.

Site Observation and Administration of Construction   123

All observations and comments in the reports should be true and complete to the best of the architect's knowledge and belief. Many architectural firms have forms which they have designed for this purpose. The AIA has a standard form (Architect's Field Report, Document AIA Document G711) which is quite suitable and is widely used. The report should be written and issued promptly, preferably on the day of the visit, and certainly no later than 1 or 2 days following.

Pending Claims. On the occasion of each site visit, the architect should specifically inquire of the contractor or the superintendent if there are any pending or unreported claims under consideration. The question and answer should be recorded in the observation report. This will serve the purpose of bringing all claims promptly to the forefront, where they can be immediately dealt with and quickly resolved, thereby preventing the accumulation of unresolved claims. It will also help to prevent old issues from being resurrected at a later time.

Architect's Versus Contractor's Field Function. The architect's observation in the field is in the nature of a periodic examination or viewing of the work in process or completed as contrasted to the contractor's continuous daily superintendence and supervision of the trade workers and artisans involved in the day-to-day execution of the work. The architect should be observing and evaluating, whereas the contractor is controlling and directing the work.

The term inspection is used only twice in the architectural agreement, to describe the architect's on-site observations for determining the date or dates of substantial completion and the date of final completion. (2.6.14)

Architect's Personnel in the Field

Recent architectural graduates, trainees, or interns are assigned office tasks in design, drafting, shop drawing review, specification writing, and construction administration to fill out the architectural teams. This is important for the professional development of the individuals involved as well as for the advancement and propagation of the profession. These activities are performed under the direct supervision of experienced architects and are checked by them. Senior personnel are always available in the office for consultation and guidance of the less experienced. However, in the case of site observation duties, it is not appropriate to assign inexperienced or unqualified staff members unless accompanied by and under the close supervision of fully qualified and experienced architects.

Record Keeping During Construction Administration

The architect's basic record keeping during the construction period will primarily consist of keeping track of documentation, accounting for contract funds and time, and recording communications among the affected parties. These actions should result in a memorialization of all significant transactions growing out of the contract and the construction process. The following items should be a part of the architect's construction administration record keeping:

1. Field Observation Reports; follow-ups on all unresolved items
2. Shop drawings, samples, product data: checklist of all required submissions
3. Submittals; checklist of all required submittals
4. Correspondence
5. Payment requests, schedules of contract sum and time, change orders, construction change directives, bidding alternates, and reconciliation of cash allowances
6. Change orders and construction change directives
7. Drawings and revisions
8. Specifications and revisions
9. Status of requests for information and requests for quotations
10. Memos on delays caused by weather, strikes, unavailability of materials, and other causes

Keeping the Owner Informed

Construction Period Conditions. Architects' exposure to professional liability claims is at its peak during the construction period. This is understandable considering the conditions that usually prevail. The contract documents, particularly the drawings and specifications, will be tested and often found to be imperfect. The contractor's performance is not always in strict conformance with the contract requirements for workmanship and materials. The owner is often under great financial and emotional pressures due to the magnitude of the undertaking and the uncertainty of the construction environment. The owner is relying on the architect and contractor to deliver the specified project for the contract price and no later than the projected completion date. Many owners do not have the reserve financial capacity to undertake any extra costs nor the flexibility to accept late completion.

This is the stage of the architect's involvement with the owner where there is the greatest possibility of a major misunderstanding which could escalate to an irreconcilable difference. During the previous phases of architectural service, the architect and owner would have worked together very closely and without much outside interference. However, previously established rapport could be seriously undermined when the owner receives conflicting information from the contractor, workers on the jobsite, and well-meaning but uninformed friends. Confidence and security in the architect may dwindle.

After the construction contract is signed the owner will not always be completely aware of all of the architect's numerous activities being performed on the owner's behalf unless the architect makes them known. The architect should send copies to the client of all correspondence with contractors, testing laboratories, consultants, utility companies, and governmental agencies. The owner should also be regularly receiving the field reports, insurance certificates, bonds, schedules, warranties, interpretations and decisions, change orders, construction change directives, certificates for payment, certificate of substantial completion, and certificate for final payment.

Another possible source for owner dissatisfaction during the construction phase is the architect's obligation to render fair decisions and interpretations on contentions of the owner or contractor. The architect should be meticulous in obtaining the viewpoints of both parties before deciding such issues, and should be open, fair, and prompt in all decision making. Some owners feel that their architect should side with them against the contractor even when their position cannot be justified. This could lead to a client/architect problem.

The architect administering the contract in behalf of the owner can go a long way toward controlling and calming the situation by competent contract administration.

Advising and Informing the Owner. One of the architect's most important duties, required by B141 and promised by A201, is keeping the owner informed. It is an owner's basic right to know what is going on. Using almost identical language, the two documents both provide that the architect shall advise the owner during construction. (B141, 2.6.4. and A201, 4.2.1) This is a general requirement for the architect to keep the owner informed, to answer any questions raised by the owner, to provide counsel on the advantages and disadvantages of various options and courses of action, and to make appropriate recommendations. The architect is thus in a position of trust and confidence and is clearly serving in a fiduciary capacity. The owner proceeds in reliance on the architect's knowledge, technical skill, and honesty of purpose.

The architectural and construction contracts also require, "On the basis of on-site observations as an Architect, the Architect shall keep the owner informed of the progress and quality of the work...." (B141, 2.6.5 and A201, 4.2.2)

Neither the owner-architect agreement nor the general conditions contain any guidance or details as to how the advisory and informative functions are to be accomplished. However, architects over time have devised practical procedures for carrying out these responsibilities.

Certificate for Contractor's Payment. The architect's issuance of a certification for the contractor's payment is a form of notification to the owner "... that the work has progressed to the point indicated and that, to the best of the Architect's knowledge, information and belief, quality of the Work is in accordance with the Contract Documents." (B141, 2.6.10 and A201, 9.4.2)

Architect's Field Report. The architect's periodic site visitation should always be reported to the owner by means of a promptly issued written report.

Although the field report is addressed to the owner, copies should be sent to the contractor and all other interested parties. The primary purpose of the report is to keep the owner informed of the progress and quality of the construction. Legitimate secondary purposes are to instruct and inform the contractor and to create a credible record of on-site activities, inquiries, decisions, and actions.

All observations and comments in the field report should be candid and complete, devoid of all editorial bias. The report should not be used as a means of ordering changes to the contract work or time. If additional work is to be required or authorized, the established change order procedure should be followed.

The field report is a good place to record any pending changes which might affect the construction so that contractors can avoid work that would have to be altered or removed when the proposed change is ultimately authorized.

At each site visit, the architect should specifically inquire of the contractor if any claims are pending and, if so, they should be noted in the field report. If not, this should also be noted. This will serve to bring to light all pending issues involving money or contract time extensions and thereby prevent insidious accumulations of suppressed dissatisfactions or claims which might surface at a later time. As a valid general principle for writing field reports, all relevant information which comes to the architect's attention should be reported.

Architect's Behind the Scenes Activities. The owner is not usually aware of many of the architect's office activities that are being performed in the owner's behalf during the construction period. For example, many hours are spent in the shop drawing review process. Shop drawings must be recorded when received, sent to various engineering consultants for review, compared with the contract drawings and specifications, and a record made of approvals and conditions of rejection. Many owners would have no knowledge of the necessity nor even the existence of shop drawings if not so informed by the architect. Copies of all of the letters transmitting shop drawings should be sent to the owner. Copies of all correspondence to contractors, engineers, governmental agencies, and others should also be sent to the owner.

Architect's Fiduciary Duty

As the owner's agent, the architect must not withhold any material information from the owner. Some architects have felt that they should shelter the owner from the mundane and technical details or from unpleasant situations involving the contractor, the jobsite, or errors in the contract documents. However, this is potentially a very dangerous practice, no matter how well intentioned, as it may later be construed as a violation of fiduciary duty.

The architect is the channel for communications between the owner and the contractor as provided in A201, 4.2.4. Therefore, the owner and contractor are both relying on the architect to keep the lines of communication open and to pass on all relevant information to each party. It would be a serious breach of duty for an architect to withhold or delay information from either party by design or through inadvertence.

In his book, *Architects & Engineers*, Third Edition, James Acret states on page 70:

> Silence may constitute fraud if one has a duty to disclose.
>
> When a fiduciary relationship exists between the parties, mere silence may constitute fraudulent concealment.

The architect's multiple positions as the owner's agent, administrator of the owner's contract, and quasi-judicial arbitrator of the owner's and contractor's differences, clearly impose a fiduciary responsibility on the architect. So, in addition to violating the contractual requirement to keep the owner informed, there is the distinct possibility of committing fraud when withholding information relevant to the owner's contract.

The architect's position as the owner's agent and fiduciary during the construction period should not be confused with the lawyer's position of advocacy. For the architect to assume an aggressive owner's advocacy position in dealing with the contractor would be a perversion of the architect's required quasi-judicial position. The architect must treat the owner and contractor fairly and impartially, siding with neither, while administering their construction contract. Should the owner require or desire the services of an advocate, an experienced construction lawyer would be the better choice.

Advantage of Effective Communications

Concomitant with the architect's procedures for keeping the owner aware of all that transpires relative to administration of the construction contract, the architect should strive for an overt two way communication with the owner client. This is one of the most effective ways to maintain a mutually satisfactory architect client relationship and thereby limit the possibility of a client originated professional liability claim arising out of construction phase services.

Written Communications. All advice and recommendations to the owner should be transmitted in writing when feasible. Failing that, at least all rejected advice and final decisions should be confirmed to the owner in writing. If the client will not issue all instructions in writing, the best course for the architect is to confirm in writing all instructions and information as it is received. Naturally, all written communications to the owner should be originated promptly and should be dated.

Limitations of Architect's Authority

The architectural agreement explicitly provides that the architect is to have no control over the methods, means, and techniques of construction. These are matters strictly in the contractor's province and they are so provided in the construction contract. The architect is neither in charge of nor in control of the contractor and is not responsible for the contractor's failure to properly perform the contracted duties. The contractor is solely responsible for scheduling and determining the sequences and procedures of construction. All safety precautions and programs are to be instituted and carried out by the contractor. (2.6.6) The architect should not interfere with the contractor in any of these matters and should only consult, observe, and report. The architect's authority, however, includes rejection of work which in the architect's opinion does not conform to the requirements of the contract documents and the ordering of additional inspection and testing. (B141, 2.6.11 and A201, 4.2.6) (See Chapter 21, Architect's Decisions Based on Design Concept, Aesthetic Effect, and Intent of the Contract Documents.)

Only the owner has the power to accept defective or nonconforming work. (A201, 12.3.1) If the architect were to do so without the owner's knowledge and concurrence, it would clearly go beyond the architect's contractual authority and could be considered a breach of the architect's fiduciary duty to the owner.

The architect may order minor changes in the work as long as they require no adjustment in the contract price or time and are not inconsistent with the contract documents. (2.6.13) Any such minor changes ordered should be immediately reported to the owner in the interest of maintaining open and amicable communications with the client.

Included in the particularization of responsibilities to be found in the architectural services agreement is Subparagraph 2.6.2, which describes the architect's duties and the limitations of authority set out in the AIA General Conditions. All of these duties must be performed with reasonable diligence and to the standard of care usually exercised by architects in the community. The architect is under no legal duty to perform any work not contracted for but would be liable for any activities voluntarily assumed if the work is done negligently.

The Architect's Project Representative

The architect's usual on-site services are adequately described in the AIA owner-architect agreement. (B141, 2.6.5) Basically, the normal scope of on-site services is scaled to that reasonably necessary to allow the architect to become generally familiar with the progress and quality of the work. The main objective is to determine if the work will, when completed, conform to the design concepts as delineated in the contract documents. It is definitely not within the architect's duty to supervise or superintend the day to day work of any of the craft workers or artisans on the job.

On some projects, the architect may have to spend more time at the site to adequately protect the owner's interests. This may be because of the size, importance, or complexity of the project. In such cases, an additional architect's representative may be engaged for this purpose. (B141, 3.2) Depending on the scope of the project, this may be part-time, full time, or even more than one person. This architect's on-site person is now called the Architect's Project Representative, but was formerly referred to as the Clerk of the Works.

Historical Background. The term Clerk of the Works dropped out of common usage a little over 40 years ago when the name was changed to Architect's Project Representative (APR). Along with the name change came another important difference. The clerk of the works was formerly hired by the owner, supervised by the architect, and reported to the architect. Now the APR is selected, hired, and supervised by the architect and reports to the architect. This resolved conflicts arising from confusion of authority and eliminated problems related primarily to communication. The new term better describes the position than did the former, now archaic, term.

Line of Communication. The architect's project representative is, as the title proclaims, a part of the architect's organization. As such, the communication path is the same as laid out in the AIA general conditions, requiring all communications between owner and contractor to be channeled through the architect. (A201, 4.2.4) Although the APR reports to the architect, the owner must always be kept informed. This is explicitly required in AIA's owner-architect agreement. (A201, 2.6.5) The most effective way of accomplishing this requirement is for the APR to file a daily report with the architect's office with a copy sent to the owner. In many instances the architect's project representative will keep a log book of activities at the site. It should be copied frequently and furnished to the owner to fulfill the architect's duty to keep the owner informed. Some architects, in a misguided attempt to shield the owner from day-to-day job-site problems, do not send these reports to the owner. However, there should be no job-related secrets withheld from the owner. It would not only fail to comply with the information requirement but could also be deemed a breach of fiduciary duty.

The APR's communication with the owner and contractor should be under the architect's direction. The APR should not communicate with subcontractors and suppliers without the knowledge and approval of the architect and contractor.

What Governs the APR's Activities? When an APR is to be retained for additional on-site architectural presence, the APR's normal functions and limitations are described in the AIA architectural services agreement. (B141, 3.2.2) It refers to AIA Document B352, Duties, Responsibilities and Limitations of Authority of the Architect's Project Representative, 1993 Edition. The document, just one page, should be appended as an exhibit to the architectural services agreement when an APR is to be employed. If the APR's scope is to be expanded, altered, or lessened from that described in B352, the architectural services agreement should be amended accordingly.

The APR should be stationed at the jobsite. The APR's duties are all related to furtherance of the architect's task of administration of the contract. The owner's interest in the avoidance of defects and deficiencies in the construction will be further protected by the attendance of the APR.

APR's Duties and Responsibilities. The APR's duties and responsibilities are intended to carry out the architect's field administration services under B141. All of the duties are under the direct supervision of the architect. They are described in detail in B352 and consist generally of:

> Observing of the work on-site and keeping the architect informed.
> Monitoring the contractor's construction schedules and anticipating delays.
> Responding to contractor's requests for information and providing interpretations of the contract documents.
> Processing change orders.
> Attending meetings.
> Observing testing procedures and verifying invoices.
> Maintaining records at the construction site.
> Maintaining a log book.
> Monitoring the shop drawing process.
> Observing the contractor's maintenance of jobsite records.
> Reviewing contractor's applications for payment and making recommendations to architect.
> Reviewing contractor's punch list prior to issuance of Certificate of Substantial Completion.
> Assisting the architect in determining the dates of substantial completion and final completion.
> Assisting in acquiring contractor's terminal documentation to be submitted to the owner.

Site Observation and Administration of Construction 131

APR's Limitations of Authority. The APR on the jobsite must be careful not to exceed the authority invested in the architect in the architectural services agreement. Accordingly, B352 itemizes several actions that the APR should *not* be involved in:

> Authorizing deviations in the contract documents.
> Approving substitute materials or equipment.
> Personally conducting or participating in any tests.
> Assuming any of the responsibilities of the contractor's superintendent or subcontractors.
> Expediting the work for the contractor.
> Assuming control over construction means or methods or safety programs.
> Authorizing or suggesting that the owner occupy the project in whole or in part.
> Issuing a Certificate for Payment or Certificate of Substantial Completion.
> Preparing or certifying to the preparation of a record copy of the drawings, specifications, addenda, change orders and other modifications.
> Rejecting work or requiring special inspection or testing.
> Accepting contractor's submittals not required by the contract documents.
> Ordering the contractor to stop the work or any portion of it.

The architect should supervise the work of the APR to assure that the duties are performed skillfully and timely and that the APR understands and abides by the limitations of authority. The APR is potentially a useful tool for reducing or eliminating problems that could otherwise escalate into extra construction costs, erroneous construction, or professional liability claims. Conversely, the APR, if not properly qualified, instructed, and supervised, could create problems too profound to contemplate. The APR, being present on the jobsite, can be tempted to get excessively involved in the contractor's duties and responsibilities. This will unnecessarily expose the architect to uninsured liability for construction error and safety infractions.

11

Construction Administration When the Contract Is Bonded

Construction Bonds

The existence of a bond on a construction project will have significant effect on the architect's administration of the contract during the construction phase.

It is the owner's prerogative, and it is so stated in the AIA General Conditions, to require the contractor to furnish a bond covering faithful performance of the contract and payment of obligations arising from the contract. (Subparagraph 11.4.1 of General Conditions of the Contract for Construction, Fourteenth Edition, AIA Document A201, 1987)

If the bond requirement is stipulated in the contract documents or bidding instructions, the contractor should include the bond cost in the bid. However, if the owner imposes the bond requirement after bidding or negotiation of the contract price, the owner must pay the cost in addition to the contract price. The bond premium must be paid in advance.

A construction contract bond guarantees that the contractor will perform the contract in full, including the specified warranty periods, and that the owner will receive the work free and clear of all liens and encumbrances. A bond is usually issued in the form of two separate documents, a performance bond and a labor and material payment bond, each in the full amount of the contract and referred to collectively as a bond. Bonds for construction are normally issued by corporate sureties, usually insurance companies, which are licensed and regulated by the departments of insurance in each state in which they operate. Financially capable individuals can be sureties, but this is no longer common.

Obligations of Surety

A bond is a three party instrument where the surety or guarantor (the bonding company) guarantees the performance of its principal (the contractor) for the benefit of the obligee (the owner) up to the face amount of the bond (the penalty or penal sum). The bonding company guarantees that the contractor will do all that is required of it by the contract documents and that all of the subcontractors, suppliers, and workers will be paid up to the face amount of the bond. The surety's obligations are basically the same as those of its principal. But in addition, it must conduct itself in accordance with the insurance statutes of the various states which generally require that they make prompt investigations and that they settle claims promptly and fairly. Surety companies can be required to pay punitive damages if they fail to follow the statutory requirements. This has resulted in a dramatic change in the ways that some sureties handle claims. They can no longer stonewall without incurring a serious risk of liability for punitive damages.

If the contractor defaults, the surety may use its own judgment as to how it will complete the contract. It may elect to continue with the defaulting contractor, to hire a new contractor, to hire subcontractors and suppliers directly, or to pay money to the owner.

Cost of a Bond

Many owners consider that payment of the bond premium is good value for money in limiting or eliminating some of the financial risk of construction. Bonds cost approximately 3/4 of 1 percent up to 1 1/2 percent of the construction contract price, depending on the contractor's financial stability, building experience, business history, and size of contract. At the end of the contract the total premium will be recalculated on the final contract price including all changes. The fee paid for a bond is not based on the same concept as a casualty insurance premium where the rates are set to create a pool of funds from which to pay losses. Bonding companies do not expect to pay losses from their own funds but expect to reimburse themselves for all costs and legal expenses from the guarantees and assurances they receive from their principal. The surety will not pay any of its own funds until after the principal's assets are exhausted. The surety industry considers a bond premium as a fee for extension of credit.

Contractors normally maintain a close and continuous relationship with their bonding company. Initially they must submit an extensive application giving detailed information on their business organization, experience in various types of construction, equipment, personnel, status of work in process, and financial position. The contractor is required to keep its bonding company informed on a current basis as to all relevant changes in operating and financial conditions and status of contracts, bonded and unbonded.

Contractor Prequalification

The surety's ability to acquire relevant information in respect to the contractor's financial stability far surpasses the usual owner's resources. Thus the requirement for a bond is an excellent contractor prequalification service.

When a contractor is unable to obtain a surety bond, it indicates the surety's opinion that the contractor is undercapitalized, underequipped, improperly experienced, has too much work in process, or has underbid the job. Considerable risk is undertaken by an owner who decides to go ahead with a contractor who cannot obtain a surety bond.

Some owners instruct their architects to specify in the bidding instructions that all bidders must be bondable and that a bond will be required but then do not require a bond from the successful contractor. Although the owner will save the cost of the bond, there will be no financial shoring in the event of the contractor's default. Other owners will require that bid bonds be submitted with all proposals under the theory that only bidders with a continuing satisfactory relationship with a bonding company will submit a bid. Most bonding companies make only a nominal charge or no charge for bid bonds for their regular customers. Owners who dispense with bonds altogether should be especially careful in reviewing prospective contractors' qualifications and should obtain reliable financial data, credit information, and legal advice before proceeding into a construction contract.

Surety as Adversary

Bonding an otherwise unsuitable contractor is not a good idea because the bond will not improve the contractor's capabilities or performance. A bonding company will not step into a faltering contractor's shoes or pay money to the owner until the contractor has actually become incapable or insolvent. It is usually necessary for the owner to file a law suit or arbitration demand against the contractor and surety and obtain a court judgment or arbitrator's award before receiving any monetary relief from the surety.

When a claim is made against a bond the surety will usually take every legal means possible to be exonerated of its obligation or to find some other party to assume some or all of the liability costs. Prime targets in this search for scapegoats are the owner and architect who might have compromised the surety's security by their departure from established procedures set out in the contract documents. The architect is particularly vulnerable if there has been any deviation from the professional standard of care. The owner will have difficulty in perfecting a bond claim if it is in default of any aspect of the construction contract.

Surety as Ally

The bonding company makes an excellent ally in situations where the contractor is exhibiting danger signs of possible impending default. When a contractor appears to be in financial difficulty, the owner is usually reluctant to continue paying the contractor's invoices. To simply stop paying would seem to be financially prudent, but would be a breach of the contract. One solution to this dilemma lies in the owner's making all future payments jointly to the contractor and its surety. Although the AIA agreements do not provide for this procedure, the contractor would have difficulty objecting to it. It is in the surety's financial interest to prevent the contractor's default and it has the legal leverage to apply appropriate pressure on the contractor. In some cases the surety will intervene, providing financial assistance and counsel to the contractor to facilitate successful completion of the contract.

Owner's Right to Approve Contractor's Surety

A surety is only as effective as its financial strength and ability to discharge its bonded obligations, so it is necessary that the owner reserve its right of approval of the contractor's bonding company. The owner should exercise its right to reject a contractor's surety only when information becomes available indicating the surety is unacceptable in some specific way such as financial incapacity or unfavorable reputation in respect to claims management. However, the architect should not specify use of a particular bonding company in the bidding instructions or elsewhere because of the close relationship that usually exists between a contractor and its surety. For a contractor to change bonding companies would necessitate filing of a new application and financial statement and a waiting period.

AIA Standard Bond Forms

In the interest of promoting uniformity and familiarity of procedures in the construction industry, the AIA has issued a standard form for a two-part

Construction Administration When the Contract Is Bonded

Performance Bond and Payment Bond, December 1984 Edition, Document A312. This form is acceptable to and is used by most corporate sureties. It is considered good practice to specify that bonds be submitted on the standard AIA form.

Use of the AIA bond forms is particularly desirable, since some sureties would otherwise include provisions in their bonds that would severely limit the owner's rights. For example, some sureties would include a provision that the bond is exonerated if the owner and contractor agree to any change orders without the prior written consent of the surety.

Transmitting Bond to Owner

When a bond is required by the contract documents and is submitted by the contractor to the architect for transmission to the owner, the architect should refrain from expressing any opinion as to the suitability of the surety or the adequacy of the bond. To express such opinions would be in the province of legal or insurance experts and beyond the expertise of architects. Should an owner rely to its detriment on erroneous opinions of its architect, the architect could become liable for resultant damages.

Commencing Work Before Issuance of Bond

If a bond is specified in the contract requirements, the design professional should make sure that the bond is received from the contractor, approved by the owner, and in effect before any work is commenced on the site. Bonding companies are extremely reluctant to issue bonds after the contract work has been started. If the contractor commences work explaining that the bond has been applied for and will be issued in due course, there is good reason to believe that the contractor is having difficulty in procuring a bond and will never succeed in obtaining it. If the contractor in this case should default, there will be no bond protection for the owner and other aggrieved claimants.

Keeping Surety Informed

Presumably, a surety has reviewed the contract documents and evaluated its risks before setting the rate and providing the bond. Thus, if the conditions of the contract are later changed in any material way without previous notification of the surety, this could release the bonding company from its obligation. The architect administering the contract should therefore keep the surety informed of any changes in conditions which would affect the surety's exposure to risk and in most cases should notify the surety or obtain its consent before proceeding.

It is a very simple matter to send a copy of all change orders and construction change directives to the bonding company. It is not necessary to obtain the surety's approval for changes made according to the procedures provided in the contract. If the contract time is altered or extended by the owner, the surety need not be notified as this notice is specifically waived in the AIA standard bond form. However, there is no harm in giving this information to the surety.

Should the owner or architect learn directly or indirectly, through rumor or otherwise, of the contractor's failure to pay subcontractors or suppliers or other evidence of the contractor's economic distress, the owner should promptly notify the bonding company. This would allow the bonding company to take whatever steps it deems appropriate to assist the contractor and thereby prevent or limit its own losses and possibly those of the contractor. The owner should simultaneously alert and confer with its own legal advisors.

Termination of the Construction Contract

Whether or not the job is bonded, should the owner find it necessary to terminate the contract for cause as provided for in the AIA General Conditions (Subparagraph 14.2.2), the owner, upon certification by the architect that sufficient cause exists to justify such action, must give the contractor and the surety, if any, 7 days' written notice prior to terminating employment of the contractor. The architect's certification must be well founded on the reasons stated in the contract and should be based on solid documentable evidence capable of withstanding the inevitable legal challenge of the surety. (See Chapter 20, Termination of the Construction Contract.)

If the job is bonded, the previously described termination and certification procedure must be preceded by additional requirements found in the bond: The owner must give the contractor and surety written notice of a conference to be held within 15 days of the notice in which to discuss methods of performing the contract. If the owner, contractor, and surety agree, the contractor will be given a reasonable time to perform. The termination cannot be declared sooner than 20 days after the conference notice. According to bond provisions, the surety's obligations are greatly reduced when the owner is also in default. The most common owner default would be failure to pay the contract amounts certified by the architect. However, the owner could be unintentionally in default by any failure to perform and complete or comply with any other terms of the contract. The architect must be very careful in performing any of the owner's obligations, or in acting for the owner, that bond protection is not inadvertently reduced or lost.

Consent of Surety

The surety will need to know of any changes which lessen its financial security in the contract. In the event of a default, the surety will undoubtedly attack an architect who has approved the final payment and an owner who has paid it when the defaulting contractor has used all of the money, lost incentive, or is insolvent. To prevent this eventuality, it is important to obtain the surety's unconditional consent to the making of the final payment to the contractor. Some bonding companies have their own forms for this consent but they usually contain objectionable conditional language such as, "if in the architect's judgment." It would be more appropriate to require use of AIA Document G707, Consent of Surety to Final Payment, 1994 Edition, which is worded as an unconditional approval.

A similar problem is associated with any reduction in or partial release of the retainage. The AIA has a standard form for this purpose, AIA Document G707A, Consent of Surety to Reduction in or Partial Release of Retainage, 1994 Edition.

Overpaying the Contractor

Whether the contract is bonded or not, all contractor's payment applications certified by the architect should always be carefully and fairly judged as to percentage of completion on a line by line basis, not allowing the amounts of understated items to offset overstated items. But in the case of a bonded contract, the bonding company will be very sensitive to any negligent overcertifying or overpaying of contractors when there later has been a default. Approving payment for materials or equipment on the site but not yet incorporated into the work, or for materials or equipment stored or being fabricated elsewhere will be considered by the surety to have lessened its security. However, if the owner and contractor had previously agreed to such payments and the surety had been informed, the surety cannot complain.

Sometimes contractors include in their payment request work which has not yet been done, with the rationale that it will soon be started and will be done by the time the payment has been received from the owner. The architect cannot approve this procedure, as the certificate is supposedly based on completion percentages in place no later than the date of the architect's site verification.

Often during the course of construction of a bonded contract, the architect will receive a form letter from the bonding company requesting such information as the percentage of completion of the work, the amount previously paid to the contractor, and the quality of workmanship. The architect owes no duty to the bonding company to fill in these forms and assumes some liability without compensation by doing so.

Owner's Claim Against Bond

In the event of default by the contractor, it would be advisable for the owner to direct a prompt claim to the bonding company. Considering the widespread legal repercussions of a contractor in default, an incomplete or defective construction project, and a claim against a surety, the owner should without delay consult with its own legal counsel, preferably one with construction industry orientation and experience.

12

Shop Drawing Procedures

Shop Drawings : Friend or Foe?

To the construction industry, shop drawings seem to be a necessary evil. Contractors find them expensive to produce and architects find them unappealing to review. Both find them time-consuming and costly to administer. We seemingly cannot construct buildings without them; but they have become a perennial source of annoyance and confusion and more importantly, a significant source of professional liability claims against architects. Undiscovered mistakes in shop drawings will often lead to unexpected or undesired construction results as well as high-ticket economic claims against architects, engineers, and contractors. Some shop drawing anomalies have resulted in costly construction defects, tragic personal injuries, and catastrophic loss of life.

In decades long past, shop drawings were treated in a fairly casual, offhand manner by contractors and architects alike. Often, to save time and trouble, contractors would instruct the subcontractor or supplier to submit the shop drawings or samples directly to the architect. The shop drawings would then appear unexpectedly in the architect's office, and they would have to sit and wait until someone got around to looking them over. If the contractor started complaining that the job was being held up, the architect would get one of the junior drafters to browse through them, mark any obvious errors, stamp them "approved" and call the subcontractor to pick them up. But it is not anything like that any more.

Shop Drawing Procedures in AIA Documents

This analysis of shop drawing problems and procedures will be based on the situations that would prevail if the owner, architect, consultants, and contractor contracted with each other using the standard form agreements issued by the American Institute of Architects. Use of the following documents will be assumed: Owner-Architect Agreement, Fourteenth Edition, AIA Document B141, 1987; Architect-Consultant Agreement, Sixth Edition, AIA Document C141, 1987; General Conditions of the Contract for Construction, Fourteenth Edition, AIA Document A201, 1987 (incorporated as part of the Owner-Contractor Agreement).

The contractor is obligated by the contract documents to submit shop drawings, product data, and samples for certain parts of the work. The architect is obligated by the owner-architect agreement to "review and approve or take other appropriate action upon Contractor's submittals such as Shop Drawings, Product Data and Samples...." (Subparagraph 2.6.12). This is included among the architect's duties during the Construction Phase -- Administration of the Construction Contract.

Submittals Defined

The AIA General Conditions provides definitions for each of the contractor's submittals:

> "Shop drawings are drawings, diagrams, schedules and other data specially prepared for the Work by the Contractor or a Subcontractor, Sub-subcontractor, manufacturer, supplier or distributor to illustrate some portion of the Work." (A201, 3.12.1)

> "Product Data are illustrations, standard schedules, performance charts, instructions, brochures, diagrams and other information furnished by the Contractor to illustrate materials or equipment for some portion of the Work." (3.12.2)

> "Samples are physical examples which illustrate materials, equipment or workmanship and establish standards by which the Work will be judged." (3.12.3)

They may be additionally described:

Shop Drawings show how components are to be made and assembled. They are usually specified to be submitted by the manufacturers, suppliers, and custom fabricators of such components as structural steel, steel reinforcing bars, hollow metal doors and frames, cabinets and casework, millwork, toilet partitions, doors, windows, electrical switchgear, sheet metal, air conditioning ductwork, building signs, miscellaneous metals, elevators, kitchen and laboratory equipment, and similar special fabrications and equipment.

Product Data, usually in the form of product and equipment lists, schedules, brochures, catalog cuts, performance data, and other descriptive materials, are often specified to be submitted for finish hardware, fastenings, electrical devices and fixtures, plumbing fixtures and equipment, air conditioning controls and equipment, roofing, insulation, flooring, paints and wall finishes, and similar manufactured products.

Samples are often specified to be submitted to illustrate the colors, finishes, and textures available in the materials specified for flooring, wall finishes, roofing, paints and coatings, woods, glass, metals, plastics, ceramics, brick, stone, fabrics, and furnishings. Samples are generally required in all cases where a choice is available. The color and texture samples are also used in the architect's office in designing, refining, and presenting the overall color scheme.

(These lists of examples of shop drawings, product data, and samples are not intended to be exhaustive, but merely to illustrate the general range of trades usually specified to make such submittals.)

These and similar submittals are not considered contract documents. They are submitted for the purpose of illustrating how the contractor proposes to conform to the requirements and the design concepts expressed in the construction drawings and specifications. (3.12.4)

Are Shop Drawings Really Needed?

The drawings and specifications prepared by architects and engineers will show the general design concept of the project and each of the major components and their relationships to each other. Some of the subcontractors and suppliers must prepare additional drawings, diagrams, schedules, and other data to illustrate the specific way in which their particular company or shop will undertake to fabricate, assemble, or install their product.

Shop drawings are needed by the fabrication shops for their own use in instructing their own personnel how to carry out the requirements of the contract documents. Fabricators will produce the shop drawings even if they are not required to submit them for architect's approval. In many cases, the building could have been built satisfactorily even if the architect had not reviewed the shop drawings. The principal reason architects and engineers need to review the shop drawings is to ascertain that the contractor understands the architectural and engineering design concepts and to correct any misapprehensions before they are carried out in the shop or field.

The principal reason architects review shop drawings of any particular trade or component is to determine if the contract drawings have been properly understood and interpreted by the producers and suppliers. The shop drawings should prove to the architect's satisfaction that the work of the contract will be fulfilled. If the shop drawings indicate that the work depicted will not comply with the intent of the contract drawings and specifications, the architect has an opportunity to notify the contractor before the costs of fabrication, purchase, or installation have been incurred.

The costly and wasteful alternative to this procedure would be simply to wait until the work is in place and then examine it and condemn or reject it. It is much more economical to review and correct the shop drawings than to remove and replace erroneous construction. Proper use of the shop drawing review system should prevent costly errors caused by misunderstanding of the contract requirements.

As long as it remains the practice of a majority of architects in the vicinity to specify the submittal of shop drawings for certain trades, then for architects it is a matter of complying with the professional standard of care. The architect's position would be difficult to justify if shop drawings had not been required for the usual trades if some party has been injured or suffered economic disadvantage where the checking of shop drawings could have prevented the loss.

Not requiring shop drawings for architectural review is conceding some degree of design prerogative to the subcontractor. Appearance problems and other construction difficulties that could have been discovered in the shop drawing review will be solely in the hands and discretion of the subcontractor. By the time it is noticed in the field it could be too costly to rectify.

Unspecified Shop Drawings. If the architect does not specify submission of shop drawings for a specific trade, they usually will not be submitted for review. In fact, most architects will not accept unspecified shop drawings for review. (A201, 3.12.5) To accept them would merely create the duty to review them. This would not only increase the architect's uncompensated work load but would also unnecessarily increase the possibility of erroneous approval, carrying with it potential liability to anyone who might suffer injury or incur loss.

Specifying Unneeded Shop Drawings

One way of lessening the exposure to risk of error in reviewing shop drawings is to refrain from specifying them in any case where the contract documents are sufficiently explicit to adequately depict the product or assemblage.

If shop drawings of a certain trade have been specified, but not submitted and therefore not reviewed, the architect could be found negligent if mistakes were carried out in the construction which could have been prevented if the shop drawings had been checked. The architect's liability position would have been better if submission of the shop drawings had not been specified at all. Architects should be attentive that all specified submittals are actually received from the contractor. It is a good idea to prepare a check list of all specified submittals at the beginning of the construction period, so that each may be checked off as received. The contractor should then be reminded to submit any missing submittals.

A procedure for processing shop drawings may be inferred from various related provisions of the AIA General Conditions. Subparagraph 3.10.2 requires the contractor to prepare and keep current a submittal schedule which is to be coordinated with the construction progress schedule. The submittal schedule should show when each shop drawing, product data, and sample is to be submitted to the architect and when the architect is expected to return it to the contractor. The schedule should allow the architect sufficient time to properly review the submittal and to refer it if necessary to the appropriate engineering consultant. Time should also be allowed in the schedule for redrawing and resubmission of shop drawings which have to be revised or replaced.

Specifying Shop Drawings

There is no standard list of trades that must have their shop drawings reviewed. It is still a matter for the professional judgment and discretion of the individual architect or engineer in each unique situation.

The sensitive trades are those involving structural stability, safety, or appearance. However, any trade's shop drawings could involve large sums of money even if these significant elements are absent.

The general understandings as to the definition and purpose of shop drawings and limitations of the architect's approval appear in the AIA General Conditions. (3.11, 3.12, and 4.2.7)

The specific details of shop drawing submittal, such as the number of copies required, routing, and time limitations will usually be specified in the supplementary general conditions or special conditions written for the particular project at hand.

Identification of the particular trades that must submit shop drawings will appear in the various trade sections of the specification, in one of the 16 standard CSI divisions.

Figure 12.1 Shop Drawing Procedure

	Activities	Time	Procedural Milestones
		------s------	Subcontract signed
ADMINISTRATION		s	
	Shop drawing	s	
	production by	s	
	subcontractor	s	Shop drawings
		s -------	submitted to general
		g	contractor
	General contractor	g	
	review and	g	
	approval	g	
		g--------	Shop drawings
		a	submitted to architect
	Review by architect	a	
	& engineering	a	
	consultants	a	Architect rejection,
		-----a--------	shop drawings must
		s	be resubmitted
	Subcontractor	s	
	correction of	s	
This section might	shop drawings	s	
not occur at all but	for resubmission	s	
in some cases		s--------	Shop drawings
would have to be		g	resubmitted to
repeated several		g	general contractor
times until shop	General contractor	g	
drawings are	review & approval	g	
finally approved		g--------	Shop drawings
		a	resubmitted to architect
	Review by architect	a	
	& engineering	a	
	consultants	a	Architect approval
		-----a--------	of shop drawings
PURCHASING & SHOP FABRICATION		s	
		s	
	Subcontractor	s	
	issues purchase	s	
	orders and starts	s	
	fabrication of	s	
	materials	s	
		----s--------	Delivery to jobsite
JOBSITE CONSTRUCTION		s	
	Installation at	s	
	jobsite by	s	Completion of the work
	subcontractor	s	depicted in the shop
		----s--------	drawings

s = Subcontractor processing time
g = General Contractor processing
a = Architect review time

Scheduling Shop Drawings

The schedule depicted in Figure 12.1 shows the usual routine of processing shop drawings from inception to final use on the jobsite.

It takes time, sometimes several weeks for the major trades, to produce their shop drawings. No responsible subcontractor or supplier would commence the costly production of shop drawings without a signed contract in hand, or at least a credible verbal assurance from a trusted general contractor.

The general contractor's proposed construction schedule from inception to completion will reflect the principle that no individual task can be commenced on the jobsite until after its shop drawings have been approved and the work fabricated. This sometimes lengthy pre-jobsite process must be realistically reflected in the schedule.

Considering the long lead time from contract signing to jobsite installation, sometimes months, each subcontractor must prepare its own time schedule that meshes with all other trades and jobsite activities.

The Normal Submission Routine

After receiving the contract, the subcontractor must reanalyze the drawings and specifications. Previous review of the documents had been for the limited purpose of preparing a competitive bid proposal and not necessarily to solve all of the minor technical and logistical problems. The shop drawings must then be prepared to reflect available materials and proposed procedures. Simultaneously, provisional purchase arrangements would be entered into with suppliers and sub-subcontractors. Upon the subcontractor's completion of the shop drawings, they are submitted to the general contractor.

The shop drawings of one fabricator will differ from those of another in the same trade on account of different machinery, processes, personnel, and preference for or access to different materials.

The general contractor must review the submittals to ascertain that field dimensions and conditions are properly depicted and that the specified materials and procedures are being used. The contractor, after approving them, submits the shop drawings to the architect. (A201, 3.12.5 and 3.12.7)

When the architect receives the submittals, it must then be determined whether they will be checked in-house or sent on to the consulting engineer or designer who prepared that section of the contract documents. The architect still must review such referred shop drawings in respect to their intended interface with the architectural requirements and to determine if they properly coordinate with adjoining and related trades.

Shop drawings reviewed solely by the architect's own staff must be carefully compared with the contract drawings and specifications.

The architect's approval of the shop drawings will often be conditioned on the correction of various misinterpretations of the contract documents. In the event that the corrections are extensive, the architect will usually completely disapprove them and require correction and resubmission. They are then sent back to the general contractor, approved, conditionally approved, or disapproved.

The general contractor then routes the shop drawings back to the subcontractor or supplier who originally submitted them. Shop drawings that have been rejected must be revised and resubmitted through the same process until finally approved by the contractor and architect.

On a building project of modest size, there could be 20 to 30 shop drawing submissions. Larger projects often entail considerably more. Keeping straight the status and orderly movement of submittals, rejections, and resubmittals to and from all of the parties involved requires the systematic logging in and out in architect's, engineer's, and contractor's offices.

Contractor's Review of Shop Drawings

The contractor is obligated to review and approve all submittals before conveying them to the architect. (A201, 3.12.5) The contractor's review should be for compliance with all information given in the contract documents as well as for suitability to field conditions and dimensions. The contractor certainly cannot review for conformity with the design concept or the intent of the documents. The contractor is required to make such submissions to the architect with reasonable promptness, in such sequence as to cause no delay in the work, and in accordance with the submission schedule. Architects should be very strict in enforcing the requirements of Subparagraph 3.12.5, AIA Document A201. If the submittals do not exhibit a contractor's review stamp showing "approved", they should be immediately returned to the contractor. The same subparagraph also states that the architect may return, without action, any submittals which are not specified in the contract documents.

The general contractor's review, in addition to assuring that realistic field conditions and dimensions are reflected, is to make sure that all contract requirements are being met. The contractor is in a much better position than the architect to make determinations relating to physical field conditions.

Shop Drawing Procedures

The contractor should not proceed with any field work governed by shop drawings until after they have been approved.

Having reviewed the shop drawings of all of the trades enables the contractor to exercise realistic scheduling controls and practical coordination between the trades. It is the contractor's responsibility to assure that all work on the job is in conformance with approved shop drawings. (A201, 3.12.6 and 3.12.7)

Consultants' Review of Shop Drawings

All submittals which further illustrate or describe work originally designed by consultants such as civil, structural, electrical, or mechanical engineers should be referred to the original designer for review. The architect should also check them to the extent of coordination requirements such as physical interrelating or meshing with work of other disciplines. Some examples:

> Doors should not bump into lighting fixtures on the ceiling.
> An electrical switchboard 90 inches high will not fit in a space with a 7-foot ceiling height.

In the case of highly technical matters, the architect should provide leadership by encouraging coordination among consultants. Some examples:

> Low voltage control wiring should not be specified both in electrical and air conditioning sections.
> Lighting fixtures cannot be in the same place as air conditioning registers.
> Ducts, conduits, pipes, and structural members cannot occupy the same space.

Architects, when making their agreements with consultants, should be sure that shop drawing review and coordination is included in the consultants' duties. The AIA Standard Form of Agreement Between Architect and Consultant, Sixth Edition, Document C141, 1987, provides in Subparagraph 2.6.11 for the consultant to review the contractor's submittals in respect to the portion of the work entrusted to the consultant. This subparagraph is harmonious with the comparable provision in the AIA Owner-Architect Agreement (B141, 2.6.12). However, if AIA Document C142, Abbreviated Architect-Consultant Agreement, First Edition, 1987, is used, a paragraph must be added to cover submittal review as it is not included in the basic form.

Monitoring Progress of Submissions

Monitoring the progress of the contractor's submissions of shop drawings, product data, and samples and the review status of each can be a complicated process if not approached in a systematic manner. Many architects have custom-designed schedules which are used for this purpose. Also, the AIA has designed and issued Document G712, Shop Drawing and Sample Record, October 1972 Edition, which is suitable for the purpose. This is also a good use for the desk top personal computer. All shop drawings, product data, and samples should be registered and date stamped each time they are received, sent out, or acted upon. The important information which must be accounted for is:

1. Receipt date
2. Title, trade, or item
3. Shop drawing number assigned
4. Specification section number
5. Contractor, subcontractor, or supplier
6. Number of copies received
7. Date, number of copies sent, and to whom referred for consultant review
8. Date, number of copies received from consultant, and consultant's recommendation for action
9. Architect's action as to approval, rejection, conditional approval, or request for resubmission
10. Distribution of shop drawings and copies of letters of transmittal. (To contractor, jobsite, owner)

Each resubmission of a shop drawing will require a new set of entries to track its progression through the review process. Faithful attention to the schedule will not only help keep the situation straight but will yield a permanent record of the chronology. Needless to say, all submittals to and from the architect should be accompanied by letters of transmittal which should be preserved as a further record of the process.

Keeping the Client Informed

Most clients of architects are not fully aware of the importance and role of shop drawings nor of the large amount of time and effort expended in the review process and its related administration. In fact, many inexperienced clients have no reason to know of the existence of the shop drawing system or of the necessity of professional review and comment. This should be completely explained to the client. It is also appropriate to send the client copies of all shop drawing letters of transmittal to keep the client currently informed and aware of this significant behind-the-scenes process.

Some experienced owners not only know the value of the shop drawings but wisely require a complete file set of all construction submittals to be assembled as an aid in the future maintenance of their buildings.

Qualified Personnel

In the architect's office, it is crucial that the usually unpopular and uninspiring task of shop drawing review be assigned to a qualified person, one who is intimately acquainted with the contract documents and the design concept or intent of the documents. Otherwise, how could the reviewer comply with the requirements of the Owner-Architect Agreement (Subparagraph 2.6.12) and of the AIA General Conditions (Subparagraph 4.2.7), which have similar language and both of which promise that the architect will be reviewing submittals "only for the limited purpose of checking for conformance with information given and the design concept expressed in the Contract Documents."?

Shop Drawing Stamps

Architects, in most cases, express their opinion of the shop drawings, product data, and samples by use of the rubber stamp which usually has some exculpatory language in fine print plus some options which can be exercised by use of check marks. Often the stamp says something like "Review is for general compliance with Contract Documents. No responsibility is assumed for correctness of dimensions or details."

The various options to be selected include

> Reviewed
> Rejected
> Revise and Resubmit
> Furnish as Corrected
> No Exception Taken
> Make Corrections Noted
> Submit Specified Item

Space is also usually provided for the date of review and action and the shop drawing number. The wording on a shop drawing stamp will not serve to change or extend the meaning of Subparagraph 4.2.7 of the AIA General Conditions. Therefore, the following words "Submittals have been reviewed and action taken in accordance with Subparagraph 4.2.7 of AIA General Conditions" could be used, with the appropriate options available for checking. The accompanying letter of transmittal should have additional comments which are needed to explain the reviewer's action or conditions of approval.

Architect's Approval of a Shop Drawing

The checking of shop drawings is a mundane and uninspiring task in a design office, so there is the temptation to assign it to someone low in the pecking order. This is a serious mistake, because the checker should be intimately aware of the contract requirements and capable of recognizing potential construction difficulties and design problems.

Architects must be alert to the possibility of holding up construction progress by taking too much time in shop drawing review and processing. Accordingly, many architects insist that all shop drawing processing time be shown on the general contractor's construction schedule. This would allow the architect to object in advance to an inadequate time allowance for review procedures.

The AIA General Conditions make it clear that the architect's approval of a shop drawing does not relieve the contractor of responsibility for requirements of the contract documents. The contractor is unquestionably responsible for errors or omissions in shop drawings.

A serious underlying concern of all architects is the possibility of inadvertently sanctioning hidden errors or unidentified revisions in a shop drawing. Architects rely on the AIA General Conditions provisions that require the contractor to disclose all deviations from the contract documents and to obtain the architect's written approval of specific deviations. (A201, 3.12.8 and 3.12.9)

The contractor is relieved of responsibility for deviations from contract requirements only if the contractor specifically informs the architect of the deviations in writing and the architect has given specific approval of the deviation in writing. At the same time, however, the architect should be extremely careful and thorough in checking shop drawings to minimize the possibility of error.

For years we have been reading in the professional literature that we should avoid using the word "approved" when describing the outcome of our review of a submittal. However, arbitrators and the courts have consistently rejected the idea that an architect or engineer could avoid responsibility for reviewing shop drawings merely by using some other word or an enigmatic expression such as "no exception taken". The AIA documents now accept the reality that architects really do approve (with or without conditions) or disapprove shop drawings. The AIA Owner-Architect Agreement (Subparagraph 2.6.12) and the AIA General Conditions (Subparagraph 4.2.7), using identical language, state: "The Architect shall *review* and *approve* or take other appropriate action..." (my italics). Therefore, it is my opinion that there should be no problem with having "approved" included as one of the options on the shop drawing stamp.

Whatever words the architect uses in trying to avoid saying approved, there comes a time in the process when the contractor needs a final decision on whether or not the work can proceed in accordance with the shop drawings. Thus in the practical world of construction, all words that do not reject the shop drawings will be taken as approving them.

The architect's review is not to be taken as an approval of any safety precautions, because these are the contractor's responsibility. The review also is not intended to interfere in any way with the contractor's prerogative of determining and controlling construction means, methods, techniques, sequences, or procedures. The architect's review is limited to determining if the requirements of the contract documents are being met and that the completed work will be in compliance with the contract documents.

The most important principle to be followed at this juncture is to make sure that whatever words are used on the shop drawing stamp and the accompanying letter of transmittal are an accurate portrayal of the intended action. If the approval is in any way conditional, choose the words carefully so no one is led astray. The contractor is required to identify specifically in writing any shop drawing revisions other than those requested by the architect on previous submittals. (3.12.9)

Improper Use of Shop Drawings

Architect Misuse. The architect should not use the shop drawings as a medium for making changes in the contract requirements. If the architect needs or wants to make a change, it is proper to initiate a change order or a construction change directive or to order a minor change (all provided in Article 7 of AIA Document A201). The only corrections architects and engineers should make on submittals are to bring them into conformance with the requirements of the contract documents.

Contractor Misuse. Contractors, subcontractors, and suppliers should not use the shop drawings as a means of suggesting substitutions from the contract requirements. Should it become advisable to recommend changes, the contractor should make specific requests of the architect, explaining the particulars and the reasons. If it is necessary to deviate from the contract requirements in the shop drawings, then the contractor must point out the deviations at the time of submittal. If the architect is misled into approving shop drawings containing unlabeled deviations, the approval will be void. The architect's approval of a contract deviation is valid only when the architect has approved the specific deviation. (A201, 3.12.8 and 3.12.9)

Various provisions of AIA standard form documents (A201, B141, C141, C142 and G712) have been quoted briefly and should be reviewed in their entirety for their complete language and context to avoid possible misinterpretation.

13

Payment Certifications

Contractor's Application for Payment

When an owner and a contractor enter into a construction agreement using the standard AIA agreement and General Conditions, they thereby designate the architect to validate the contractor's periodic requisitions for funds to be paid by the owner. Both parties thereby signify their confidence in the integrity of the architect's certificate. They both rely on the architect's technical competence and integrity, with the expectation of fairness and impartiality. The contractor's surety, employees, subcontractors, and suppliers as well as the owner's insurers and lenders proceed in reliance on the certainty of a fair and objective payment system.

It is essential that payment certifications be issued strictly in conformance with all applicable provisions of the construction contract. The architect's power does not include changing the payment procedures of the contract. The payment and certification system described here is based on the standard General Conditions of the Contract for Construction, AIA Document A201, Fourteenth Edition, 1987. (See Appendix D.)

Typically, construction agreements call for monthly payments for the value of all work in place or suitably stored on the site as of the last day of the month, with the payment due on or before the tenth day of the following month. The parties can agree to more or less frequent payments and any due dates they find convenient. According to the AIA General Conditions, each payment applied for by the contractor should be in the format of a schedule of values which had previously been submitted to and approved by the architect. (A201, 9.2.1 and 9.3.1) The value of each item of work is assessed as to its percentage of completion as of the last day of the month, the date of the application. The total amount of each application will be the value of all of the work completed to the application date reduced by the amount of the agreed retainage, with a credit allowed for all previous payments made by the owner.

The application for payment may be made in any format convenient to the contractor. Many contractors prefer to use the AIA standard forms or a similar format. The applicable forms are Application and Certificate for Payment, Document G702, and the Continuation Sheet, Document G703. These forms are quite convenient as they are arranged in logical order and contain an affidavit for the contractor's signature as well as the Architect's Certificate for Payment. If these forms are not used by the contractor, the architect must prepare a certificate form to be issued.

Upon receipt of the application, the architect must make a site examination to determine that the percentage of completion stated for each item does not exceed reality on the jobsite. (A201, 4.2.2 and 4.5.5) Each item must stand on its own with no averaging of "overs" with "unders" and without anticipating of completion of any item a few days thereafter.

Representations and Limitations of Certificates

The architect's issuance of a certificate for payment is a representation to the owner that, to the architect's knowledge, information, and belief, the work has progressed to the point indicated. This must be based on the architect's physical presence and visual observation at the jobsite. The architect also represents that the quality of the work is in accordance with the contract documents and that the contractor is entitled to the sum of money stated in the certificate.

The issuance of a certificate is not a representation that the architect has:

1. Made exhaustive or continuous on-site inspections to check the quality or quantity of the work
2. Reviewed construction means, methods, techniques, sequences, or procedures
3. Reviewed copies of requisitions received from subcontractors and material suppliers and other data requested by the owner to substantiate the contractor's right to payment
4. Made examination to ascertain how or for what purpose the contractor has used money previously paid on account of the contract sum. (A201, 9.4.2)

Payment Certifications 157

Issuance of a certificate does not constitute acceptance of defective or nonconforming work. (A201, 9.6.6)

Neither the architect nor the owner have any duty under the contract to pay the subcontractors and suppliers or to see that they get paid. However, the architect may, if they request, release information to them regarding the amounts which have been paid to or withheld from the contractor for their portion of the work. (A201, 9.6.3 and 9.6.4)

During the time when a dispute has arisen or arbitration is pending, the architect must continue administering the payment procedure and issuing certificates when due. The owner must pay any amounts certified by the architect and the contractor is obligated to continue the work of the contract. (A201, 4.3.4)

Decisions to Withhold Certificate

If it is the architect's opinion that the required representations cannot be made, the architect may decide to withhold certification in whole or in part to whatever extent is reasonably necessary to protect the owner from loss. If the architect is unable to issue a certificate in the amount applied for, the architect must notify the contractor and owner within 7 days after the date of the application. If the contractor and architect cannot agree on a revised amount, the architect will issue a certificate for an amount for which the necessary representations to the owner can be made. Additionally, the architect may decline to issue a certificate altogether or may nullify in whole or in part previously issued certificates if necessary to protect the owner from loss because of :

1. Defective work not remedied
2. Third party claims filed or reasonable evidence indicating probable filing of such claims
3. Failure of the contractor to make payments properly to subcontractors or for labor, materials, or equipment
4. Reasonable evidence that the work cannot be completed for the unpaid balance of the contract sum
5. Damage to the owner or another contractor
6. Reasonable evidence that the work will not be completed within the contract time and that the unpaid balance would not be adequate to cover actual or liquidated damages for the anticipated delay
7. Persistent failure to carry out the work in accordance with the contract documents.

Whenever the causes for withholding certification are removed, the architect may issue a certificate for the amounts withheld. (A201, 9.5.1)

Overcertification

The danger of an architect's overcertification of funds can arise from various sources, not the least common being a schedule of values which has been "front end loaded" by the contractor. This is the process of attributing larger than realistic amounts to the operations which will be completed first and reducing the sums for operations which will be accomplished later in the construction period. By artful adjustments in the breakdown, the contractor can be paid more than the completed work justifies and, because of this, the retention can become insufficient to protect the owner from overpayment. The owner will have completely and effectively lost the advantage of the retention. A willfully distorted schedule of values is fraudulent and when skillfully contrived is not easy to detect. If the architect has any suspicion that this phenomenon is taking place, the contractor should be asked to submit substantiating data to validate the schedule of values.

A second major source of possible overcertification is in the estimation of the percentage of completion of each item. If the operation is one which lends itself to counting or calculating units or areas, then that should be done. It would not be unreasonable to ask the contractor how the figures were derived and to require evidence of the reasoning and computation. Each line item on the payment application should be separately weighed and considered and should be individually justifiable. An underestimated item cannot be used to offset an overstated item. Sometimes, at the time of compilation of the billing, certain items might have been optimistically estimated but not subsequently completed to the expected level, thereby resulting in overstated items. The architect has no choice but to mark them down to reality. Contractors often object to this treatment, arguing that the work is progressing and will be up to the stated completion by the time the bill is paid. The architect, being in a trusted fiduciary position, cannot allow this procedure. All completion percentages stated must be correct as of the date of the architect's site examination and the certificate cannot predate the site visit.

Certificate Must Be Fair

The architect cannot arbitrarily reduce or refuse to issue a certificate. The owner and surety are relying on the architect to protect the owner from overpaying. At the same time, the contractor and all who have a monetary interest in the owner's payment are relying on the architect for fair treatment. The architect must take this duty very seriously and carry it out skillfully, carefully, and honestly. It should be done strictly within the time periods set out in the contract.

It is generally accepted that an architect, when certifying payments due to a contractor, is acting in a quasi-judicial or arbitral capacity rather than in an exclusively ministerial role and thus has immunity from suit. There is little likelihood of a contractor prevailing in a legal claim against an architect for

Payment Certifications

reduction in or refusal to issue a certificate or nullification of a previous certificate. However, the architect could lose immunity if the alleged professional transgressions are due to negligence, collusion, fraud, bad faith, or malicious intent. Although autocratic and inflexible architects are an eternal pain to contractors and do not necessarily best serve their clients in this way, they are within the bounds of allowable professional behavior.

If the contractor is not given the opportunity to present its viewpoint before decisions are finalized, or is otherwise unfairly treated, the architect may lose immunity and be liable to the contractor for any monetary loss. The surety is rightful in expecting the architect to exercise ordinary care to assure that the contractor will not be overpaid since contract provisions for periodic payments, retentions, and other safeguards are as much for the surety's protection as the owner's. The architect may be liable to the surety if any losses are attributable to issuance of erroneous certificates if the architect is proved to have acted negligently or fraudulently.

Final Payment

When the architect has made the final inspection, if the work is found acceptable and the contract fully performed, the final certificate for payment should be issued. The certificate is a further representation to the owner that all conditions listed in the contract as precedent to the contractor's entitlement to final payment have been fulfilled. (A201, 9.10.1 and 9.10.2)

In those contracts which include provision for retainage, a further certificate for payment of the retainage will follow, usually five days after expiration of the lien period for employees, subcontractors, and suppliers. Substantiating data for release of retainage will include

1. A release of the contractor's own lien rights
2. Title company lien guarantee
3. Consent of surety, if any

Contract provisions will vary, as mechanics' lien laws are not uniform in the various states.

The architect's final certificate should include reconciliation of all outstanding matters, including all pending change orders and construction change directives, allowances, determination of liquidated damages, retention, and previous payments and credits.

Accord and Satisfaction

The final payment constitutes an accord and satisfaction; that is, by paying it the owner is thereby acknowledging that there are no outstanding claims against the contractor other than those previously claimed and unsettled, those which might develop during the warranty period, and possible latent defects. (A201, 4.3.5 and 9.10.3) Similarly, the contractor, by accepting the final payment, acknowledges that there are no outstanding claims against the owner, other than any previously claimed and remaining unsettled. (A201, 9.10.4)

The architect's rendering of the certificate for final payment is the final element of service under the Owner-Architect agreement. (AIA Document B141, Subparagraph 2.6.1) Any further services are classified as additional services subject to additional agreement and compensation. (B141, Article 3)

Substantiation for Payment Requests

The AIA General Conditions requires that each interim payment request be accompanied by such data substantiating the contractor's right to payment as the owner or architect may require, such as requisitions from subcontractors and material suppliers. (9.3.1) Substantiation for the final payment is more extensive and includes an affidavit that payrolls, bills for materials and equipment, and any other indebtedness connected with the work have been paid or otherwise satisfied, and certificates of any insurance which is required to remain in force after completion. Consent of surety, if any, to the final payment will also be required. Other substantiating data such as receipts, releases and waivers of liens, claims, security interests, or encumbrances arising out of the contract, must be submitted if required by the owner. (A201, 9.3.1 and 9.10.2)

Some of the substantiating data may be submitted by the contractor by use of these standard forms of the AIA: Contractor's Affidavit of Payment of Debts and Claims, Document G706; Contractor's Affidavit of Release of Liens, Document G706A; Consent of Surety to Final Payment, Document G707. Consent of Surety to Reduction In or Partial Release of Retainage, Document G707A.

The contractor should be made aware at the outset of the contract of the extent of substantiating data which will be required for all payment applications. It must be remembered that the certificate is not a representation that the architect has reviewed any of the owner-required substantiating data. The owner should check the substantiating data submitted for adequacy and authenticity if capable or should engage qualified advisors to do so.

14

Change Orders

Who Benefits From Changes During Construction?

Contrary to uninformed popular opinion, no one benefits from changes ordered during the construction period. They are generally disruptive to the orderly progress of the work and are usually an economic burden on both the owner and contractor. They are often symptomatic of someone's failure to properly fulfill their function in the construction process.

Changes in the scope or details of construction originate from various sources. Owners will have second thoughts or will embark on excursions of economic downgrading. Contractors will offer specification substitutions for various reasons, some more honorable than others. Sometimes, faulty construction documents will generate the unexpected need for alternative materials or processes. Some changes, such as those caused by unavailability of specified materials, unforeseen conditions, or changes in governmental requirements usually cannot be avoided. Practically every change in contract conditions will cost more than the same items would have cost if included in the original contract and the full value of deducted items will not be credited to the owner.

Changes in the contract and work schedule cause confusion, and on projects with numerous changes, the job superintendent and subcontractors sometimes become uncertain of the exact state of the contract at any point in time. Add to this the uncertainty of prospective changes which have been discussed and price-quoted but not yet ordered. Sometimes, they are never definitely confirmed or canceled. It would be in the best interest of all concerned if changes could be firmly controlled and severely limited in number or in rare circumstances even eliminated.

Owner's Right to Make Changes

Neither owner nor contractor has an inherent right to unilaterally change any of the terms of a validly executed contract unless the contract itself contains a provision which specifically allows for changes. A construction contract would be very impractical indeed if an owner could not make necessary changes as the construction progresses; therefore, nearly all construction contracts recognize that some degree of flexibility is a practical necessity and will include a change order procedure. The owner will have the right to order changes and the contractor will be required to carry them out in return for an equitable adjustment in the contract price and time.

Change Orders and Construction Change Directives

Construction contracts that include the AIA General Conditions, AIA Document A201, 1987, have the advantage of a practical change procedure which properly reflects the roles, rights, and obligations of owner, contractor, and architect. The AIA General Conditions defines a change order as a written instrument prepared by the architect and signed by the owner, contractor, and architect stating their agreement on a change in the work and the amounts of adjustment, if any, in the contract sum and time. (A201, 7.2.1). Although the contract requires that change orders be in writing, judges and arbitrators are generally reluctant to strictly enforce this against contractors when the extra work has been done with the owner's or architect's knowledge and consent.

In those occasional circumstances where the owner and architect have signed the change order but the contractor will not, the contractor is required to proceed with the work, reserving resolution of the disagreed portions to a later time. A change order not agreed to and not signed by the contractor is called a construction change directive, a term coined by the AIA and first used in the 1987 AIA General Conditions. (Paragraph 7.3) At any time that the contractor later agrees to its terms or mutual agreement is obtained by adjustment of its terms, it is then deemed to be a change order.

The disputed portions of a construction change directive, if not resolved by negotiation, will be decided by the architect. If the architect's decision is not acceptable to the parties, it is subject to arbitration. The contractor is not obligated to proceed without consent with construction change directives involving work relating to asbestos or polychlorinated biphenyl (PCB). (A201, 10.1.3) Change orders and construction change directives, upon their execution, become contract documents. (A201, 1.1.1)

Change Order Procedure

When a modification in the work is contemplated by the architect or owner contingent upon an acceptable price and time quotation, a request is sent to the contractor describing the proposed change. A standard form of the AIA, Proposal Request, Document G709, is suitable and may be used for this purpose. The request need be signed only by the architect, as it is simply a request for information and not an order to change the work of the contract in any way. If the quotation tendered by the contractor is acceptable to the owner, a change order may then be written and circulated for signatures. Change orders may be administered using the standard AIA Change Order form, Document G701. If the contractor is not agreeable to any aspect of the proposed change, the AIA standard form, Construction Change Directive, Document G714, should be used. If the construction change directive is later found to be acceptable to the contractor, an appropriate change order should be issued. If the construction change directive continues to be unacceptable to the contractor, it becomes a claim to be decided by the architect.

Any unsolicited written demand for additional compensation or time for changed conditions received from the contractor is treated as though it were a proposal request. To whatever extent it is acceptable to the owner, as advised by the architect, it may be incorporated into a change order and submitted to the contractor for final acceptance as a change order or, if rejected in whole or in part, as a construction change directive.

A proposal request as well as the resulting change order must contain a description of the work sufficiently detailed to enable the contractor to price it accurately and to estimate the effect on the time schedule. The work description and drawings which would have been adequate for the pricing proposal may not be sufficiently detailed for construction, so additional drawings and specifications may have to be prepared after the change order is signed.

Supplementary drawings and specifications needed for the construction of changes will take the form of specifications, new drawings, or amendments to existing drawings. New specifications and drawings should be consistent with the proposal description, appropriately identified, and dated. Existing drawings amended to show changes should be carefully annotated and dated to assure that changes can be easily distinguished from the original contract. The usual practice of identifying different changes by consecutively numbered delta symbols should not be abused by the inclusion of unmarked changes or improperly described changes. For example, corrections of errors should not be labeled as clarifications. The amended or supplementary drawings and specifications are a part of the change order documentation and therefore become contract documents. The state of the drawings before the amendments should be recorded by preserving a print or a reproducible tracing. Owners, contractors, and subcontractors should always preserve all superseded drawings until after all contract billings have been rendered and all payments made. The architect should promptly notify the contractor whenever change order proposals are rejected by the owner so that any work which was being held up to accommodate the impending change can then proceed.

Pricing and Billing of Change Orders

The potential for owner misunderstanding and dissatisfaction is very high in the pricing of changes during construction. Most experienced owners realize that the contractor's costs for making changes will often be higher than the same or similar work would have cost if included in the original bid. Change order work will often cost 10 to 15 percent more than if it had been included in the original bid. This is for several reasons:

1. Lack of a competitive environment
2. Inability to easily fit the change into the existing schedule
3. The inordinate amount of paperwork and distraction experienced by the contractor.

Basically, it is difficult for anyone involved to make money on change orders, so the fewer there are, the more cost effective the project.

Although all contracts include the implied covenant of fair dealing, it is possible that some contractors would take advantage of the lack of competition and exaggerate the costs. Here, the use of estimating costbooks will provide a valid "third party" look at the costs and serve as an objective source by an impartial organization.

One way of partially controlling this problem is to require the general contract bidders to state their and their subcontractors' overhead and profit markups at the time of submitting their bids. It is also possible for the architect to specify the contractor's and subcontractors' acceptable markups in the supplementary general conditions so the contract price will be predicated on the specified markups being applied to all change orders. Another element in controlling or limiting the costs of changes is to request all bidders to quote unit costs for adding or deducting certain specified relevant materials and operations such as concrete, excavating, or painting.

Unit prices for adding many materials or operations will be different from those for deducting on account of the effect of associated fixed costs. It is quite common that unit prices for adding will be more than for deducting and higher for small quantities than for large quantities. This is a function of actual construction cost accounting and not a device for victimizing owners. For similar reasons, it is customary for contractors to attribute profit and overhead to additive but not deductive change orders.

Change Order

Change order quotations submitted by contractors should be itemized to enable the architect and owner to properly review and evaluate the costs. Unless a lump sum has been negotiated and agreed upon, the detailed breakdown should include:

1. Labor
2. Materials
3. Transportation
4. Subcontracts
5. Bond
6. Insurance
7. Permits
8. Testing
9. Coordination and superintendence
10. Overhead and profit

The subcontract amounts should be similarly detailed. Materials estimates should include delivery and sales taxes. Knowledgeable contractors, protective of their own economic interest, will impose an expiration date on all change quotations. This will cover their exposure to price rises, but more importantly, prevents uncontrolled open-ended disturbance to their construction scheduling.

All change orders should include the agreed adjustment to the contract time, and if there is to be no adjustment, then that should be stated. Some change orders are similar to cost-plus contracts when the scope of work cannot be ascertained in advance. In such cases, it is common to quote unit prices, labor rates, and overhead and profit markups in the approved change order. After the change work is completed, the total charge can be determined using the agreed rates.

Contractors should not proceed with any extra work for which they expect to be paid without suffering the formalities of a written change order. One notable exception to this advice is emergency work which must be commenced immediately to mitigate further damage. When the emergency work is completed it should be billed promptly in such detail and substantiation as to enable the owner and architect to audit and approve it. Rushed changes, those which must proceed immediately to avoid scheduling disruptions will also be ahead of the paperwork. Diligent contractors will follow quickly with the change order information so the work can be promptly billed. Customarily, change order work is billed monthly for payment in the same manner as the original contract on the basis of percentage completed less the agreed retention.

Changes in Time

Most changes during construction will have some direct effect on the time of the contract, either deferring or advancing the estimated time of completion. Changes to non-critical path work should not affect completion, unless the change prolongs the activity sufficiently that it is then on the critical path.

When completion will be delayed by a change, the contractor will logically expect to be granted the extra time in the change order as well as the additional costs of extended overhead charges in the field and in the home office operation. Some changes will have the effect of extending work of the basic contract into future times of higher labor costs. This would be a proper charge on the change order even when the more costly labor is not related to the work of the change order.

Acceleration and Impact Claims

If the owner delays the contractor by failure to make portions of the work area available on time or in some other manner deprives the contractor of contract time, it constitutes acceleration of the contract. This will cause the contractor extra expense for more crews, overtime wages, or damages for late completion. In all fairness, rightful acceleration claims should be recognized immediately and the agreement formalized in a change order.

The cumulative effect of excessive or confusing changes leads to the disruption of carefully planned schedules and a loss of construction momentum. This is difficult to identify, quantify, and attribute to individual change orders as they occur. The effect will generally become apparent at or near the end of the job, when the contractor is trying to account for the unexpected and mysterious schedule slippage and failure to meet the completion deadline. In most cases, when the contractor submits a claim for additional time due to this so-called "impact" effect, it is immediately rejected by the owner and architect as being vague, baseless, and unprovable. However, contractors who appeal their impact and acceleration claims to knowledgeable arbitrators and judges in an effective manner have a fairly good chance of prevailing.

Owners and architects could lessen the burdensome effect of these claims by initially evaluating them more realistically and by negotiating a reasonable settlement. Moderate compromises on each side will usually prove to be more economical than the costs of arbitration or litigation and the risk of a harsh award.

Architect and Surety

The architect has the basic administrative burden of keeping the contract straight and monitoring the progress and status of all changes. This is a very sensitive duty as it occurs simultaneously with construction progress and if improperly performed could cause unnecessary confusion and construction delay.

Architectural construction administrators should devise and use methodical systems and procedures for continuous monitoring of the status and progress of change order proposal requests, quotations, approvals, and disapprovals. Contractors' demands for extra compensation or time should be processed in a timely fashion to avoid the accumulation of unresolved claims and the inevitable unpleasant confrontation with the owner.

When a project is bonded, the surety could be relieved of its responsibility if extensive or costly changes materially alter its originally assumed risk. Therefore, it is essential to keep the surety informed of change orders as they occur. Insignificant change orders could be accumulated and sent as a group with later important changes or when the aggregate of changes becomes significant.

Contractor's Claims for Extra Compensation

When contractors submit claims for extra compensation, the architect should immediately react by researching the contract documents and the circumstances giving rise to the claim and responding promptly. If "extra" work is already in the contract, no change order can be issued, and the reason should be furnished to the contractor in writing. If the claim arises from alleged imperfections in the architect's actions or documents, the architect must be deeply introspective and scrupulously fair in evaluating the claim. If the decision is in favor of the contractor, a change order must be issued charging the costs and time to the owner. The architect and owner must then resolve the issue between them.

Architect's Minor Changes

According to the General Conditions (7.4.1), the architect is entitled to order minor changes in the work not involving adjustment in the contract sum or time. Such changes must be consistent with the intent of the contract documents and must be in writing. A change order should be issued and executed to preclude later claims by the contractor or owner in respect to cost or time. (See Chapter 21, Architect's Decisions Based on Design Concept, Aesthetic Effect, and Intent of the Contract Documents.)

Use Change Orders for all Contract Changes

The change order process should be used to formalize all adjustments in the contract work, cost, or time, such as:

> Owner's carrying out of contractor's work (2.4.1)
>
> Resolution and accounting of contract allowances (A201, 3.8.2.4)
>
> Cost differences resulting from owner's rejection of subcontractors (A201, 5.2.3)

Cost of property insurance ordered by contractor and charged to owner (A201, 11.3.1.2)

Cost of insurance charged to contractor (A201, 11.3.4)

Cost of replacing insured damaged property (A201, 11.3.9)

Cost of uncovering and replacing work which was not required to be inspected and which after uncovering proved to be in accordance with the contract documents (A201, 12.1.2)

Owner's acceptance of uncorrected defective or nonconforming work, with or without a credit, should be formalized with a change order. (A201, 12.3.1) Architectural services provided under the Owner-Architect Agreement, Fourteenth Edition, AIA Document B141, revised in 1987, include, as a basic service, administration of the construction contract, although services connected with the change order process are classified as contingent additional services and therefore merit additional compensation.

15

Responsibilities of The Owner in a Construction Contract

The Owner as a Member of the Construction Team

Construction accomplished through written contracts requires efficient and effective teamwork. The main components of the usual team are the owner, contractor, and architect. In addition, the main players have extensive teams of their own, consisting of consultants, advisors, suppliers, and subcontractors. Altogether, dozens of persons and entities are required to combine their efforts and cooperation to achieve their mutual objective. (See the chart of contractual relationships, Figure 1.1.) The main players usually comprise a new untried team, the combination having never before worked together on a single project, even though they may be highly experienced individually. Owners are often the weakest link in the team because they are not always as experienced and expert in their roles as are the contractors and architects they employ.

An effective team is heavily dependent on each member's appropriate conduct during the construction period. For guidance, contractors and architects can look to the customs and standards of their trade or profession and to the contract. However, the owner, not being organized as a part of the construction industry, is limited to reliance on the contract and on the advice of its advisors. Owners do not necessarily know how other owners conduct themselves in a given situation. The architect is always available to counsel the owner at any time it is needed.

Some owners find themselves in this unique contractual position only once or twice in a lifetime while others are constantly in the construction marketplace. Some experienced owners administer their roles very efficiently and effectively through their own construction departments often staffed with architects, engineers, lawyers, accountants, estimators, contract administrators, and other highly skilled construction and real estate experts.

Owners have the highest stake in the physical development of their property and thus have the most to gain or lose by their own cooperation and appropriate behavior or the lack of it. The owner will usually pay for its bizarre, untimely, or inappropriate responses or any other unusual behavior. Inexperienced owners who choose to remain uninformed and uninvolved in the construction process will not obtain optimum results and could easily lose in the process.

AIA General Conditions

The latest edition of the AIA standard form, General Conditions of the Contract for Construction, is a good source of information relating to the owner's rights, duties, and responsibilities under a standard construction contract. The General Conditions (or some edited version of it) is made a part of most construction contracts in the United States and therefore has become the main practical standard for judging the customary duties and practices of architects and contractors during construction. Likewise, usual owners' obligations are presented and described in the document. (See Appendix D.)

Owner's Responsibilities

In analyzing the owner's activities under the AIA General Conditions, it is immediately apparent that some of the owner's duties could be, and often are, carried out by the architect. Indeed, Subparagraph 4.2.1 states that the architect is the owner's representative and has some limited authority to act on behalf of the owner. Most of the owner's duties, however, will have to be carried out personally or by legally authorized representatives. The owner will, in some instances, confer additional authority on the architect and this must be carefully delineated in writing. The architect has no blanket power to act for the owner in any situation not previously authorized, preferably in writing.

Aside from any obligations imposed by the General Conditions, the owner and contractor are bound by the implied covenants of mutual cooperation and fair dealing. This implies that each must act promptly when required, abide by the terms of the contract, treat each other fairly, and avoid overreaching and other forms of sharp or deceptive practice.

Although the contract documents are technical and voluminous, the owner, in furtherance of its own interest, should reserve the necessary time and effort to read and examine them in their entirety. Anything not completely understood should be taken up with the architect for explanation and discussion. The owner should have its legal counsel review the entire contract and answer any questions remaining in the owner's comprehension of the contract. The owner should not execute the agreement without a thorough understanding of and concurrence with its contents.

Signing the Contract

After the owner has signed the agreement, an act not usually entrusted to the architect, the owner and contractor should sign all the rest of the contract documents. This is important, as it identifies the documents which comprise the contract. Drawings and specifications, in particular, undergo many revisions, and a substantial dispute could develop as to which edition of a document is the basis of the contract. In the event that the owner or contractor or both fail to sign the documents for identification, the architect is required to identify such unsigned documents upon request. (A201, 1.2.1) (See Chapter 2, The Contract Documents.)

After the Contract is Signed

Prior to start of construction, the contractor will submit a list of intended subcontractors and suppliers for consideration by the owner and architect. This should be carefully reviewed and the contractor immediately notified if, after due investigation, the owner or architect has reasonable objection to any proposed person or entity. The contractor should then propose other names acceptable to the owner and architect. In the event the substitution requires a change in the contract sum, an appropriate change order should be issued. (A201, 5.2)

It is important that the work of the contract not be commenced prior to the effective date of all insurance and bonds or prior to the imposition of lenders' liens. All governmental approvals and permits must also have been obtained. When all these preliminary essentials have been accomplished, the owner should give written notice to the contractor that construction may commence. In the event that the owner neglects to issue the notice to proceed, the contractor must give 5 days' written notice to the owner before proceeding with work on the site. This would alert the owner in time that the contractor could still be stopped if all the preliminary technicalities have not been completed. (A201, 8.2.2)

The owner and architect must also receive the contractor's time schedule, which should indicate completion on or before the agreed completion date. (A201, 3.10) The purpose of the submittal is for receipt of scheduling information only and not for critical review and approval, as the contractor is solely responsible for construction procedures, techniques, and sequences.

The contractor will have to engage a testing laboratory acceptable to the owner to render certain specified testing, inspecting, and approval services. The owner should promptly accept or reject the nominated testing laboratory in writing. The contractor will pay for all specified testing, while the owner must pay for any additional testing. (A201, 13.5)

The various administrative notices required of the owner, such as a notice to proceed or to correct work, are usually taken care of by the architect and often are in the form of an entry in minutes of a meeting and distributed to all interested parties. Legal notices such as a demand for arbitration, notification of surety, or declaration of termination should be prepared by the owner's legal counsel.

During the construction period, the architect is required to visit the site from time to time to carry out its obligations under the owner-architect agreement and the construction contract. The architect should notify the owner in advance of all site visitations to enable the owner or its representative to be present if so desired. The owner should attend the preconstruction jobsite meeting if one has been arranged, even though the owner has no contractual duty to visit the site at any time. (See Chapter 7, The Preconstruction Jobsite Conference.)

Owner Must Furnish Information

If requested in writing, the owner is required to furnish to the contractor information about the project site needed for giving notice and filing or enforcing mechanics' lien rights. (A201, 2.1.2) The owner is also required to furnish promptly, if requested, reasonable evidence that the owner will be financially capable of meeting its obligations under the contract. (A201, 2.2.1)

The owner must also provide land surveys, descriptions of legal limitations, utility locations, and a legal description of the site. (A201, 2.2.2) It is the owner's responsibility to secure and pay for necessary approvals, easements, and assessments exacted by governmental regulation, except for those fees and permits specified to be the contractor's responsibility. (A201, 2.2.3) Any information or services which are the owner's responsibility must be furnished with reasonable speed to avoid causing delay in the work. (A201, 2.2.4)

Responsibilities of The Owner in a Construction Contract

Insurance and Bonds

The owner is responsible for purchasing and maintaining owner's liability insurance, boiler and machinery insurance, and property insurance. (A201, 11.2 and 11.3) Property insurance shall include fire and extended coverage and physical loss or damage including theft, vandalism, malicious mischief, collapse, falsework, temporary buildings, and debris removal including demolition caused by enforcement of any applicable requirements, and covers reasonable compensation for architect's services and expenses required as a result of such insured loss. (A201, 11.3.1.1) Although Document A201 requires copies of all specified owner's insurance policies be provided to the contractor before start of construction (11.3.6), common practice is to submit certificates of such insurance.

If the owner does not intend to purchase property insurance as specified, the contractor must be notified of this prior to start of work. The contractor can then arrange for the necessary insurance to cover the interests of the contractor, subcontractors, and sub-subcontractors. The architect should prepare a change order charging the costs to the owner. (A201, 11.3.1.2.)

Insurance certificates, as evidence of the contractor's compliance with the insurance requirements of the contract, are collected by the architect and submitted to the owner for approval. (A201, 11.1.3) The owner should have its insurance counsel review the certificates before approving or disapproving. This should be done promptly, because all insurance and bonds should be in proper form and in effect before any work of the contract is undertaken on the site.

The owner has the right to require the contractor to furnish a performance bond and payment bond, both in the full amount of the contract. If this requirement is in the bidding instructions or contract documents prior to execution of the contract, the bonds must be provided within the contract sum. If the bond requirement is imposed after the contract sum has been established, the premium will be added to the contract sum. (A201, 11.4.1) (See Chapter 9, Construction Insurance.)

Payments

The owner's most evident obligation is to provide the financial means for the mobilization and application of resources by the contractor. A steady, timely, and reliable flow of funds is essential to keep the personnel and materials supplied to the construction process.

Before signing the contract, if requested by the contractor, the owner must provide reasonable evidence of the financial arrangements which the owner has made to fulfill its obligations under the contract. If this is not forthcoming, the contractor is not obligated to sign the contract or proceed with the work. After the work is started, the contractor can from time to time request this information and, if it is not promptly furnished by the owner, may stop the work. (A201, 2.2.1 and 14.1.1.5) (See Chapter 20, Termination of the Construction Contract.)

Prior to the first application for payment, the contractor must submit a schedule of values allocated to various portions of the work. The schedule, to be reviewed and approved by the architect, will be used as the basis for reviewing all payment applications. (A201, 9.2.1)

According to the contract, as each payment to the contractor becomes due, the contractor is to apply for it using a form similar to Application and Certificate for Payment, AIA Document G702, May 1983 Edition, and the Continuation Sheet, Document G703. The application should be made to the architect, who is obligated to make a site examination to verify the percentages of completion of all items. If the architect agrees that the stated degree of completion of each item is correct, the architect will execute the certificate on the contractor's application form. If the architect certifies a lesser amount than was applied for, an explanation should be attached. The owner is then required to pay the contractor the amount within the agreed time period. (A201,9.3 and 9.4)

If any payments are paid after the due date, they should bear interest at a rate to be agreed by the parties, otherwise at the legal rate prevailing from time to time at the site of the project. (A201, 13.6.1) While any claim, including arbitration, is pending, the owner is obligated to continue making payments when due and the contractor is required to continue with the work. (A201, 4.3.4 and 4.5.3)

The contractor's payment application may include the value of materials and equipment suitably stored on the site unless the contract specifically provides otherwise. Materials and equipment stored or being fabricated off the site may be billed to the owner only if the owner and contractor have previously agreed to this. Such an agreement should include location, protection of owner's title, security from fire and theft, transportation, and insurance. (A201, 9.3.2 and 10.2.1.2) Applications for payment may include amounts for the percentage of work completed on change orders and construction change directives. (A201, 9.1.1 and 9.3.1.1)

Owner's Failure to Pay

Should the architect fail to issue a certificate within 7 days after the contractor's application, through no fault of the contractor, or if the owner fails to pay the certified amount within 7 days after the date specified in the contract, then the contractor upon 7 days' additional written notice, may stop the work until payment has been received. When the delinquent payment has been made and the work resumed, the architect will issue a change order charging the owner with the costs of shutdown, delay, and start-up and a corresponding extension in the contract time. (A201, 9.7.1) (See Chapter 20, Termination of the Construction Contract.)

Owner's Right to Stop the Work

When the contractor persistently fails to carry out the work in accordance with the contract documents or fails to correct improper or defective work, the owner may order the contractor to stop the work until the cause for the order has been eliminated. The stop work order must be in writing and signed personally by the owner or by an agent specifically so empowered in writing. The owner's power to stop the work does not impose a duty on the owner to exercise this power for the benefit of the contractor or anyone else. (A201, 2.3.1) The architect does not possess the power to stop the work of the contract. If in the architect's judgment the work should be stopped, it will be necessary to obtain the owner's concurrence and action. (See Chapter 20, Termination of the Construction Contract.)

Owner's Right to Carry Out the Work

If the contractor defaults or does not carry out the work in accordance with the contract documents, the owner may, with prior certified approval of the architect, give the contractor notice to correct the default or defective work within 7 days. If the contractor fails to proceed promptly and diligently, the owner must give a second written notice and 7 additional days. If the contractor after the second notice and during the second 7-day period fails to proceed and continue work diligently, the owner may correct the deficiencies and the architect will issue a change order charging the costs, including additional architectural fees, back to the contractor. If the remainder of the contract sum is insufficient to cover the costs, the contractor must pay the difference to the owner. (A201, 2.4.1)

If the contractor's performance has declined to a level so low that the owner must consider taking such drastic measures as stopping the work or carrying out any of the work, the owner should be in close consultation with the architect and legal counsel. In case of later legal action, which is almost certain to follow, the architect's prior approval of the owner's actions will have to be proved. The architect's approval should be based upon independent opinion and in the form of a written certificate.

If the contractor fails to keep the work site clean as required in the specifications, the owner is entitled to order the work done by others and charge it to the contractor. If a dispute develops among the owner, contractor, and owner's separate contractors as to relative responsibility for cleaning, the owner may clean up and allocate the charge to the various parties in proportions determined by the architect. (A201, 3.15 and 6.3.1)

The owner is entitled to carry out construction work or operations related to the work of the contractor utilizing its own forces or by contracting with separate contractors, and the contractor is required to cooperate in respect to use of the site and mutual scheduling. If the contractor determines that it has been caused extra expense or delay, a claim should be made which will be decided by the architect. (A201, 6.1 and 6.2)

Separate Contractors

The owner is responsible for communications and coordination among its own forces, separate contractors, and the contractor. However, the owner is required to channel all its own communications to the contractor through the architect. (A201, 4.2.4) Communication through less formal channels may lead to confusion, delay, and extra expense.

The contractor is responsible for whatever cutting, fitting, or patching is necessary to complete the work and to make the various parts of the work fit together properly. The contractor is not allowed to damage or endanger by cutting or patching the work of the owner's forces or separate contractors, and conversely they cannot cut or patch the contractor's work without prior written permission of the owner and contractor or other separate contractors. (A201, 3.14)

Owner's Claims Against the Contractor

When unknown concealed conditions are uncovered during excavation or unusual physical conditions are otherwise encountered, if they are more costly than shown in the contract documents, the contractor is entitled to claim for the resulting additional expenses and time. Similarly, when the unknown conditions prove to be easier or less costly than depicted in the contract documents, the owner in some cases could be entitled to a credit on the contract price and time. In either event, the architect should investigate the situation, make a determination, and issue an appropriate change order. (A201, 4.3.6)

The architect is authorized to order minor changes in the work providing there is no change in the contract sum or time. (A201, 7.4.1) If either owner or contractor disagrees with the architect's order for a minor change, the only relief is to arbitration, unless the architect is willing to revise the order.

Any other owner claims against the contractor should first be submitted to the architect for investigation and determination. The architect, after obtaining the contractor's viewpoint, will make a decision binding on both parties but subject to arbitration if either party wishes to contest it. All claims for the architect's decision or interpretation should be in writing.

Responsibilities of The Owner in a Construction Contract

Defective Work

The architect is the final judge of the acceptability of the contractor's workmanship and materials. The contractor is obligated to remove and replace any work which the architect deems not in compliance with requirements of the contract. However, if the owner desires to accept nonconforming work rather than have it removed and replaced, an appropriate and equitable adjustment will be made in the contract price. (A201, 12.2.3 and 12.3.1) Construction defects coming to the attention of the owner during the warranty period should be promptly reported to the contractor. (A201, 12.2.2)

Owner and Architect

By the time the project is ready for construction, the owner and architect will have been working closely together for several months or longer. During this time the architect has been dealing with the owner on the basis of an objective professional advisor. Although the client will consider the architect's opinions and advice, the client has the last word in all matters and makes the final decisions. However, after the construction contract is signed, the relationship must change. The architect will still be the owner's professional advisor but must also be an independent judge of the performance of the owner and contractor under the contract and must make fair rulings on the claims of owner and contractor alike. (A201, 4.2.12) Sometimes owners feel that their architect should side with them because they are paying the fees, but contractors would not freely enter into this form of construction agreement if the architect could not be trusted to render objective and impartial rulings, fair to both sides.

In the event of the architect being fired, deceased, or having resigned, the owner is required to appoint a new architect, acceptable to the contractor, to fulfill the duties of architect in the administration of the construction contract. (A201, 4.1.3)

The Ideal Owner

The owner will receive the most from the construction process by reviewing all submittals promptly when received, making all decisions promptly, reporting all noted defects and deficiencies promptly, conferring with retained experts when necessary, and paying all payment certificates when due. Owners, particularly of larger, more costly projects, should confer with insurance counsel on all insurance and bond matters, construction industry lawyers on legal concerns, and experienced accountants on accounting subjects. In general, the owner should be responsive, cooperative, and reasonable and should not interfere with the contractor's progress.

If the owner lacks confidence in its expertise or finds these important duties onerous or excessively time-consuming, serious consideration should be given to extension of the architect's administrative duties or, alternatively, the engagement of a construction manager, either independent or on staff, to carry out some or all of the diverse owner functions required during the construction period.

Custom Construction Is a Complex Venture

Construction contracts are not self-executing. They must be administered in behalf of both parties. Many prospective construction owners casually enter into construction contracts as though they were buying a standard commodity off the shelf, utilizing a cut-and-dried procedure.

Entering into such a complex and economically significant venture with this casual attitude will frequently produce poor results and is likely to leave the unenlightened owner in a state of profound disappointment, possible financial losses, and with a terminal case of distrust of the construction industry.

Many owners labor under the naive delusion that there must be some governmental entity to look after their interests so they have nothing to be concerned about. They do not realize that most construction inspection agencies, such as the building department, perform only cursory spot checks to test general compliance with the building code. They do not assure specific compliance with all of the requirements. Building regulations are minimum standards of health and safety, usually considerably less restrictive than the typical construction contract.

Other governmental agencies are involved in policing the provisions of their particular area of law enforcement. For example, the contractors' licensing boards of each state will respond to complaints alleging violations by their licensees. It is not their function to monitor contract compliance by contractors for individual owners.

Lenders and sureties sometimes send their representatives to the building site for the purpose of protecting their own interests. Most of the time, they do not even report these visits to the owner.

Essentially, the owner's position is one of having to watch out for and protect its own interests.

Owners are Unique

Of the two principal parties in a construction contract, one is an experienced general contractor who presumably knows what to do, knows what is expected of it in the agreement, and knows how to protect its interests.

The other party is the owner, a nonstandard, one of a kind, unique entity. Owners run the gamut from the novice beginner to the extremely capable and well seasoned veteran. Many owner entities are governed by lay boards or committees comprised of usually competent management people, although possibly unschooled in the construction process. Whereas contractors and all of their personnel are constantly in the construction business, many owners are not. Contractors learn from their mistakes and, after considerable experience, generally know how to avoid problems. On the other hand, unseasoned owners do not have the educational advantage of past mistakes. Every routine event will be a new experience and elementary mistakes will undoubtedly be made.

The owner entity is legally empowered to enter into any contract and is free to conduct its affairs in any manner that in its sole judgment it deems advisable. If the owner is knowledgeable it will be able to make its decisions rationally and reasonably. Those owners who have the necessary judgment and expertise, or have it available in-house, will know when it is capable of acting in its own behalf and when it should seek outside advice and assistance.

The novice owner will probably react to each demand made upon it as a separate crisis to be met as it occurs, unrelated to the complete process, and with no idea of what is yet to come. It will often respond to ordinary problems in an unexpected manner, sometimes considered bizarre by the contractor. It will not be able to look ahead and to anticipate each demand that will be imposed on it for appropriate decisions and provision of pertinent information. It will not necessarily know when it is time to seek outside advice and counsel.

The contractor is perfectly capable of looking out for its own interests, whereas the unsophisticated and untested owner enters the fray as an inadequate and unequipped adversary.

Owner's Rejection of Work

Typically, the inexperienced owner imagines the process as one where the owner's main duty is to pay the contractor's invoice periodically during the construction period, with few, if any, other duties. This is usually accompanied by the mistaken belief that the contractor's sole aim is to please the owner. The owner is astounded when the contractor objects to changing completed work rejected by the owner because the owner does not like it or it was not what was expected. The owner is stunned when the contractor demands extra compensation. The owner does not always realize that the basis for rejection of work must be on objective grounds such as what was specified, the building regulations, and upon industry standards, but certainly not on the unrealistic standard of an owner's faulty visualization.

Owner's Untimely Decision Making

The neophyte owner seldom realizes the disastrous effect that delayed, imprecise, or piecemeal decisions can have on the construction schedule. All owner options such as items to be purchased from allowances, color choices, and model selections must be decided and transmitted to the contractor well in advance of long lead ordering deadlines. Late decisions have the dual effects of lessening the available choices and delaying construction.

Normally, the owner is responsible for obtaining all governmental authorizations for the use of the land and for the construction. This includes zoning approval and the general building permit. It is also the owner's responsibility to furnish the building site and all rights of way, easements, and access to public utilities. If all of these approvals and facilities are not provided when needed and the contractor is caused any delay, it must be charged to the owner.

Effects of Delayed Completion

When construction is delayed, the unknowing owner will customarily assume that the primary effect will be deferment of building occupancy. If the contract included a liquidated damages provision, the owner's conclusion is that the contractor will simply reduce the contract price by the agreed amount per day of delayed occupancy and that this will cover the owner's extra costs. The uninformed owner would perhaps not realize that the contractor will be seriously expecting to recover all increased operating costs incurred by any construction delay caused by the owner or the owner's agents. In addition, the contractor will expect an equitable extension of the agreed completion date for all time lost due to inclement weather as well as by other factors, such as strikes and unavailability of materials.

Unexpected Charges for Extra Work

Inexperienced owners usually have difficulty in objectively evaluating the contractor's requests for additional compensation for extra work. Often the owner will ask for additions to the contract, seemingly of little or no apparent economic consequence, and is surprised later when a change order billing arrives. The contractor also usually requests a corresponding deferral of the completion date.

This type of misunderstanding often occurs when extra costs and time are not completely discussed and agreed upon at the same time extra work is authorized. Usually, the owner expects that there will be no charge because the contractor said nothing about money, while the contractor assumes that any intelligent owner should have known that nothing is free and that an appropriate charge would be processed in due course. Often the extra work would not have been ordered if the owner had known the complete cost and time consequences.

Even in those cases where owners recognize the fairness of an extra charge for additional work, they are often staggered by the large number of ancillary add-ons that comprise the complete charge. It is an edifying revelation to discover how a bill for labor and materials can grow insidiously by the customary addition of wage mark-up, transportation, supervision, insurance, bond, tool rental, scaffolding rental, overhead, and profit. It might also include the work of subcontractors with their own overhead and profit. These are all legitimate costs if incurred.

It is even more mystifying to the owner to review credit invoices. When work is eliminated from the contract, the credit is often reduced by charges for re-creating, transportation, and restocking. Deductive change orders seldom provide credit for the contractor's or subcontractors' overhead and profit. (See Chapter 14, Change Orders.)

Completion Formalities

At the end of construction, the untrained owner, after taking control of the building and moving in, feels there is only one formality left to accomplish, and that is to pay the contractor's final invoice. Most, if not all, of the usual closing out procedures which are normally taken care of by the expert, are usually neglected or overlooked by an untrained owner. The notice of completion may not be filed, written guarantees not collected, operating instructions for mechanical equipment not secured, spare parts not obtained, lien releases forgotten, records not kept of the names and addresses of subcontractors and suppliers, and as-built record drawings not assembled and preserved.

Prior to expiration of the contractor's one year warranty period, the neophyte contract administrator seldom thinks about arranging for a comprehensive examination of the building to have the original contractors fine tune their work. When defects surface shortly after expiration of the warranty period, the owner is resentful that contractors are reluctant to return without additional compensation. Unversed owners do not always realize that the contractor and subcontractors have no further obligation under the contract after the warranty period has passed. (See Chapter 19, Closing Out the Job.)

Solving the Problem with a Tough Contract

The amateur do-it-yourself owner-administrator is sometimes aided and abetted by the general practice lawyer who is not experienced in the practicalities of construction law and the peculiar customs and practices of the construction industry.

The inexperienced owner, teamed up with the uninitiated lawyer reason that a really tough construction contract will whip the errant contractor into line and protect the owner's interests. So they obtain an AIA standard form contract and go through it very carefully, paragraph by paragraph, removing all provisions that in any way could possibly be construed as being favorable to the contractor. They then add very stringent provisions giving the owner every imaginable advantage over the contractor. If the owner succeeds in finding an unsuspecting contractor willing to sign the resulting agreement, then the construction process and the owner-contractor relationship are off to a rough start. (See Chapter 4, Construction Contracts.)

16

The Contractor's Responsibilities

What is Expected of the General Contractor

Contractors are generally aware of the main features of a standard building contract and only experienced owners are as well informed. Essentially, the contractor is required to furnish all of the necessary labor and materials needed to erect the building and site improvements and the owner is obligated to pay for it. Time is an important element in most construction contracts, both as to the contractor's completion of the work and when payments must be made by the owner. The additional terms and conditions of the contract necessary to make it workable will vary widely depending on which of various available standard forms of agreement are used and how they are amended to suit the requirements of the situation at hand.

Most of the standard form contracts issued by contractor organizations tend to favor the contractor. They also tend to be completely silent on the duties and status of the architect. It is not expected that the contract would be administered by an architect. The standard contract forms of the AIA are thought to be fairly written to give equal weight to the interests of owners and contractors.

The two main owner-contractor agreements issued by the American Institute of Architects are:

> Standard Form of Agreement Between Owner and Contractor (where the basis of payment is a Stipulated Sum), Twelfth Edition, AIA Document A101, 1987, (Appendix B)
>
> Standard Form of Agreement Between Owner and Contractor (where the basis of payment is Cost of the Work Plus a Fee with or without a Guaranteed Maximum Price), Tenth Edition, AIA Document A111, 1987 (Appendix C)

When either agreement is used, depending on the payment method agreed, the companion must be:

> General Conditions of the Contract for Construction, Fourteenth Edition, AIA Document A201, 1987 (Appendix D)

These documents were written to respect the rights and obligations of contractors and owners alike and reflect the presence and role of the architect during the construction process. (For other standard AIA Construction Contract Forms, See Chapter 4, Construction Contracts.)

The responsibilities of the contractor under the AIA General Conditions are in addition to the duties imposed by the usual customs and practices of contractors in the same community. Also, both parties are bound by the implied covenants of fair dealing and reasonableness present in any contract. Additionally, the architect and owner will naturally expect the contractor to provide all of the unwritten behind-the-scenes services and adjuncts customarily a part of any competent contracting business. The contracting firm should have a complete organizational structure, appropriately equipped and staffed to provide estimating, purchasing, contracting, accounting, engineering, and construction services for all projects undertaken.

The firm should have secretarial and clerical personnel and procedures sufficient to initiate and respond to all written communication requirements of the contract. The operation should be financially capable of performing any contract it undertakes and should be appropriately licensed to do business in any jurisdiction in which it operates. All of these diverse functions might be performed by a single individual in a one or two person firm. The firm should be able to carry out all necessary bookkeeping functions, to account for all funds received and disbursed, and to render bills and accounts when necessary.

Signing of the Contract

The contractor should sign the contract in the proper name of the contracting firm, and the person signing should have the legal power to bind the firm. Some corporations and partnerships require signatures of two officers or partners on all agreements. The contractor, for its own protection should make sure that the owner entity is also signing properly. In addition, owner and contractor should sign all of the contract documents for identification and each party should retain a complete set of signed documents for their own files. The architect should maintain a complete identical duplicate set, marked for identification but not necessarily signed by the parties, for use in administration of the contract. Should the parties neglect to sign all of the documents, the architect must be able to unequivocally identify the contract set if later requested to do so. (AIA Document A201, Subparagraph 1.2.1)

The contractor's signing of the contract is a representation that the contractor has visited the site, is familiar with local conditions, and has correlated visual observations with requirements of the contract documents. (A201, 1.2.1) It is also the contractor's confirmation that the specified contract time is a reasonable period for performing the work. (A201, 8.2.1)

Drawings and Specifications

When the work of the contract is finally completed, the contractor must return or suitably account for all drawings and specifications to the architect except for one record set. The contractor is not entitled to use the drawings or specifications on other projects. (A201, 1.3.1)

The contractor is required to study and compare the contract drawings and specifications and the land survey and to report any noted errors, discrepancies, or building code violations to the architect. Before starting work, the contractor must also take field measurements and compare these with the contract documents, reporting all discovered errors and discrepancies to the architect. (A201, 3.2.1, 3.2.2 and 3.7.3) The contractor has no obligation to find any errors and is not responsible for the existence of any errors or code violations in the documents, discovered or undiscovered, but could be held responsible for proceeding with work known to be erroneous or in violation of building codes. (A201, 3.7.4) The contractor could be held proportionately liable for failing to notice errors which should have been recognized, given the average contractor's knowledge and skill. The contractor should never proceed in uncertainty.

Errors, discrepancies, and ambiguities discovered and reported should be resolved by the architect. This might require redesign, issuance of revised drawings or specifications, and an appropriate adjustment in the contract sum and time.

Contractor Responsible for Results

The contractor has the general responsibility:

> To perform the work in compliance with the contract documents and approved submittals (A201, 3.2.3)
> To supervise and direct the work (A201, 3.3.1)
> To be responsible for the acts of its employees and subcontractors (A201, 3.3.2)
> To enforce discipline on the job and not employ unfit persons (A201, 3.4.2)
> To inspect the work of each trade to assure that it is suitable for subsequent work (A201, 3.3.4)
> To pay for all labor and materials, tools, equipment, water, heat, utilities, and services necessary for and reasonably incidental to the proper execution and completion of the work (A201, 3.4.1)

The contractor is solely responsible for and should have control of the construction means, methods, techniques, sequences, and procedures for coordinating the work of the contract. (A201, 3.3.1) Neither the architect nor owner should interpose its wishes or interfere with the contractor's performance of any of these functions as this would jeopardize the objective of giving the contractor full and unrestricted control over these areas of primary responsibility. It is the contractor's prerogative to let subcontracts and purchase materials without respect to the architect's choice of arrangement of specification subdivisions or construction drawings. (A201, 1.2.4)

Construction Quality

The ultimate quality of the construction product is primarily in the control of the contractor. The contractor establishes the standard of workmanship integrity for its organization. This is enforced in many direct as well as subtle or indirect ways. The contract administrators, in negotiating subcontract agreements and purchase orders, can enforce the quality standard or can undermine it by bargaining it away. The contractor's superintendent on the job as well as the subcontractors' supervisors can insist on high workmanship standards or they can conspire to slide by on the minimums barely acceptable under the specifications. The general contractor can hold the subcontractors to the specified quality or higher, or can act as the agent of mediocre subcontractors or suppliers in persuading the architect to approve marginally acceptable work.

Taxes and Licenses

The contractor must:

> Pay all sales, consumer, or use taxes (A201, 3.6.1)
>
> Secure and pay for building permits and governmental fees, licenses, and inspections (A201, 3.7.1)
>
> Pay all royalties and license fees (A201, 3.17.1)
>
> Comply with and give notices required by all laws, building codes, and ordinances (A201, 3.7.2)

In addition, the contractor must pay for any taxes and licenses that are specified in the supplementary or other conditions or in the trade sections of the specifications. The owner is ordinarily responsible for taxes and assessments that relate to the land or its use, such as property taxes and assessments for sewers, water, street lighting, schools, and similar fees and charges.

Safety and Accident Prevention

Safety on the construction site, provision of safeguards and warnings, and the giving of safety notices are all in the province of the contractor. (A201, 10.1.1 and 10.2.1 through 10.2.3) The contractor is required to appoint a person to be responsible for accident prevention on the jobsite, and it is understood to be the contractor's superintendent unless the contractor designates otherwise. (A201, 10.2.6)

The contractor is obligated to proceed with the work expeditiously, to assign adequate personnel, materials, and supervision, and to achieve substantial completion within the contract time. (A201, 8.2.3)

Separate Contractors

When the owner has contracted directly with more than one contractor to work simultaneously or serially on the same site, the contractors are known as separate contractors. This would be the case, for example, when one contractor has been engaged to construct the building while separate contractors have been hired for installation of utility lines or for installation of laboratory equipment or painting. In such situations, it is essential that the contractors cooperate with each other. So it is up to the owner when contracting with the separate contractors to anticipate the cooperation problems and to provide for the necessary coordination and communication. (A201, 4.2.4 and 6.1.3)

The owner should contract with each of the separate contractors on substantially the same basis to avoid conflicting provisions which would inhibit or prevent the necessary coordination and cooperation. Each separate contract should incorporate the same or similar general conditions including complementary insurance requirements. The AIA General Conditions contains appropriate provisions requiring cooperation among separate contractors and the owner's employees. (Article 6) These provisions are necessary to establish the owner's right to do work with its own forces and to award separate contracts without encountering unexpected or unreasonable objections from other contractors working simultaneously on the same site. The site and its facilities or limitations must be shared on a fair basis. Such necessities as parking, electrical, telephone and water services, storage areas, loading and unloading areas, office spaces, temporary toilets, scaffolding, shoring, and hoisting facilities must often be provided on a cooperative basis. Complete duplication of these services would be needlessly extravagant, so it is to the owner's advantage to specify which contractor is to provide each element and the basis on which the other contractors will be allowed to participate in its use.

Although each separate contractor is responsible for its own scheduling and completion dates, it is the owner's responsibility to schedule the separate contractors and its own employees and to coordinate the schedules. Where the work of the separate contractors comes together, each of the contractors is obligated to report discrepancies which prevent their work from properly fitting. It is the owner's responsibility to resolve all differences between separate contractors. The separate contractors are mutually responsible to refrain from damaging each other's work in process or the completed work and to pay for any damage which they cause. (A201, 3.14.2)

Communications

Otherwise well-managed construction and administration processes will deteriorate into misunderstandings when logical communication channels are not established and consistently utilized. Communications between the owner and contractor should always be routed through the architect. Similarly, the contractor must be the conduit for all communications with subcontractors and suppliers. Communications with the architect's consultants should be through the architect, and the owner is the medium for communications with or among the separate contractors. (A201, 4.2.4) The owner must also provide the communication channel for all of the owner's consultants and advisors.

Subcontractors and Suppliers

The reputation of a general contractor in respect to meeting progress schedules and promised completion dates, providing superior workmanship, and quoting fair and competitive prices is directly related to its assemblage of subcontractors and suppliers with which it regularly does business. Well-managed contracting firms endeavor to establish and cultivate long-standing relationships with their principal subcontractors and suppliers. Thus they are assured of continuing availability of a cadre of high quality skilled craftspeople and competent supervision under the control of financially capable subcontractors.

The general contract does not create any contractual relationship between the owner or architect and the subcontractors or suppliers. The contractual relationships between the general contractor and the subcontractors and suppliers will be created by the subcontracts and purchase orders issued by the contractor. (See Chapter 18, Architect's Relationship With Subcontractors.)

The contractor is required to submit a list of the proposed principal subcontractors and suppliers to the architect for transmission to the owner. The architect should respond in writing stating whether or not the architect or owner, after appropriate investigation, has any reasonable objection to the contractor's list. Should the architect fail to reply promptly, this is deemed to be acceptance of the list. (A201, 5.2.1)

The contractor cannot contract with any subcontractors or suppliers to which the architect or owner has made timely objection, and the contractor is not required to contract with any entities to which it has made reasonable objection. (A201, 5.2.2) Should the substitution by a mutually acceptable entity require a revision in the contract sum, an appropriate change order should be issued charging or crediting the difference to the owner. (A201, 5.2.3) After the subcontractor list has been established, the contractor cannot make a change in it if the owner or architect lodges a reasonable objection to such change. (A201, 5.2.4)

Hazardous Waste Materials

In the past, certain substances were used in construction which have since been identified as "harmful to your health". Prominent among these materials are asbestos fibers, often used in acoustical walls and ceilings, flooring, siding, and insulation, and polychlorinated biphenyl (PCBs), used in high capacity electrical transformers. These toxic substances must be removed or rendered harmless.

If hazardous materials such as asbestos or PCB which have not been rendered harmless are encountered on the project, the contractor is required to cease working in the affected area and to notify the owner and architect of the condition in writing. It is necessary then to determine definitely whether or not the suspected materials have been accurately identified. This requires testing which would be at the owner's expense. (A201, 13.5.1) If the suspected materials are proved not to be hazardous, the contractor should promptly resume work. (A201, 10.1.2 through 10.1.4) However, should the materials be confirmed as hazardous, the contractor is justified in discontinuing work in the affected area until it has been made safe. The owner should then make arrangements with an experienced hazardous materials contractor competent to remove or render harmless the offensive materials. Only when the work area is considered safe and harmless should the original contractor resume the work of the contract.

Most professional liability insurance policies carried by architects exclude all activities involving asbestos or PCB. Therefore, upon the discovery or suspicion of the presence of these or other toxic materials on the site, architects should decline to render any services involving their discovery, removal, or disposal. Those architects who have made a specialty of problems dealing with toxic substances must obtain an appropriate special endorsement from their liability insurance carriers.

According to Paragraph 9.8. of AIA Document B141, neither architects nor their consultants are required to assume any responsibility in respect to the discovery, presence, handling, removal, or disposal of hazardous materials or the exposure of persons to them in any form at the project site.

Contractor's Warranty

A warranty is a written statement guaranteeing the integrity of a product and promising the purchaser that the provider will be responsible for repairing or replacing the product for a period of time. In the General Conditions, the contractor warrants that the materials and equipment furnished under the contract will be new and of good quality unless otherwise specified, and that the work will be free of defects and in conformance to the contract documents. The contractor's warranty naturally excludes any defects caused by the owner's neglect such as inadequate maintenance or improper operation. Normal wear and tear is also excluded. (A201, 3.5.1)

Warranty Period

The customary construction industry warranty is one year. There is no reason a shorter or longer period could not be specified if desired. However, substantial deviation from the norm would be likely to create strong resistance from most building contractors as well as an addition to the contract sum to cover the extended warranty time period.

Upon the owner's discovery of conditions which the contractor must repair or replace during the warranty period, the owner must give a timely written notice to the contractor. The 1-year warranty period starts to run as of the date of substantial completion. For work first performed after substantial completion, the start of the 1-year period is extended to one year after the actual performance of the work. (A201, 12.2.2)

Warranties are not required to be in writing, but it is common practice for architects to specify that the contractor's 1-year warranty be submitted in writing. This should

The Contractor's Responsibilities

be submitted by the contractor along with all other specified warranties as part of the contract closing out procedures.

Architects, when transmitting the contractor's written warranties, should refrain from rendering any opinions as to the sufficiency or language of the warranties as this would constitute a legal opinion. The transmittal letter should merely identify the warranties as to the specification section and contractor and contain a recommendation that the owner seek legal review of all warranties.

Warranties are only as good as the contractors who have executed them. If the contractor has gone bankrupt or is otherwise financially incapable of providing the warranty service at the time it is needed, the warranty is worthless. To provide a backup to the general contractor's warranty, it is standard practice to specify written warranties from the major subcontractors, manufacturers, and suppliers such as electrical, air conditioning, roofing, and plumbing. This yields a greater protection to the owner, as the general contractor's resources would then be supplemented by those of the subcontractor.

It is quite common to require a longer warranty period on roofing, such as two years. Should the owner require a longer warranty, a roofing bond of 10, 15 or 20 years is available through some manufacturers upon payment of a bond premium and the involvement of a surety that will issue a roofing bond. Only roofs of a specified higher quality are eligible for roofing bonds.

The contractor's customer service during the warranty period is of extreme importance to most owners, and this fact is recognized by high-quality contracting firms. Although the contractor has no obligation to continue rendering warranty services after expiration of the specified warranty period, many firms voluntarily follow through on repair work for a longer time as a matter of public relations and as a way of cultivating prospects for future business.

Some contractors are not cooperative in performing corrective work during the warranty period, and in those cases, the owner would have had to withhold contract funds to be disbursed only after satisfactory completion of the 1-year warranty period. This lengthy retention of funds cannot be done unless the contract so provides.

When the job is bonded, the work of the warranty period is included in the bond. Should the contractor refuse to respond on warranty work, the bonding company should be notified.

Architect's Services During Warranty Period

The architect's services under AIA Document B141 terminates at the earlier of issuance to the owner of the final certificate for payment or 60 days after the date of substantial completion. (B141, 2.6.1) Thus, most of the services rendered in the owner's interest during the warranty period will have to be self-performed unless the architect is specifically retained for these functions. Among the services of value to the owner at this time would be the receiving, evaluating, and channeling of the owner's service requests to the contractor and a comprehensive examination of the premises just prior to expiration of the warranty period. Any list of warranty corrections submitted to the contractor should be carefully qualified to eliminate items which are not the contractor's responsibility. The owner is responsible for defects caused by normal wear and tear, abuse, insufficient maintenance, improper operation, and alterations to the contractor's work. (A201, 3.5.1) (See Chapter 26, Analyzing Liability for Construction Defects.)

Contractor's Indemnification of Owner and Architect

The AIA General Conditions contains a broad indemnification clause in which the contractor undertakes to pay any losses or claims against the owner, architect, and architect's consultants and hold them harmless from certain claims and losses arising out of the contractor's and subcontractors' negligent acts. Although the indemnification applies even when the owner, architect, or architect's consultants are partly at fault, the contractor's obligations do not extend to the liability of the design professionals arising out of :

1. The preparation or approval of maps, drawings, opinions, reports, surveys, change orders and designs
2. The giving of or failure to give directions or instructions by the design team, provided such giving or failure to give is the primary cause of the injury or damage. (A201, 3.18.1 through 3.18.3)

Limitation on Owner's Rights to Appoint New Architect or Change Architect's Duties

Should the owner and architect wish to agree in writing to restrict, modify, or extend the architect's construction phase duties, it will be necessary to obtain the contractor's consent, which consent must not be unreasonably withheld. (B141, 2.6.1, and A201, 4.1.2)

If the architect's employment is terminated or if the architect dies or becomes incapacitated, the owner is required to engage a replacement architect against whom the contractor has no reasonable objection. The new architect's duties and status under the contract are the same as those of the former architect. (A201, 4.1.3)

The Building Contractor's Dilemma When Confronted with Defective Construction Documents

There are very few general contractors, subcontractors, or suppliers who, at one time or another, for one seemingly compelling reason or another, have not made the deliberate decision to deviate from the contract drawings or specifications. Clearly such departure from the agreement, made without the formality of obtaining the owner's prior approval, is usually perceived by the contractor as a practical business necessity which will maintain the project momentum, eliminate excessive administration and avoid delay. Sometimes it is purely for economic motives.

One of the main reasons contractors give for defection from the contract requirements is that the plans or specifications are incorrect, impractical, or ambiguous or at least they have that appearance. Experienced contractors have always known that it is unrealistic to expect a complex set of construction documents to be completely free from error or uncertainty. Architects and engineers are painfully aware that there is no such thing, literally, as a perfect set of plans and specifications.

The Contractor's Legal Responsibility. The contractor who faithfully executes a fixed price contract in accordance with erroneous contract documents will probably be safe from attack by the owner in the event that the completed work fails to live up to the owner's expectations.

However, the situation is different in the case of a cost plus contract. A contractor under a cost plus contract is under a legal duty to correct such errors in the contract documents as would be detected by a reasonably competent contractor when the correction of such errors would achieve substantial cost savings.

Where the contract includes the AIA General Conditions, the contractor is required to carefully study the contract documents and at once report to the architect any error, inconsistency, or omission. (A201, 3.,2.1 and 3.2.2) Even if a contract does not contain this provision, or a similar one, it would be prudent to notify the architect or the owner upon discovery of any errors in the contract documents. This is especially true if the defect might be injurious to a third party.

The practical advantages to the contractor in reporting errors discovered in the documents are twofold: to avoid liability for execution of defective work, and to capitalize on the opportunity to make appropriate monetary and time schedule adjustments to the contract.

A contractor who knowingly executes work in accordance with erroneous contract documents will be liable to a third party if it was foreseeable that a third party might sustain injury as a result of the execution of the work.

Correction of Errors Can Effect Contract Price and Time. There is no question that the contractor is entitled to an increase in compensation if the amended contract requirements result in greater costs. Similarly, the time for performance of the contract might also be affected. Conversely, in some cases the corrected documents will result in a reduction of cost or construction time and the contract should be adjusted accordingly.

Contractor's Voluntary Correction of Errors. One possibility when the contractor is aware of defects in the documents, is to voluntarily correct the perceived errors and execute the work in the manner which the contractor thinks should have been originally required. In some cases this could be very helpful to the owner and architect and the contractor's action would be deemed appropriate if not heroic. However, in many instances, the contractor's gratuitous assumption of the architect's duty will produce a result unsatisfactory or unacceptable to the owner or architect. The adverse consequences, usually unexpected by the contractor who had expected gratitude, could range from having to shoulder the entire liability for any shortcoming in the new design in addition to any workmanship defects in the construction.

Deliberate Breach of the Contract. A contractor who willfully breaches the contract with the owner, even with good motives, occupies the least favored position in the law. A contractor who unintentionally breaches the contract is given the benefit of the doctrine of *substantial performance*. This doctrine holds that the contractor is still entitled to be paid the contract price, minus an amount that would compensate the owner for the difference between the value of the project as constructed and the value as planned.

However, a contractor who willfully breaches the contract will be charged with the cost of rebuilding the project to conform with the contract documents, even though rebuilding might be uneconomic. For example, a contractor who willfully substituted steel for copper pipe was liable for the cost of removing and replacing the pipe rather than for the difference in value between the project as planned and the project as built.

Moreover, even if the contractor's redesign and construction were perfect and beneficial to the project, it would be very difficult, if not impossible, to collect compensation for any additional costs for extra work not approved by the owner in advance of execution.

Contractor's Expertise. General contractors and subcontractors are expected to be skilled in their respective fields and many times are included in certain bidding lists only because of their unique expertise and specialized experience. On occasion, a contractor will execute construction required in contract documents which contain errors, inconsistencies, or omissions of which the contractor may not be aware. If the contractor follows these erroneous plans to the detriment of the owner or injured third parties, the claim would likely arise that the contractor should have known that the plans were defective or inappropriate. This occurs particularly in those commonplace situations where the error should be apparent to any qualified contractor engaged in that type of construction work.

If a judge or arbitrator agrees with this line of reasoning, the contractor could be held proportionately liable along with the architect who prepared the faulty documents. When erroneously executing the type of error not normally considered within the area of experience of contractors, the contractor would most likely be held not legally responsible.

Errors Caused By Owner's Faulty Directions to Contractor. A misdirected owner, for reasons of economy or otherwise, might order a contractor to deviate from the contract drawings or specifications for the purpose of obtaining a contract credit or other supposed advantage. In the event that the constructed result is considered defective, the disappointed owner might be sustained in the claim that the contractor should have possessed sufficient expertise to predict the improper or poor result and should have advised the owner against the change.

If the contractor had the forethought to advise the owner not to make the design change, it certainly should have been in writing with the reasons stated. A better procedure would have been to refer the matter back to the original architect or engineer for a recommendation and a change order. This would then relieve the contractor of responsibility for this type of design defect.

Contractors' License Laws. In many states, the state contractor's license law imposes the obligation on all contractors to carry out the requirements of the plans or specifications. The license law also prohibits willful or deliberate disregard of the building codes. Thus, a contractor who carries out requirements of the contract documents that violate the building codes may risk suspension or revocation of the license, and might also be disciplined for failure to carry out the requirements of the contract documents. In such a situation, the contractor has no option but to report any noted building code discrepancy to the owner or the architect, and insist on a satisfactory resolution before proceeding with the work.

A contractor's best course is to faithfully follow the plans and specifications unless they require construction known to be erroneous or would violate the building codes. The contractor should carefully check the documents for obvious errors and apply to the owner or architect for additional instructions reconciling all detected anomalies. By insisting on all known errors being clarified by the architect, the contractor can avoid being held legally liable for design defects. It provides the added advantage of being able to adjust the contract price and performance time if these are also affected.

17

Role of the Construction Superintendent

Contractor's Superintendent

The AIA General Conditions, Document A201, 1987 edition, has only two explicit references to duties and responsibilities of the contractor's superintendent. Subparagraph 3.9.1. requires that

> "The Contractor shall employ a competent superintendent and necessary assistants who shall be in attendance at the Project site during performance of the Work. The superintendent shall represent the Contractor, and communications given to the superintendent shall be as binding as if given to the Contractor."

Subparagraph 10.2.6 requires that

> "The Contractor shall designate a responsible member of the Contractor's organization at the site whose duty shall be the prevention of accidents. This person shall be the Contractor's superintendent unless otherwise designated by the Contractor in writing to the Owner and Architect."

However, in analyzing the contractor's duties and responsibilities, it becomes apparent that some of those responsibilities could most conveniently be carried out by the superintendent at the job site, while others would more easily be carried out by the contractor's home office personnel. Obviously it is the contractor's prerogative to assign the contracted responsibilities in whatever way is deemed most expedient under the circumstances, apportioning duties variously to field and office personnel in the most strategic and advantageous manner.

The assignment of some of the contractor's responsibilities to the field superintendent is based mainly on the expectation that the superintendent will always be present at the jobsite. The general conditions specifically require the superintendent to be "in attendance at the Project site during performance of the Work." (A201, 3.9.1) Also, the General Conditions mandate that the superintendent "represent" the contractor. (A201, 3.9.1) The term Contractor is defined as, "the Contractor or the Contractor's authorized representative." (A201, 3.1.1) Therefore, the superintendent is clearly empowered to carry out all of the contractor's functions at the jobsite. In addition, the General Conditions designate the superintendent as a receiver of communications "as binding as if given to the Contractor." (A201, 3.9.1)

Except for very large and costly projects, the superintendent will not usually have a secretary or other clerical personnel present on the jobsite. Some contracting firms make a home office secretary available by telephone for the superintendent's necessary correspondence. However, in most cases the field superintendent will not become directly involved in the processes of letter writing and the giving of written notices other than oral notification of home office personnel of the circumstances giving rise to the need for written communications. The primary paperwork function will be performed by contract administrators or clerical employees, depending on size of the project, resources of the contractor, and organizational structure of the contracting firm.

The superintendent has all of the traditional duties on the jobsite usually assigned to contracting firms. These include being responsible for dimensional control, scheduling of subcontractors, labor and materials, scheduling of all field operations including looking ahead, coordinating trades, quality control, maintaining job site records, keeping the contract documents and all modifications "straight", and in general managing the project at the site. This is a very large order and requires a person of superior technical and interpersonal skills including an intimate knowledge of the construction trades and processes, an organized mind, leadership ability, and diplomacy.

Supplementing this full agenda will be all of the contractor's responsibilities spelled out in the General Conditions which cannot be conveniently performed by the home office staff and which become by default an additional assortment of duties tacitly assigned to the superintendent in charge of the jobsite. While many of these functions are not expressly assigned to the superintendent, the contracting firm's leadership and staff will naturally expect them to be a part of the field operation.

Any of the duties and responsibilities of the superintendent deriving from the General Conditions can easily be eliminated or added to if desired by owner and contractor, merely by modification of the General Conditions. Be aware, however, that the AIA documents are coordinated, so it might be necessary to make appropriate changes in other companion documents to maintain consistency.

Preliminary Site Visit

Subparagraph 1.2.2. states that

> "Execution of the Contract by the Contractor is a representation that the Contractor has visited the site, become familiar with local conditions under which the work is to be performed and correlated personal observations with requirements of the Contract Documents."

This site visitation could have been made by the superintendent but more likely would have been made by the contractor's estimating department personnel during the process of preparing the contract proposal. If the superintendent did not personally visit the site before the contract was executed, the site should be investigated before any work is started for familiarization with local conditions.

Review of Contract Documents and Field Conditions

The superintendent, before starting any work, should carefully study and compare the contract documents, property survey, and soil tests and correlate them with information obtained during field observations. Any errors, omissions, inconsistencies, or anomalies discovered must be reported to the architect. The contractor is under no contractual obligation to find any discrepancies, but would be held financially responsible for carrying out any work knowing that it is based on erroneous information. Any damage or loss which is caused by the contractor's deliberate failure to report recognized errors or inconsistencies will be charged to the contractor. It would be extremely difficult to prove that a contractor knew about an unreported defect in the contract documents, but in certain situations a competent contractor would be assumed to have sufficient knowledge to recognize certain types of errors. At all times during the construction period, the contractor should report recognized errors to the architect in writing, even though the general conditions do not specifically require written notice.

During the course of construction, the contractor is obliged to report to the architect any violations of building codes or other governmental regulations which are discovered in the contract documents. However the contractor is under no contractual duty to determine that the contract documents are in compliance with applicable laws and ordinances. If the contractor knowingly proceeds with work in violation of building regulations without notice to the architect, the contractor will be held financially responsible for the consequent costs. (A201, 3.2.1, 3.7.3 and 3.7.4)

Before any construction activities are initiated at the site, the superintendent should take appropriate field measurements and compare them with the contract documents in an attempt to discover any errors or discrepancies. Any apparent mistakes should be reported to the architect as previously discussed. (A201, 3.2.2) It is the architect's responsibility to resolve any such inconsistencies or errors discovered in the documents. The contractor should not proceed with construction in uncertainty. If correction of errors in the documents results in additional or less building cost or time, the architect should prepare the necessary change order documentation for the owner's and contractor's approvals.

Preconstruction Jobsite Conference

If a preconstruction jobsite conference has been scheduled by the architect, the superintendent should attend. This will serve as a valuable orientation to the architect's personality and procedures as well as to the peculiarities and particulars of the project and site. (See Chapter 7, The Preconstruction Jobsite Conference.)

Daily Construction Log

Most organized construction companies require their superintendents to maintain a written journal of all significant events which occur on a jobsite on a daily basis. The daily log should have entries to record the following types of information:

1. Day and date
2. Weather conditions
3. Trades and personnel on the job
4. Earth, materials, and equipment received
5. Earth, debris, and excess materials removed
6. Names and roles of visitors
7. Delays and impending delays
8. Anticipated progress during next 7 days
9. Brief summary of accomplishments and difficulties
10. Inspections, rejections, and approvals received
11. Architect's instructions or interpretations received
12. Changed conditions
13. Unexpected or unforeseen conditions
14. Accidents, injuries, and property damage
15. Safety procedures initiated

Role of the Construction Superintendent

The log should have an entry for every day. This should also include days not worked because of labor disputes, adverse weather, lack of governmental approvals, owner's orders, lack of materials or any other reason. All days should be accounted for between commencement and completion or termination of the work.

Some contracting firms allow the superintendent to telephone one of the main office secretaries each day to dictate the information for a neat typewritten daily log. The daily log should be carefully preserved at the end of the project as a source of authentic information. It will be valuable if needed to substantiate the contractor's billing for extra work and in the event of future lawsuits or arbitrations.

Construction of the Work

The fundamental and by far the most all-encompassing duty of the contractor which will be carried out by the superintendent is imposed by Subparagraph 3.2.3, of the General Conditions which bureaucratically states that "The Contractor shall perform the Work in accordance with the Contract Documents and submittals approved pursuant to Paragraph 3.12." In simple language, this means, "build the building".

Supervision and Construction Procedures

The superintendent must skillfully and diligently supervise the construction and direct the work. An important general principle permeates the General Conditions, that it is the contractor's prerogative to determine, control, and be responsible for construction means, methods, techniques, sequences, and procedures and for coordinating all portions of the work. Many of these principles will be determined by the superintendent and all will be carried out in the field operation. (3.3.1)

On a continuous basis, the superintendent must inspect the work of each trade to determine that the work is being done by skilled workers and that it will be suitable to receive subsequent operations and trades. The superintendent must enforce strict discipline and good order among all persons on the job site. (A201, 3.3.4, 3.4.2, 3.13.1 and 3.15.1)

The superintendent is also responsible for keeping the site neat and orderly and in conformance with applicable laws and contract requirements. The site must be kept clean and free of waste accumulations. At completion of the project, all work and the site must be cleaned to the degree specified and all excess materials, equipment, and temporary structures removed. (A201, 3.13.1 and 3.15.1)

The contractor must obtain and pay for all required governmental permits and approvals, unless otherwise specified, and arrange for all governmental inspections. (A201, 3.7.1) These significant details are usually organized by the superintendent.

Progress Schedule

The contractor is required to prepare and submit a construction progress schedule based on the agreed time limits and predicated on expeditious and practicable execution of the work. A schedule of all specified submittals, coordinated and consistent with the construction schedule, must also be submitted. The superintendent's input and assistance with the scheduling logic, proposed construction sequences, and time allowances estimated for each of the operations would be invaluable to the scheduling process and should increase the possibility of producing a realistic and attainable schedule. The construction schedule is submitted for the architect's and owner's information only, while the submittal schedule is subject to the architect's review and approval. The construction and submittal schedules must be revised from time to time as necessary to reflect the actual construction progress. (A201, 3.10.1 and 3.10.2) The work is required to conform to the most recent schedules. (A201, 3.10.3)

Documents and Samples at the Site

The superintendent is expected to maintain at the site a complete up-to-date set of all drawings, specifications, and other contract documents, change orders, addenda, and modifications as well as all approved shop drawings and samples for reference by the owner or architect. (A201, 3.11.1) Conforming to this contract requirement will enable the superintendent more easily to keep the current status of the contract straight in mind.

Shop Drawings, Product Data, and Samples

All shop drawings and other submittals from subcontractors and suppliers should be carefully checked and approved by the superintendent at the site. This is to ensure that realistic job site conditions and actual field measurements are reflected and that they comply with the contract documents, before they are submitted for the architect's review. (A201, 3.12.5 and 3.12.7) Work requiring submittal of shop drawings or samples should be done in conformance with the approved submittals. (A201, 3.12.6)

Cutting and Patching

Cutting, fitting, and patching necessary to make the various parts of the work fit together properly is the contractor's responsibility. The superintendent will have to take care of all necessary coordination among the subcontractors to assure harmonious interconnections of the various trades. (A201, 3.14.1) Any cutting and patching which causes damage to the work of the owner or separate contractors will be the responsibility of the contractor. (A201, 3.14.2)

Role of the Construction Superintendent

Communications Facilitating Contract Administration

Although the contract is between the owner and contractor, the line of communication is through the architect to induce orderly administration. All communications with the architect's consultants should be through the architect, and similarly, all communications directed to subcontractors should be through the contractor. (A201, 4.2.4) The superintendent, being the representative of the contractor on the jobsite, is the proper recipient of communications directed to the contractor. (A201, 3.9.1) It is up to the contracting firm to establish its own two-way communication system between its home office and its superintendents at the various jobsites.

Continuing Contract Performance

While awaiting final resolution of contractor's or owner's claims against each other, whether submitted to arbitration or not, the contractor should continue with the work and the owner is obligated to continue making payments as contracted. (A201, 4.3.4 and 4.5.3) For either to discontinue performance could be deemed a breach of the contract thereby possibly excusing the other from any further performance.

Claims for Concealed or Unknown Conditions

Should physical conditions be encountered on the site which were concealed or unexpected, and which are materially different from those represented in the contract documents, the party discovering the conditions should notify the other party before conditions are disturbed. Notice must be given within 21 days of the first observance of the conditions. The architect will then investigate and determine whether the contract sum should be increased, decreased, or remain the same. A similar determination should be made in respect to extension of the time of performance. (A201, 4.3.6)

The superintendent should document the discovery and circumstances of all unexpected conditions and occurrences by suitable entry in the daily log as well as by photographs, notes, observations, and measurements when applicable. This will assist the contractor in properly invoicing the job or in making claims for extra compensation. It will also help the architect to make fair and appropriate determinations.

Claims for Additional Time

Record keeping by the superintendent on the site is extremely important to the smooth-running job and to enable the contractor to be properly remunerated for changes in contract conditions and for additional contract time caused by various factors including adverse weather. All claims for additional time due to weather must be documented as to days of occurrence and the adverse effect the weather had on the scheduled construction. This information must be accumulated on a daily basis and should be faithfully entered in the daily job log. Only abnormal weather which could not have been reasonably anticipated and which had an adverse effect on the construction can be claimed for time extension. (A201, 4.3.8.2)

Cooperation with Owner and Architect and with Separate Contractors

The contractor is in full charge of the construction site and is required to provide access by the owner and architect to all parts of the work in all stages of progress of construction. (A201, 3.16.1)

The owner reserves the right to do work on the project site with its own forces and to award separate contracts for portions of the work. The contractor is obligated to cooperate with the owner and the separate contractors and to allow them reasonable use of the site and storage areas. The owner is required when contracting with separate contractors to use contract provisions similar to those used in the original owner-contractor agreement. The owner is also obligated to coordinate the work of its own forces and the work of separate contractors. The contractor is required to cooperate in the coordination of its time schedules with those of the owner and separate contractors. (A201, 6.1.1, 6.1.3 and 6.2.1)

Where any part of the contractor's work interfaces with the work done by the owner or separate contractors, the contractor should report to the architect any observed discrepancies or defects which would render the work of others unsuitable for continuation of construction. The contractor's failure to notify the architect will constitute acceptance of the owner's or separate contractor's work, except for defects not then reasonably discoverable. (A201, 6.2.2)

Changes in the Work

The owner has the right under the contract to order changes without invalidating the contract, and the contractor is required to carry out the changes. Changes, where the scope of work, price, and time are agreeable to the parties, are provided for by a written change order. Those changes that are not acceptable in whole or in part to the contractor are provided for by construction change directive, and the remaining disagreements are left for later resolution by the architect or by negotiation or arbitration. An order for a minor change in the work, not involving any change in contract sum or time schedule, may be ordered by the architect. The contractor must promptly carry out all changes ordered by the owner or architect. (A201, 7.1.1, 7.1.3 and 7.4.1) (See Chapter 14, Change Orders.)

Time of Construction

It is important to the owner's interests that no construction on the site is commenced before the mortgage liens, if any, are in place or prior to the effective dates on all insurance specified to be carried by the contractor or furnished by the owner. If the notice to the contractor to proceed with construction, preferably written, is not provided by the owner, the contractor is required to give the owner written notice no less than five days prior to starting work. (A201, 8.2.2) The contractor is required to pursue the work diligently and efficiently, using adequate forces, and to complete the work to substantial completion within the contract time. (A201, 8.2.3)

The superintendent should have all of the pre-construction procedures organized before construction is commenced on the site. These will consist of reviewing documents and site conditions, obtaining of necessary governmental permits, preparation of construction schedule and submittal schedule, submission of insurance certificates, and provisional mobilization of initial labor, materials and subcontractors. The superintendent should be present and accompany the architect at all scheduled site visitations. This is for the purpose of answering any questions posed by the architect and to receive the architect's instructions and determinations.

Substantial Completion

Substantial Completion is the stage of completion when the owner may occupy or make use of the work for its intended purpose even though there may be various items of incomplete or defective work. When the contractor considers that the work is substantially complete, the contractor must prepare a punch list (referred to in the AIA General Conditions as an inspection list) tabulating all remaining incomplete and defective work. The architect will review and possibly add to the contractor's list. The superintendent must promptly organize and schedule the completion of all punch list work. The architect will then make a determination as to completion, binding on owner and contractor, by issuance of a certificate of substantial completion. The owner and contractor both sign the architect's certificate as an indication of their mutual acceptance of the responsibilities assigned to each of them in respect to such matters as utilities, insurance, security, maintenance, and damage to the work. (A201, 9.8.1 and 9.8.2)

The superintendent must then start making arrangements with the subcontractors and suppliers to obtain all specified written warranties, operating manuals and instructions, spare parts, and record drawings for submission to the architect. (See Chapter 19, Closing Out the Job.)

Protection of Persons and Property

One of the most significant responsibilities of the construction superintendent is the duty to safeguard persons and property on or about the site. The contractor's duty to provide safety and to prevent accidents is specifically assigned to the superintendent by Subparagraph 10.2.6 of the General Conditions which states, "The Contractor shall designate a responsible member of the Contractor's organization at the site whose duty shall be the prevention of accidents. This person shall be the Contractor's superintendent unless otherwise designated by the Contractor in writing to the Owner and Architect."

The superintendent must initiate, maintain, and supervise all safety precautions and programs to be carried out on the work. (A201, 10.1.1) The superintendent must take reasonable precautions for safety and provide reasonable protection to prevent damage, injury, or loss to persons on or about the site, work or materials, or adjacent property. (A201, 10.2.1) The superintendent must give any required notices and comply with all safety laws. (A201, 10.2.2)

The superintendent must erect and maintain all reasonable safeguards needed for safety and protection and post all necessary warning notices. Owners and users of adjacent properties and utilities must be warned of possible hazards. (A201, 10.2.3)

Use and storage of explosives and other hazardous materials or methods must be with the utmost of care and be supervised by competent personnel. (A201, 10.2.4) It is the superintendent's ongoing duty to take responsible managerial charge of the premises and not allow the site or structure to be damaged during construction by overloading or other misuse. (A201, 10.2.7)

In the event of emergency, it is the superintendent's responsibility to exercise discretion in acting to prevent or mitigate threatened damage or injury to persons or property. Claims for additional compensation or time extension arising out of emergency action will be considered and decided by the architect. (A201, 10.3.1) The superintendent should keep complete, detailed, and current records, including photographs where appropriate, of all emergency activities to enable the contractor to compile and prove all incurred costs.

Asbestos And PCB

In alteration work in existing buildings, it is not uncommon for the contractor unexpectedly to find materials reasonably believed to be asbestos or polychlorinated biphenyl (PCB) which have not been rendered harmless. If this happens, the superintendent should immediately stop work in the area affected and give written notice of the condition to the owner and architect. If the materials prove to be asbestos or PCB which have not been rendered harmless, the contractor is not required to continue work in the affected area unless it is further agreed in writing by the owner and contractor, or determined by the architect or by arbitration if demanded by either party. The contractor cannot be required to perform any work related to asbestos or PCB without consent, notwithstanding provisions of Article 7 of the General Conditions relating to construction change directives. (A201, 10.1.2 and 10.1.3)

It is now very difficult for architects and contractors to obtain insurance coverage for operations involving asbestos or PCB. The General Conditions require the owner to indemnify and hold harmless the contractor and architect from losses arising out of work in an area affected by asbestos or PCB. (A201, 10.1.4)

Uncovering and Correction of Work

If work has been covered contrary to the specifications or architect's request before examination by the architect, it must be uncovered for observation and replaced at the contractor's expense, with no time extension allowable.

In the event work not specified to be observed before covering is requested by the architect to be uncovered for examination, the contractor is required to comply with the architect's request. However, if the uncovered work proves to have been satisfactory, the contractor will be paid for uncovering and replacing and be granted a reasonable time extension.

If the uncovered work is not in accordance with the contract documents, the contractor will be required to pay all the costs of uncovering and correction, unless the unsatisfactory condition was caused by the owner or a separate contractor, in which case the costs will be charged to the owner. (A201, 12.1.1 and 12.1.2)

All work rejected by the architect or failing to comply with contract requirements shall be properly corrected by the contractor. All defective work shall be removed from the site. (A201, 12.2.1 and 12.2.3)

Defective work which becomes apparent within 1 year after the date of substantial completion should be corrected after receipt of owner's written notice. (A201, 12.2.2) During the 1-year warranty period, any correction notices received by the contractor will generally be referred to the superintendent for investigation, coordination, and appropriate action if the superintendent is still on the contractor's staff. Otherwise the contractor will have to assign it to other field personnel, preferably someone who is familiar with the project.

Tests and Inspections

Testing and inspections of portions of the work specified by the architect or required by public authorities must be arranged for and facilitated by the contractor to be performed by a competent independent testing agency acceptable to the owner. The superintendent must make portions of the work accessible for testing or observation by testing personnel and the architect. All required certificates or reports of testing, certification, inspection, or approval are required to be acquired by the contractor and promptly delivered to the architect. (A201, 13.5.1, 13.5.2 and 13.5.4)

Replacement of Superintendent

When a contractor replaces the superintendent on the project, a certain amount of disorder and confusion will result. This is because a lot of the superintendent's organization and administration is kept in mind rather than on paper. The previous edition of the AIA General Conditions (1976 Edition) prohibited the contractor from changing superintendents without the architect's approval. However, writers of the 1987 Edition deemed this to be interference in the contractor's responsibility for means, methods, and techniques of construction. Thus it would be a mistake for an architect to specify an approval requirement for changing superintendents or any other limitation on the contractor's control of the work.

18

The Architect's Relationship with Subcontractors

Contractual Relationships

It is well understood that there is no contractual relationship between the architect or owner and any of the entities controlled by the contractor. This general principle is stated in most construction contracts.

The general contractor contracts directly with subcontractors and suppliers. They in turn enter into contractual relationships with their sub-subcontractors and suppliers and so on. Many of these second and third tier subcontractors and suppliers are completely unknown to the architect or owner and sometimes even to the general contractor. Nevertheless, these are the people and organizations who actually supply a great percentage of the materials and equipment that architects specify.

These entities are too important to the success of any project to be completely ignored by the architect. Subcontractors account for 3 percent of the claims made against architects' insurers. It is clear that these second and third tier contractors feel strongly that architects should have some duty to consider their position.

Construction conditions controlled by the architect frequently cause frustration, dissatisfaction, and anger among subcontractors and suppliers. In some situations, it is simply a matter of the subcontractors' suppressed desire to express relevant opinions based on superior knowledge of their specialized portion of the work. In other cases, it is of deeper significance, that of economic survival.

Whenever an architect makes a harsh or insensitive decision on the contractor's questions, suggestions, or claims, it is usually passed on to the subcontractors or suppliers who must actually bear the burden of inconvenience and economic loss. The architect usually does not hear about these practical difficulties, being insulated by the general contractor. Often the complete viewpoint of the subcontractor is not effectively expressed and advocated by the general contractor.

Relationships among the Parties Governed by Contracts

The interrelationships among architect, owner, contractor, subcontractors, sub-subcontractors, material suppliers, and manufacturers of materials and equipment are governed by the contract documents of each specific construction project. It will be assumed that all of the contracts will be on AIA standard form documents.

> Between Owner and Architect:
> AIA Document B141 (Appendix A)
>
> Between Owner and Contractor:
> AIA Document A101 (Appendix B) or
> AIA Document A111 (Appendix C) and
> AIA Document A201 (Appendix D)
>
> Between Contractor and Subcontractor:
> Standard Form of Agreement Between Contractor and
> Subcontractor, Twelfth Edition, AIA Document A401, 1987

The construction contract is between the owner and the contractor and there are no other parties to that contract. The AIA General Conditions, AIA Document A201, makes it abundantly clear that the contract creates no contractual relationship between the owner and any subcontractor or sub-subcontractor. (A201, 1.1.2.) Logically this also extends to exclude suppliers and manufacturers of materials and equipment from any contractual relationship. It is obvious that there can be no contractual relationship between the architect and the contractor, subcontractors, sub-subcontractors, suppliers, or manufacturers.

Architects should thus condition their behavior and if possible that of the owner to avoid interfering with the relationships between the contractor and those contractually subordinate. They also should avoid inappropriate conduct which could create the appearance of relationships where none exist.

Communication

During the design period and when construction documents are being prepared, architects need to consult with various authentic sources of technical information. They often call upon friendly contractors and subcontractors to discuss pragmatic solutions to practical problems.

Representatives of building material and equipment suppliers and manufacturers regularly call on architects to point out the favorable characteristics of their products and to urge their use. After reviewing their catalogues and promotional literature, architects often contact suppliers and manufacturers for additional data on their products and to discuss suitability, details of application, installation, availability, costs, and warranties. Conferring with these sales agents and technical representatives is an invaluable source of relevant information. At the time of these discussions and solicitations of advice, the construction contract has not yet come into existence so there is no restriction on the architect's communication with these sources. However, after the construction contract is signed, the architect's interactions with these second and third tier contractors will be governed by the contract.

The architect should respect provisions of the AIA General Conditions which require all communications with subcontractors and material suppliers to be channeled through the contractor. (A201, 4.2.4) Subcontractors are similarly bound by the same provisions of the contract. Sometimes it would seem very convenient for the architect to continue the friendly direct relationship with these cooperative sources of information, but it would be better practice to at least inform the contractor of these informal communications. The architect must be careful not to interfere with the contractor's contractual relationships. It is very easy to create false impressions and unrealistic expectations. What might seem an innocuous statement by the architect could be construed by a subcontractor or supplier as approval for changes in specification, price, construction time, or other conditions.

It would be safer, and less likely to cause misunderstandings, to communicate in writing and through the channel provided by the contractor. Any of the architect's consultants who must communicate with those controlled by the contractor should similarly route their inquiries through the architect and contractor as required by the AIA general conditions. As a practical matter, this can be accomplished by sending copies of all communications to all interested parties. For example, when the architect's structural engineer must communicate with the contractor's steel fabricator on a routine matter, a direct letter from engineer to fabricator with copies mailed to architect and contractor will be more efficient than separate letters from engineer to architect, architect to contractor, contractor to fabricator and back again. By this procedure, the architect and contractor can monitor the transaction and can take timely steps to preserve the contract if necessary.

Architects' on-site conferences with subcontractors should always be in the presence of the contractor or superintendent and any instructions given, decisions made, or agreements reached should be reported in the architect's written observation report. Office conferences with subcontractors should be similarly monitored by the contractor and duly reported to all concerned.

Shop Drawings

The majority of shop drawings, product data, and samples originate with subcontractors or suppliers. They should be submitted to the architect only through the contractor who must have reviewed and approved them first. (A201, 3.12.5) Sometimes in the spirit of cooperation and as a time saving expedient, an architect will accept submittals directly from a subcontractor. The architect should not violate the contract procedures in this or any other manner, as it creates an embarrassing precedent making it difficult to hold the contractor to any of the administrative provisions in the contract. It also negates whatever little liability protection was inherent in the specified procedure.

All submittals should similarly be returned through the contractor. If, occasionally, expedience dictates a shortcutting of the specified procedures, the architect should at the very least see to it that the contractor is informed of every step in the progress of submittals as well as given the opportunity of independently reviewing and approving them. (See Chapter 12, Shop Drawing Procedures.)

Contractor's Right to Subcontract

The term subcontractor means any person or entity who has contracted directly with the contractor to provide labor or materials or both. It does not include any separate contractors who have contracted directly with the owner nor any subcontractors of separate contractors. A sub-subcontractor is any person or entity who has contracted with a subcontractor. Suppliers of materials or equipment can be subcontractors or sub-subcontractors.

Consistent with the contractor's duty to maintain control over the means, methods, techniques, sequences, and procedures of construction, it is the contractor's prerogative to let subcontracts, purchase materials, and assign work in any desired combination without regard to the architect's division of work in the specifications or on the drawings. (A201, 1.2.4) However, architects should endeavor to keep abreast of the logical flow of the separate building trades and write their specifications accordingly. Otherwise, avoidable labor disputes can unknowingly happen.

Additionally, inappropriate specifying of some work will cause the item to be missed by the usual supplier. For example, if wood blocking needed for electrical fixture installation, normally installed by the carpentry contractor, is specified under electrical work, the electrical estimator might not include it as it is not normal electrical work, whereas the carpentry estimator is not likely to examine the electrical section of the specifications and thus will miss the item. The result is that the general contractor will lack bid coverage on the item. Neither subcontractor is likely to willingly agree if the value is sizable.

Subcontractor List

As soon as is practicable after the contract is signed, the contractor must submit a list of all proposed subcontractors and suppliers to the architect. (A201, 5.2.1) If after due investigation the owner and architect have any reasonable objection to any on the list, the contractor must replace that person or entity. If the replacement bid is higher than the contractor's original sub bid then the owner will have to pay the difference or receive a credit if it is lower. The contractor is entitled to assume there is no objection to the list if no immediate response is forthcoming.

Usually, a disappointed subcontractor will not quietly accept rejection, so the architect and owner must have solid reasons for their action. An architect's recommendation to the owner that any subcontractor or supplier be rejected must be based on proper research and substantial grounds not motivated by personal reasons. An architect who acts in the owner's best interest and without malice is not likely to be successfully sued for libel or slander.

After the list of subcontractors and suppliers has been approved, the contractor cannot make any changes if the owner or architect have any reasonable objection. (A201. 5.2.4) Although the owner and architect may suggest subcontractors and suppliers, the contractor is not obligated to contract with anyone to whom the contractor has made a reasonable objection.

Subcontract Agreement

AIA's standard form of subcontract agreement (A401) used by some contractors is in harmony with AIA owner-contractor agreements (A101 and A111) as well as the AIA General Conditions (A201) and the owner-architect agreement (B141).

AIA's subcontract agreement has been balanced as fairly as possible to respect the rights and interests of contractors as well as subcontractors. The form has been approved and endorsed by the American Subcontractors Association and the Associated Specialty Contractors, Inc. The Associated General Contractors (AGC) has not sanctioned A401 although it has approved the AIA owner-contractor agreements and the AIA General Conditions.

Most contractors are inclined to use AGC's subcontract form, one similar to it, or one which each firm has perfected over many years of mending and patching to solve practical problems and close loopholes. They often favor the contractor's position over that of the subcontractor.

An architect cannot require the contractor to use any particular subcontracting form, unless it has been specifically required beforehand in the bidding instructions or supplementary general conditions.

Unwritten Expectations

Regardless of the agreement forms used, subcontractors have certain basic unwritten expectations from the owner, contractor, and architect. Underlying all is the fundamental expectation of fair play.

During the bidding period, subcontractors expect contractors to keep their proposals confidential and not to use them as leverage to obtain lower bids from competitors. The tacit understanding is that the lowest responsible bidder gets the job at the price bid without having to extend further discounts or other concessions.

Subcontractors would like to be consulted when the construction schedule is being compiled and whenever it is amended. Bid proposals are usually based on the work being performed in normal sequence with no more than the usual interference from other trades. Subcontractors expect reasonable advance notice for start of their part of the work. They expect to receive a reasonable supply of the contract documents so they can run their office, estimating, purchasing, and field operations efficiently.

They expect to be provided with reasonably located parking, material and trash storage, and reasonable access to water, electricity, hoisting, telephone, and toilets. They rely on the contractor to provide skilled superintendence, accurate measurements and quantities, fair allocation of space for piping, conduit, and ducts of various trades, proper scheduling of all the other subcontractors, and keeping of order on the job. They expect the contractor to promptly relay all questions and suggestions to the architect or other subcontractors and to return quickly with answers. Above all, they expect to be paid promptly in accordance with the agreement.

All of these expectations are reasonable and are based on the normal situation on a well-run job presided over by a skilled superintendent. All of these wishes and more can be, and usually are, provided by competent general contractors.

In addition, there are many unspoken aspirations and understandings involving the architect. The architect's basic duty is to design all subcontract work properly and respect the normal work sequences and usual interfaces between adjoining trades.

Subcontractors trust that the architect has properly researched the specifications so that all catalog numbers and technicalities are complete and up-to-date. They expect the architect to process submittals promptly and to expedite options and selections which must be made by the owner or architect. They rely on the architect to pitch in and assist when difficulties arise with governmental authority questioning the design or specifications.

They also rely on the architect's technical ability, reasonableness, honesty of purpose in judging materials and workmanship, and interpreting the contract documents. They hope the architect will recognize their superior knowledge of their trade and will at least listen when suggestions are being offered. Some architects assume that the subcontractor is always motivated by self-interest and is only trying to lower the quality or cost to the owner's detriment.

Architect's Authority to Reject Work

Although the architect has the power to reject work that is not in conformity with the contract documents, the architect is under no legal obligation to the contractor or subcontractors either to exercise or not to exercise this authority. (A201, 4.2.6)

General contractors often find themselves in an awkward position when a subcontractor's work is not acceptable to the architect but the subcontractor is steadfast in claiming that the architect is unreasonable and is demanding a higher standard than was specified. The general contractor is then obligated to assist the subcontractor in appealing the architect's decision. The contractor loses credibility with the subcontractor when unable to sway the architect and will lose the respect of the architect and owner when trying to sell substandard subcontractor's work.

Architect's Administration of Payments

Many subcontractors do not take advantage of a significant right given to them in the AIA General Conditions. They are allowed to inquire of the architect of the status of the contractor's applications for payment. The architect is authorized to release information to a subcontractor pertaining to its portion of the work in respect to dates and amounts paid, percentage of completion approved, amounts disallowed, and amounts of retainage. (A201, 9.6.3 and A401, 12.4.2) Material suppliers have the right to similar treatment. (A201, 9.6.5)

Neither the architect nor owner are obligated to pay or to see to it that subcontractors and suppliers are paid. (A201, 9.6.4)

However, the contract requires the contractor to pay the subcontractors and suppliers from funds that have previously been received on account of their work. (A201 9.6.2) Also, the contractor may not include in any payment request money that is not intended to be paid to subcontractors or suppliers on account of disputes or any other reason. (A201, 9.3.2)

Should the owner or architect happen to hear of any complaints from subcontractors or suppliers, or rumors in the industry that any have not been paid for previously approved work, this should serve as a serious warning. Concern about the contractor's incipient financial instability should cause the architect to advise the owner to confer with legal and accounting counsel. The surety, if any, may have to be put on notice, and changes in the contract payment procedure may have to be negotiated to protect the owner's interest. This will also usually have a side effect of protecting the interests of subcontractors and suppliers.

The owner may terminate the contract should the contractor fail to pay subcontractors or suppliers their allocated share from the funds advanced in periodic payments. (A201, 14.2.12)

Should the general contract be terminated by the owner for cause, all notified subcontractors are automatically assigned to the owner. If the work had been suspended for over 30 days, assigned subcontractors are entitled to an equitable adjustment in their contracts. (A201, 5.4.1 and 5.4.2) This will reimburse them for their reasonable costs of shutdown, delay, and start-up. (See Chapter 20, Termination of the Construction Contract.)

19

Closing Out the Job

Orderly Conclusion of the Construction Contract

When the construction project finally takes shape physically and the major building activity starts to taper off, the architect responsible for the contract administration should think about what is needed for the orderly conclusion of the contract.

Often the owner is thinking more about the process of leaving the old premises, moving into the new beautiful building, and occupying it forever. Similarly, the contractor and subcontractors are more concerned with rapidly completing the physical work, promptly collecting their money, and moving on to the next job.

Some projects seem to be very difficult to end gracefully. The final punch list items are often being corrected months after occupancy and the owner is never entirely pleased about paying the contractor's final bills. The paper work continues on a sporadic but gradually declining basis, and there never appears to be an actual clean finale to the whole affair. This sort of loose unraveled ending to the contract administration must not be very impressive to the client or contractor and is very time-consuming and exasperating for the architect. Additionally, some of the liability restriction offered by statutes of limitation will be diluted by sloppy or nonexistent contract termination procedures.

The purpose here is to review the normal procedures that are available to architects in the efficient conclusion of the administration of a construction contract when the architect's services have been engaged under the terms of the standard AIA Owner-Architect Agreement and the construction contract includes the AIA General Conditions. These two standard AIA documents are coordinated so that all of their related provisions are in harmony. Most importantly, all of the architect's duties described in the AIA General Conditions are anticipated and included in the AIA Owner-Architect Agreement.

When Is the Construction Completed?

How can one tell when the construction is completed? A simple question yielding an inexact, inconclusive, and sometimes complex answer. In the real world of construction, completion is not an absolute. There can be and usually are varying degrees of completion.

Even when the construction contract defines completion, the question remains unanswered for some common situations. The concept of completion has various meanings and implications depending on the specific application.

The Contract Time

The *contract time* ends at the date of *substantial completion*. The term contract time is defined in the General Conditions, Subparagraph 8.1.1:

> "Unless otherwise provided, Contract Time is the period of time, including authorized adjustments, allotted in the Contract Documents for Substantial Completion of the Work."

Contract time for the particular project at hand should be specified in the construction contract. (Standard Form of Agreement Between Owner and Contractor where the basis if payment is a Stipulated Sum, AIA Document A 101, 1987 edition, Paragraph 3.2, see Appendix B, or Standard Form of Agreement Between Owner and Contractor where the basis of payment is the Cost of the Work Plus a Fee with or without a Guaranteed Maximum Price, AIA Document A111, Tenth Edition, 1987, Paragraph 4.2, see Appendix C.)

This does not mean that there is no more work to be done. There will still be minor or trivial items of incomplete work and defects to be rectified. There could also be major items of incomplete work, but if they do not prevent the owner's use of the building the contract could still be substantially complete. When the contractor considers that the project has reached this decisive stage, the contractor is required to prepare a comprehensive list of all remaining items of work and corrections to be made to complete the work of the contract. The contractor should proceed promptly with the completion and rectification work on the list. At the same time, the list is submitted to the architect for review and further action. Failure to list an incomplete or defective item does not relieve the contractor from responsibility for its proper completion as specified.

Upon receipt of the contractor's correction list the architect is required to make an inspection to determine if the work is indeed substantially complete and in conformance with the contract requirements. If the architect's inspection discloses items, whether or not on the contractor's list, which are not in accordance with the contract documents and which would preclude substantial completion, then the contractor must rectify these items upon notification by the architect. The architect should have all design consultants examine the work which they designed such as structural, electrical, and mechanical systems and landscape installation. The architect's consultants may also find items to be added to the contractor's list. The contractor, upon completion of the items which preclude substantial completion, must again notify the architect to reinspect the work.

Determination of Substantial Completion and Final Completion

Establishing the date of substantial completion could be a major source of dispute if it is left up to the owner and contractor to decide by themselves. It is a particularly explosive situation when liquidated damages for delayed completion or bonuses tied to early completion are involved. Thus, it is practical and commonplace to assign this determination to someone else such as the architect as provided in the AIA agreements. Subparagraph 2.6.14 of the Owner-Architect Agreement, AIA Document B141, 1987, provides:

> "The Architect shall conduct inspections to determine the date or dates of Substantial Completion and the date of final completion, shall receive and forward to the Owner for the Owner's review and records written warranties and related documents required by the Contract Documents and assembled by the Contractor, and shall issue a final Certificate for Payment upon compliance with the requirements of the Contract Documents." (See also the General Conditions, Paragraph 4.2.9.)

Certificate of Substantial Completion

The architect, upon determination that the work is substantially complete, must issue a certificate of substantial completion. (A201, 9.8.2) The certificate does not have to be in any particular form but could be on the AIA's standard form, Document G704, which contains all the necessary elements to comply with the AIA General Conditions.

The certificate of substantial completion must name a specific date as the date of substantial completion. This becomes the date of commencement of all warranties except those relating to items remaining on the contractor's correction list and as supplemented by the architect. The warranty periods on these items will commence on completion of the actual performance of the work. (A201, 12.2.2) The date of substantial completion is also the ending of the contract time period for determination of liquidated damages.

The owner and contractor must also sign the architect's certificate as an indication of their acceptance of the agreed allocation of responsibility for security, maintenance, heat, utilities, damage to the work, and insurance. The certificate also fixes the time within which the contractor is to finish all items on the correction list accompanying the certificate.

The owner may now move in and occupy or utilize the project or the designated portion covered by the certificate. It may be necessary to obtain concurrence of the local building officials, the surety, and insurance carriers for the owner's partial or full occupancy of the project before final completion. The owner should be urged to make all defects and dissatisfactions immediately known to the architect. The architect should then promptly send on to the contractor all legitimate complaints, eliminating all unreasonable demands and requests not within the scope of the contract documents. It is possible to have several dates of substantial completion if the project is divided into portions to be occupied progressively as the work continues.

Notice of Completion (as Distinguished from Certificate of Substantial Completion)

Certificate of substantial completion is a term coined by drafters of AIA form documents and is uniformly used in the AIA documents to describe the date determined by the architect as the date on which the project is sufficiently complete that it may be occupied or utilized by the owner for its intended purpose.

A notice of completion, on the other hand, is a legal term and is not mentioned in the AIA documents. The notice of completion is associated with requirements of state mechanics' lien laws. In some states, such as California, a notice of completion, signed by the owner or agent, must be recorded in the office of the recorder of the county in which the work of the contract is situated, within 10 days after completion of the work. The definition of completion of the work for the purpose of complying with the mechanics' lien law is not necessarily the same as for substantial completion as defined in the AIA General Conditions, although in many situations it could be the same date.

Completion for the purpose of complying with the lien law means the work is finished and the persons who have the right to lien the owner's property have no more work left to do. Some court cases have recognized completion when minor items were left to be done while others have held that similar minor items precluded completion.

The recording of a notice of completion is very important to the owner as well as to potential mechanics' lien claimants, as it shortens the claim period and establishes a definite date for measuring the time periods in the mechanics' lien law. A notice of completion in California must be recorded within 10 days after completion in the office of the recorder of the county in which the construction site is situated.

In California, original contractors (those who deal directly with the owner or owner's agent) have 60 days after the recording of a valid notice of completion to record their lien claims while subcontractors and others who work for the original contractor have 30 days in which to file their claims. If no notice of completion, or an invalid one, has been recorded, the period for all lien claimants is extended to 90 days after cessation of labor.

Some architects and contractors include as a part of their standard service, the filling out of a notice of completion form for the owner's signature and recording with the county recorder's office. This entails the exercise of judgment as to the date to be stated as the completion date. Normally, the date of substantial completion will be the date to be used in the notice of completion. However, it is possible that a mechanics' lien claimant might later convince a judge that minor corrections on the defect list inhibited completion of their work and that the notice of completion was therefore invalid and ineffective.

Architects who do not take care of the notice of completion for their clients should at least notify their client in sufficient time that they can contact their attorney or do it for themselves.

Mechanics Lien Laws

The mechanics' lien laws of the various states provide for different time periods during which lien rights may be asserted. They also specify definite cutoff points for limiting lien claims. The prescribed notices of completion and cessation must contain specific dates of completion or cessation of labor.

It is to the advantage of owners that these notices be recorded as early as legally possible because they have the effect of reducing or limiting the time periods during which valid lien claims may be filed. There is also the advantage of providing certainty of the dates from which rights or limits can be calculated. Prompt recording of these notices is also beneficial to the general contractor, because it lessens the waiting period for final contract payments.

The twofold problem which must be faced by owners, architects, and contractors is:

1. How can the proper completion date be determined?
2. Who should make this determination?

The codes generally state that completion means actual completion of the work. If that date is unequivocally known, there is no problem. However, it is very seldom in any substantial construction contract that the construction activity just suddenly ends with no loose ends dangling.

Completion as Seen by Judges

In individual California cases where the parties were trying variously to shorten the lien period (to cut off claims) or lengthen it (to validate liens), judges have made their determinations based on the facts peculiar to each case. It seems that some courts have ruled in opposite ways on similar facts, although a fairly consistent thread appears to run through all of the cases:

Remaining work that seems to *impair* completion:

1. Minor subcontracts that are completely or mostly unperformed
2. Work that is part of the original contract but has never been done, even if fairly minor (Such as a second coat of paint which was originally specified but not applied.)
3. Work that would otherwise deprive deserving lien claimants of their rights

Remaining work that seems *not to impair* completion:

1. Minor portions of larger subcontracts which are otherwise substantially performed
2. Replacement of or corrections of minor portions of the work which had previously been performed, but imperfectly
3. Superficial punch list items such as paint touch-up
4. Minor building department corrections

These lists will give general guidance only as some cases can be found to illustrate opposite positions. However, some codes give additional guidance by prescribing the "equivalents" to completion, including occupation or acceptance of the work by the owner or a cessation of labor. For example, California Civil Code, Title 15, Section 3086, states in part:

"Any of the following shall be deemed equivalent to a completion:

(a) The occupation or use of a work of improvement by the owner, or his agent, accompanied by cessation of labor thereon
(b) The acceptance by the owner, or his agent, of the work of improvement

(c) After the commencement of a work of improvement, a cessation of labor thereon for a continuous period of 60 days, or a cessation of labor thereon for a continuous period of 30 days or more if the owner files for record a notice of cessation."

If a notice of completion or cessation is recorded it usually must be signed by the owner or its agent. Thus it will be the owner or agent who must decide what date to include in the notice. In the absence of an architect or engineer on the project, the owner should confer with the contractor in establishing the appropriate date to be used. Some construction agreements designate the contractor as the owner's agent for the purpose of signing and recording the notice of completion. This is an important function and the contractor should carefully consider the appropriate completion date to protect the owner from loss from an invalid, therefore ineffective, notice. The contractor must fight the urge to designate an unconvincingly early date intended to hasten the final payments due under the contract.

If there is an architect on the project, the owner and contractor should look to the architect for a reliable determination. The architect's good faith, arm's length decision based on a thorough independent knowledge of on-site conditions as well as the contract documents will be helpful in establishing a valid date capable of withstanding legal attack.

Certificate of Occupancy

Certificates of Occupancy are not mentioned in the AIA General Conditions, but most municipal building departments have a system of certificates of occupancy which may be issued only upon the satisfactory completion of construction pursuant to approved construction drawings. The approved drawings illustrate compliance with applicable building laws and regulations. This decision of acceptable completion will be made by a building inspector who is responsible only to the employing governmental jurisdiction.

Commonly, building code agencies accept for checking only the drawings but not the specifications booklet (project manual) or the written portions of the construction contract. Therefore, the building inspector's determination of completion is for the limited purpose of code compliance and issuance of a certificate of occupancy. The building inspector is not concerned with, and would have no practical way of knowing, whether or not the construction contract was fully performed. Many items in the contract might not appear in the construction drawings if they are not regulated by the building code. For example, colors, textures, furniture, fixtures, wall coverings, paint, coatings, floor coverings, graphics, signage, and decorative features.

The building inspector, in making the completion determination, is not required to and usually does not consult with the owner, contractor, architect, surety, lender, or insurers.

Lender's Final Disbursement

Construction lenders commonly link their final disbursement to physical completion coupled with definite proof that no liens have been filed. The principal interests of the lender are twofold:

1. To assure that the physical security for the loan is of merchantable quality
2. To assure the priority of the lender's lien

Although the owner will have to apply for the funds and provide the lender with the required assurances and proofs, the final determination remains unilaterally in the hands of the lender. The lender will declare the project complete, in accordance with terms of the loan agreement, for the limited purpose of releasing the final disbursement.

Public Work

In the case of public work projects, a different set of rules will apply in determining completion. When the work of the contract is subject to the acceptance of the public entity, that official acceptance is deemed to be the date of completion. This could be days or weeks after actual completion of the contract work. Some public bodies meet only periodically for formal acceptance of their construction projects.

On some public contracts, a cessation of labor of 30 days is deemed to be completion.

Leased Premises

When property is promised in a lease to be available for a lessee's occupancy upon completion of new construction, the lease may or may not contain a definition of completion. Without a good workable definition, the area for misunderstanding and dispute is extensive. For example, a lessee could move in upon substantial completion, but refuse to pay some or any of the rent on account of incomplete or defective work, even though minor. Some lessees, when it is to their advantage, refuse to accept occupancy on account of very minor work left to be done. Some lessors will expect to collect the entire monthly rent even though the new construction is not 100 percent complete for months after lessee's occupancy. The definition in the lease will be heavily relied upon as neither lessee nor lessor is in the superior position to successfully impose its arbitrary decision on the other.

Closing Out the Job

Similarly, agreements to buy or sell property contingent on completion of construction are ambiguous, uncertain, and subject to misinterpretation when the definition of completion is inartfully written or omitted.

So, What is Completion?

It is apparent that completion of construction is an elusive concept and that the date of completion for one purpose is not necessarily the date of completion for any other purpose.

For some purposes, establishment of mere completion without a specific date suffices. For others, a specific date must be designated. In rare cases will the date of completion be the same for most or all purposes.

Final Submissions

While the final corrections are being made in the field, the contractor should start assembling the miscellany of specified final submittals:

> All specified written warranties. When transmitting them to the owner, the architect should refrain from stating any opinions as to their sufficiency but should suggest that the owner's legal counsel review all written warranties. The architect's principal function, in respect to warranties, is to confirm that the contractor has submitted all that were specified. The word guarantee no longer appears in AIA documents, as it is synonymous with the word warranty. (A201, 4.2.9)

> All specified operating instructions, user's manuals, wiring diagrams, parts lists, and spare parts for mechanical and electrical equipment.

> Keys and keying schedules.

> Record drawings. (A201, 3.11.1) The record drawings furnished by the contractor will consist primarily of marked-up prints, drawings, and other data showing significant changes in the work made during construction. If the owner wishes the architect to prepare a set of reproducible drawings based on the contractor's record drawings, this can be done as an optional additional service in the AIA Owner-Architect Agreement, Document B141, Subparagraph 3.14.16.

> All contract drawings and specifications used by the contractor and subcontractors should be returned or suitably accounted for, except for one contract record set which the contractor may retain. (A201,, 1.3.1)

All of the contractor's final submissions should be sent to the owner accompanied by a written letter of transmittal inventorying all items in detail to create a permanent record for the architect's file. The final submittal should precede the owner's final payment.

Final Completion

When the work of the combined correction lists has been finished, the contractor should notify the architect that the work is ready for final inspection and acceptance. At the same time, the contractor should submit the final application for payment.

The architect should make a careful inspection to determine that the completed work is in compliance with the contract documents. Consulting engineers and other design consultants should make their final inspections of the portions of the work which they have designed. When the architect is satisfied that the work of the contract is complete and in conformance with the contract documents, the architect's attention should be focused on the contractor's final application for payment.

If any work has been performed not in accordance with requirements of the contract documents, it is the owner's right to accept the work instead of requiring its correction or removal. In this case, the contract price will be reduced as appropriate and equitable. If the owner and contractor cannot agree on a suitable credit, the architect will make a final determination. If the architect's opinion is not acceptable to both, it may be appealed to arbitration. (See Chapter 22, Resolution of Construction Disputes.)

Determination of Final Completion

There is no actual certificate of final completion; the architect's approval of the contractor's final application for payment constitutes the certificate of final completion. The General Conditions, Subparagraph 9.10.1, states in part:

> "Upon receipt of written notice (from the contractor) that the Work is ready for final inspection and acceptance and upon receipt of a final Application for Payment, the Architect will promptly make such inspection and, when the Architect finds the Work acceptable under the Contract Documents and the Contract fully performed, the Architect will promptly issue a final Certificate for Payment stating that to the best of the Architect's knowledge, information and belief, and on the basis of the Architect's observations and inspections, the Work has been completed in accordance with terms and conditions of the Contract Documents...."

The Final Certificate of Payment signifies that in the architect's opinion the contract is fully performed, although it does not actually determine or state a date of actual completion.

The architect's determinations of substantial completion and final completion should be preceded by the necessary field inspections and reviews of the contract documents. The architect should also obtain the viewpoints of owner and contractor before making the final determination. This will preclude later allegations that the architect neglected to take into account some relevant factor, known only to the owner or contractor.

Architect's Final Certificate and Final Payment

The contractor's final application for payment does not have to be in any particular form, but most contractors use a format similar to Application and Certificate for Payment, AIA Document G702, and the Continuation Sheet, AIA Document G703. If these forms are used, the architect's certificate which appears on G702 may be used. The wording of the certificate ties in with the wording of Subparagraph 9.10.1 of the AIA General Conditions. If the contractor does not use the AIA standard forms, the architect will have to create a form of final certificate. It could be in the form of a cover letter to the owner transmitting the contractor's payment application. The wording of the certificate could be similar to that on the AIA form. Although the payment and certificate issued at completion of the work are referred to as final, there will still be one or more additional payments to release the retainage, if any, and sums withheld to ensure completion of remaining inspection list items and unsettled claims.

As a condition precedent to the owner's making the final payment, the contractor is required by the General Conditions (9.10.2) to submit the following items:

> An affidavit that payrolls, bills for materials and equipment, and other indebtedness connected with the work for which the owner or owner's property might be responsible or encumbered (less amounts withheld by owner) have been paid or otherwise satisfied. Forms for this purpose are AIA Document G706, Contractor's Affidavit of Payment of Debts and Claims, and AIA Document G706A, Contractor's Affidavit of Release of Liens.

> A certificate evidencing that insurance required by the contract documents to remain in force after final payment is currently in effect and will not be canceled or allowed to expire until at least 30 days' prior written notice has been given to the owner. This certificate may be obtained by the contractor from its insurance agent or broker.

A written statement that the contractor knows of no substantial reason that the insurance will not be renewable to cover the period required by the contract documents.

Consent of surety, if any, to final payment. This should be an unconditional consent and could be submitted on AIA Document G707, Consent of Surety to Final Payment.

If required by the owner, other data establishing payment or satisfaction of obligations, such receipts, releases and waivers of liens, claims, security interests, or encumbrances arising out of the contract, to the extent and in such form as may be designated by the owner. The contractor must indemnify the owner against the possible liens of any subcontractors who refuse to furnish a release or waiver required by the owner. The contractor must also reimburse the owner all sums paid out in discharging liens after the final payment is made.

All the documentation submitted by the contractor to satisfy these listed requirements should be promptly transmitted to the owner without stating any opinions as to their sufficiency but with the recommendation that they be reviewed by the owner's auditors and legal counsel. The owner should be reminded that if these additional submittals are in proper accounting and legal order, the enclosed architect's certificate for final payment should be paid.

The making of the final payment constitutes a waiver of claims by the owner except those arising from unsettled liens or claims arising out of the contract, failure of the work to comply with requirements of the contract documents, or terms of special warranties required by the contract documents. (A201, 4.3.5) Acceptance of the final payment by the contractor, a subcontractor, or a supplier constitutes a waiver of claims by that payee except those previously made in writing and identified by that payee as unsettled at the time of final application for payment. (A201, 9.10.4)

Decisions to Withhold Certificate

Within 7 days after receiving the contractor's application for payment, the architect must issue to the owner a certificate for payment, with a copy to the contractor, for an amount that the architect deems proper, or must notify the owner and contractor in writing of the reasons for withholding certification in whole or in part. (A201, 9.4.1) The issuance of a certificate for payment constitutes the architect's representation to the owner that the work has progressed to the point indicated and that the quality of the work is in accordance with requirements of the contract documents, to the best of the architect's knowledge, information, and belief. These representations are based on the architect's site observations and information submitted by the contractor in the application for payment. (A201, 9.4.2)

Should the architect consider that the required representations cannot be made, the architect should withhold certification in whole or in part to the extent deemed necessary to protect the owner's interest. If this happens, the architect must so notify the owner and contractor in writing. If the contractor and architect cannot agree on a mutually acceptable amount, the architect should promptly issue a certificate in an amount for which the required representations can be made to the owner. The architect may refuse to certify any amount and may nullify previous certificates in whole or in part if, in the architect's judgment, this is necessary to protect the owner from loss on account of unremedied defective work, the filing or probable filing of third party claims, contractor's failure to pay construction bills, reasonable evidence that the work cannot be completed for the unpaid balance of the contract, damage to the owner or another contractor, evidence that the work will not be completed within the contract time and that anticipated liquidated damages will be greater than the unpaid balance of the contract, or persistent failure to comply with requirements of the contract documents. When these reasons for withholding a certificate are removed, the architect should promptly issue a certificate for the amounts previously withheld. (See Chapter 13, Payment Certifications.)

Owner's Partial Use or Occupancy

The owner may obtain partial use or occupancy of the project before completion upon the Contractor's acquiescence and upon consent of the insurers and public authorities having jurisdiction over the work. The owner and contractor must further agree upon their respective responsibilities assigned to each of them in respect to payments, retainage, security, maintenance, heat, utilities, damage to the work, and insurance.

The agreement must also cover the period for correction of the work and commencement of specified warranties. The contractor may not unreasonably withhold consent for partial occupancy. (A201, 9.9.1) A joint inspection by the owner, contractor, and architect should be prior to partial occupancy to record the condition of the work. (A201, 9.9.2) Partial occupancy does not constitute acceptance of work not in conformance to the contract documents unless otherwise agreed. (A201, 9.6.6 and 9.9.3)

Retainage

It is common in construction contracts that the owner pays 90 percent of the value of the work in place at each payment, the remaining 10 percent being withheld by the owner to be paid to the contractor at or after the completion of construction. Usually, the contract provides that the retained amount will be payable after the expiration of the period during which mechanics' liens may be claimed. This period differs from state to state. In those states, such as California, where subcontractors' and suppliers' lien rights expire 30 days after recording of a valid notice of completion, the contract will provide for payment of retainage 35 days after recording of the notice of completion provided no mechanics' lien claims have been recorded. This allows 5 days after subcontractors' lien rights have expired in which to determine that no liens have been recorded. All that remains, then, is to obtain a waiver of lien rights from the general contractor, whose lien rights would otherwise continue an additional 30 days after the subcontractors' rights have expired. Then the retainage can be paid.

The retainage does not have to be 10 percent. It could be more or less and is subject to owner's requirements and contract negotiation. On projects involving large sums of money, the retainage can become very burdensome to the contractor, and therefore it is quite common to specify reduction or elimination of further retainage after it reaches a certain amount. For example, on a $1,000,000 contract, a 10 percent retainage would accumulate by the end of the work to $100,000. The contract could provide for no further retainage after the first $50,000, if it is deemed that this sum would sufficiently protect the owner's interest. It is also possible to release some of the retention at some point such as at completion. In this example, $25,000 could be released at completion, leaving $25,000 to be retained until after expiration of the lien period. In the event that the retainage is to be reduced or partially released, if the job is bonded it is essential to obtain the surety's prior permission. This should be obtained by the contractor and should be unconditional and in writing. A standard form is available for this purpose, AIA Document G707A, Consent of Surety to Reduction In or Partial Release Of Retainage.

Liquidated Damages

If the contract provides for payment or allowance of liquidated damages to the owner in the event of late completion, the architect will have to make a computation of the sum owing, if any, or a determination that the contract has been completed on time. The contract provision for liquidated damages should be in terms of a certain number of calendar days (to comply with the definition in Subparagraph 8.1.4 of the General Conditions) or a specified date for substantial completion. The starting date for computation will be stated in the contract. (A201, 8.1.2) The contract time is terminated on the date of substantial completion. (A201, 8.1.1 and 8.1.3)

The contractually stipulated completion date must be adjusted for all agreed extensions, for inclement weather, and for any other justifiable extensions approved by the architect. Any assessed liquidated damages should be deducted from the final payment certificate.

Final Reconciliation of Cash Allowances

When the actual purchase costs of allowance items are known, the differences from the specified amounts should be adjusted by means of a change order. If the net purchase cost (including shipping and taxes as specified in A201, 3.8.2.2) exceeds the allowance, the excess is to be charged to the owner by an additive change order or, when it is less, by a deductive change order.

There should be no change in the contractor's related costs unless the owner or architect had changed the specification or quantity of the allowance item in a way that would affect those costs, up or down. In that event, it would also be appropriate to adjust the contractor's overhead and profit accordingly. If such a change affects the time of performance through no fault of the contractor, the contract time should also be adjusted in the change order.

Owner's Final Payment to the Contractor

The owner is required to pay the architect's final certificate within 30 days thereafter unless stipulated otherwise in the agreement. (A101, Article 6 or A111, Article 13)

Termination of the Contract

Either owner or contractor may terminate the contract for cause under certain specified conditions. In addition, the owner may suspend the work for convenience. (A201, Article 14) In case of termination or suspension, careful compliance with all technical provisions of the contract is very important.

Unilateral terminations, whether initiated by owner or contractor, will always be under conditions of strained relationships, financial hardship, and unrealized expectations and will undoubtedly be followed by claims and counterclaims and litigation or arbitration. Therefore, any services performed by the architect during these stressful periods must be strictly in accordance with the procedures set out in the contract documents. (See Chapter 20, Termination of the Construction Contract.)

Architect's Decisions

All unsettled claims and differences between the parties should be resolved in the process of concluding the contract. The architect should attempt to obtain the viewpoints of owner and contractor on all matters remaining in contention. A thorough examination of their positions and a complete analysis of the requirements of the contract documents should be carefully made. The architect should promptly make a competent and fair determination of each issue in controversy so that the dispute resolution process can proceed to the next step, arbitration, if necessary. (A201, 4.3, 4.4, and 4.5) (See Chapter 22, Resolution of Construction Disputes.)

Owner-Architect Relationship

As part of the contract closing process, the architect should make certain that the office files are complete and in proper order for permanent filing. The files will be of extreme importance in the event of a later legal claim against the architect. Even if the job has run smoothly and the client and contractor are both extremely pleased with the outcome of the contract, there is still the lurking possibility of claims being asserted by third parties, as yet unknown, such as later owners, occupants, lessees, and passersby. There is also the possibility of a latent construction defect becoming apparent years later.

The architect's responsibility to render basic services under the AIA Owner-Architect Agreement for the construction phase commences with the award of the construction contract and terminates at the earlier of issuance of the final certificate for payment or 60 days after the date of substantial completion unless extended. (B141, 2.6.1)

Many owners will wish to have the architect inspect the project prior to expiration of the various specified warranty periods. This service is not usually included in the basic architectural fee and would be the subject of an additional charge. A written detailed observation report should be issued. The architect should be meticulous in differentiating between maintenance or usage defects which are the owner's responsibility and construction defects which the contractor must rectify.

Amicable Conclusion of the Contract

After the owner has made the final payment, the retainage payment, and any other sums that had been held back pending completion of punch list work, the owner's and contractor's paperwork is virtually completed.

The contractor has no further duties on the jobsite other than responding to requests for adjustments, corrections, and repairs during the specified warranty periods. It has now become the owner's responsibility for day to day maintenance of the building and grounds.

After concluding the contract in this methodical, uneventful, and amicable manner, the owner and contractor should still be on good terms. The contractor should not be at all apprehensive about referring prospective customers to view the work and to talk to the owner. The owner would be more than pleased to recommend the contractor to friends and associates. This type of contract ending is good for the construction industry.

20

Termination of the Construction Contract

Construction Is a Unique Manufacturing Method

The construction process is extremely complex, depending on the cooperation of dozens of entities that may have never before worked together in the same combination on a single project. No other manufacturing process works in the same peculiar way. Most manufacturers fabricate their products under cover in clean, warm, dry, and lighted, purpose-built buildings with permanent tools, storage, security, and a stable work force. Their major suppliers remain largely unchanged for years. If automobiles, washing machines, or suits of clothes were made the same way as buildings, one at a time, no average person could ever afford these commonplace necessities.

Differing unique sites, innovative designs, changing permutations of subcontractors and suppliers, and new combinations of owners, contractors, and architects all contribute to the lack of uniformity. The construction industry attempts to standardize wherever possible to counteract the negative effects of non-standard manufacturing conditions. The principal areas of standardization have been in uniform building codes, published material standards, prefabricated components, the customs and practices of each trade, and standardization of documentation.

It is of little wonder that occasionally the delicate balance of organized chaos that exemplifies a typical construction project becomes disturbed. The most sensitive factors are usually the people involved. Imperfect communication, differing temperaments, careless craft workers, and incompetence, deception, and dishonesty all contribute their share to the irretrievable breakdown of satisfactory working relationships among the parties.

When all involved are carrying out their contractual obligations as expected, relative tranquillity will prevail and the project will progress to eventual satisfactory completion. The owner will have its project and all contributing parties will have achieved their expected rewards.

It is when any of the involved entities conducts themselves inappropriately that the normal pattern of controlled turmoil will be disrupted. It is only a matter of degree that distinguishes projects that are able to overcome the expected irregularities from those that erupt into the turbulence of irreconcilable conflict and an aborted construction contract. The usual consequence is severe financial hardship or even financial collapse of some of the participants.

Termination Provisions in the Contract

The low point of a construction administrator's life is when adverse circumstances result in the contractor's abandonment of the contract or the owner's ejection of the contractor. Such problems are the results of a period of deteriorating conditions and are seldom completely unexpected. These extreme measures have far-reaching legal implications and financial repercussions for both parties. They should not be casually considered and should be undertaken only as a last resort. Practical and equitable termination provisions, fair to both sides, are included in the AIA General Conditions of the Contract for Construction, Fourteenth Edition, Document A201, 1987 (Appendix D).

Termination of the Contract

The most disruptive of events that can occur in construction is the owner's decision to oust the contractor and terminate the contract or the contractor's decision to leave. The owner's and contractor's necessary rights to terminate and the governing terms and conditions should be included in all construction contracts. Contracts based on standard AIA documentation include a practical termination procedure that respects the interests of both the owner and contractor. The AIA General Conditions provides for three termination possibilities:

1. Termination by the contractor (Paragraph 14.1)
2. Termination by the owner for convenience (14.3)
3. Termination by the owner for cause. (14.2)

The first and second cases provide for certain conditions precedent and prescribe orderly procedures for termination, but do not require the architect's approval or certificate of opinion.

However, in the third case the contractor's interests are protected from an ill-informed or precipitate owner by requiring the architect's concurrence in the form of a certification that there is sufficient cause to justify the contractor's expulsion. (A201, 14.2.2) It also shields the owner from making a potentially costly error in judgment.

What Do Contractors Expect? The average contractor going into a construction contract expects that the owner will have the capacity and willingness to do all that is required of it by the contract. In addition, they also expect that the owner will be cooperative in all relevant matters and will not be obstructive in any way. These are the normal and reasonable contemplations that anyone would entertain when entering into any kind of business dealing with others.

It would also be reasonably anticipated that the owner would not proceed if it knew that it would be unable to perform its part of the contract. Normally the owner would have much to lose if it violated any of the contractor's expectations.

When the Owner's Behavior Deviates from the Norm

The contractor's capacity to make a profit on any contract is predicated on its ability to control money allocations and time schedules. Anything that impedes progress on the project will usually affect both factors. The contractor would then be placed in arrears on money and/or time. Most contractors will have prepared each project's cost breakdown and time schedule with the expectation that their own people, their suppliers, and subcontractors will perform efficiently in accordance with industry standards and past experience. Occasionally, their estimates will prove to be faulty or they will encounter unexpected conditions that will affect their money or time budgets. These snags and glitches are the very nature of contractors' risk taking. Contractors know that it is their responsibility to pay for their own time and money deficits. They accept this risk when they sign the contract.

Therefore, it is shocking to a contractor to discover during construction that the owner is apparently unable to hold up its end of the bargain. It means that the contractor's business has been propelled involuntarily into economic peril in proportion to the size of the endangered project. In extreme cases it will trigger the firm's bankruptcy as well as topple some of their subcontractors and suppliers.

Generally, the reverses that plague owners after the contract is signed are in the realm of unfortunate occurrences or unforeseen, although not always unforeseeable, conditions.

Sometimes owners proceed with the contract and the construction prematurely, before having in hand all such essentials as clear title to the land, lessor's permission, legal rights of way, adequate construction financing, or all necessary governmental approvals. The lack of any of these indispensable elements at the appropriate time could result in a work stoppage. It might take weeks or months to cure the problem. Meanwhile, if the construction had already commenced it would have to be curtailed or stopped.

Work might have to be halted while awaiting decisions on matters requiring a procrastinating owner's input or approval. Proposed changes in the work might be under lengthy consideration by a vacillating owner, thereby preventing the contractor from proceeding in certain work areas. When work being held in abeyance has disrupted the planned work schedule excessively, it becomes uneconomical to continue in a piecemeal manner, so field operations must be suspended until a continuous work schedule can be pursued in all work areas.

The contractor would seriously consider stopping work if the owner has ceased making prompt payments pursuant to the architect's certificates. A complete cessation of payments will undoubtedly result in an absolute work stoppage.

Reluctant Architects. Sometimes it is the architect's or engineer's procrastination or intransigence that is holding up the work. The contractor might have to stop the work if requested decisions or interpretations are not forthcoming as and when needed. Shop drawings not promptly reviewed, corrected, and returned in accordance with the approved construction schedule will prevent the construction from progressing and could be the cause of a shutdown. The construction progress could be seriously affected if the architect is not diligent in making all scheduled site visits and promptly issuing site observation reports.

Termination by the Contractor

At various times a contractor might conclude that mounting losses caused by rising costs, inefficiency, bad luck, or an uncooperative owner could be curtailed or eliminated by quitting the job and canceling the contract. However, these are not allowable reasons for termination. The contract provides that the contractor is justified in declaring the contract terminated only under certain limited specified conditions.

Termination of the Construction Contract

The contractor may terminate the contract if the work is stopped for a period of 30 days through no fault of the contractor or any of its suppliers, subcontractors, or employees, for any of the following reasons:

1. Issuance of an order of a court or other public authority having jurisdiction
2. An act of government, such as a declaration of national emergency, making material unavailable
3. Because the architect has not issued a certificate for payment and has not notified the contractor of the reason for withholding certification, or because the owner has not made payment within the agreed time
4. If repeated suspensions, delays, or interruptions by the owner, as allowed by the contract, constitute in the aggregate more than 100 percent of the total number of days scheduled for completion, or 120 days in any 365-day period, whichever is less
5. The owner has failed to furnish to the contractor promptly, upon request, reasonable evidence that financial arrangements have been made to fulfill the owner's obligations under the contract (AIA Document A201, Subparagraph 14.1.1)

If any of these reasons exist, the contractor may terminate the contract upon 7 additional days' written notice to the owner and architect. The contractor is then entitled to recover from the owner payment for work executed and for proved loss with respect to materials, equipment, tools, and construction equipment and machinery, including reasonable overhead, profit, and damages. (A201, 14.1.2)

The contractor may also terminate the contract if the owner or owner's agents (possibly the architect) cause the work to be stopped for 60 days by persistently failing to fulfill the owner's obligations in matters important to the progress of the work. The contractor, after giving an additional 7 days' written notice, may terminate the contract, provided the contractor, suppliers, subcontractors, or employees are not at fault. (A201, 14.1.3) Matters important to the progress of the work include such owner shortcomings as:

1. Failure to make the site or work areas available when needed
2. Failure to obtain necessary easements, party line agreements, utilities, and governmental approvals
3. Failure to make necessary decisions such as those relating to color and material selections and other options
4. Failure to provide owner-furnished materials, equipment, services, or separate contractors

    5. Failure to respond to shop drawing submittals or other requests for information or direction

Downside risks for the contractor who opts for termination are the negative effect on its reputation in the community, the possibility that a court or arbitrator will not agree that sufficient cause existed for termination, and the distinct likelihood of monetary loss.

The Architect's Position When the Contractor Terminates

Whereas the architect's approval and certificate is a precondition of the owner's termination of the contract, no such approval is required for the contractor to terminate. As a practical matter, in many cases of owner financial delinquency leading to the contractor's termination decision, the architect would have previously disappeared on account of not being paid. The delinquent or impecunious owner would now be self-administering the contract. This would add immeasurably to the already distressed contractor's problems.

The contractor's right to terminate is related mostly to lengthy work stoppages. Dates and lengths of stoppages are not matters of opinion as they are simple to document and to prove. The troublesome part is in documenting and proving justification for the stoppage. Contractors' terminations based on Subparagraph 14.1.3 (owner's failure to perform its obligations in matters important to the progress of the work) can become seriously contentious when the contractor claims to be delayed by the owner's lack of action and the owner maintains otherwise.

In those cases where there is a difference of opinion between the owner and contractor as to the cause of the stoppage, both parties could claim the right to terminate. This situation would require the architect, if still functioning, to make a written determination that would become binding on the parties if not appealed to arbitration within 30 days. (A201, 4.5.4.1.) The architect's determination should be based on a thorough investigation of the facts and circumstances including the viewpoints of both the owner and contractor. The opinion should be objective, siding with neither party. When the stakes are high, this type of disagreement is usually too controversial to end simply with the architect's decision. It will usually be concluded only after presentation to arbitrators. If the architect's administration had already been dispensed with by reason of the owner's financial incapacity, the dispute would have to be referred directly to arbitration.

When the Owner Ejects the Contractor and Terminates the Contract

Honorable Intentions. Most owners and contractors enter into their construction agreements honorably and with optimism and high expectations. The owner is looking forward to seeing the happily anticipated project materialize and the contractor is pleased to further its business interests by embarking on a new, hopefully profitable, construction venture. The owner is extremely pleased to have found such a seemingly qualified contractor to build the project at an acceptable contract price and with an agreeable completion date. On the strength of the contract and the contractor's assurances, the owner begins making definite plans for vacating the old premises and moving into the new on the promised date. Some owners will have made significant financial commitments for receiving newly purchased materials, furniture, and equipment and newly hired personnel based on the anticipated completion date.

When the Bubble Bursts. As the contractor starts the work on the site, both owner and contractor are brimming with optimism. Accordingly, it is with great disappointment and impending fear that the owner becomes increasingly aware of the contractor's apparent shortcomings. It seldom happens all at once. Conditions gradually deteriorate. It usually starts with small disappointments and minor violations of the contract. There will be days when no one is working on the job. There is always a rational excuse. There will be lapses in workmanship and the appearance of inferior materials on the job. Construction equipment is mysteriously removed from the jobsite. The job superintendent is working sporadically, only part-time, or is replaced. When the contractor is questioned about any of these matters, there is a glib reply and a plausible explanation, offering reassurance to the owner and often questioning the owner's lack of faith.

As time goes on, it becomes more difficult to communicate with the contractor. Telephone calls are not returned. Letters and faxes are not answered. Rumors are abroad that labor, subcontractors, and suppliers are not being paid. Invoices, statements, and demands from subcontractors and suppliers are beginning to show up in the owner's mail. The work is now starting to fall significantly behind the construction schedule.

On some projects it starts differently. The first clue of the contractor's financial slippage is seen in the monthly requisitions for payment. Most of the line items are slightly exaggerated, the change orders are overpriced, and the contractor is becoming more persistent about prompt collection of the monthly payment, usually by messenger, and early, if possible. It is suggested that the retainage be reduced or released. The contractor is becoming more strident about money matters.

Increasingly, the owner grows more seriously concerned about the competence, reputability, and possibly the solvency of the contractor.

Considering Termination. After an owner's confidence has been severely shaken by some of the illustrative symptoms, it would be extremely difficult to regain credible belief in the contractor. The owner's first reaction would be to promptly get rid of the contractor.

Termination by the Owner for Cause

When symptoms of a contractor's impending default begin to appear, owners, their architects, and legal advisors should start considering the options and whether or not to terminate the contract. It is very unsettling to an owner when the contractor is not making satisfactory progress or the workmanship is substandard, or both. Subcontractors' complaints of not being paid and the emergence of mechanics' liens are of serious concern. Often these indications are accompanied by desperate claims for extra time and compensation on flimsy pretexts or nonexistent grounds.

The owner and architect will often consider the possibility of terminating the contract, particularly if the contractor has not reacted favorably to repeated admonitions. This is allowable under the contract only under carefully defined circumstances. The owner may terminate the contract if the contractor:

1. Persistently or repeatedly refuses or fails to supply enough properly skilled workers or proper materials
2. Fails to make payment to subcontractors for materials or labor in accordance with the respective agreements between the contractor and subcontractors
3. Persistently disregards laws, ordinances, or rules, regulations, or orders of a public authority having jurisdiction
4. Otherwise is guilty of a substantial breach of a provision of the contract documents (A201, 14.2.1)

It must be the owner's decision whether or not to take the momentous step of terminating the contract. It is not a decision to take lightly, because the contractor will undoubtedly resist such a move by all legally available means. Very few contractors would respond by simply leaving quietly upon receipt of the owner's termination notice. Moreover, the owner must then face the demanding task of finding and engaging a replacement contractor to finish the job. Few contractors would be enthusiastic about stepping into a situation teeming with dissatisfaction and possible legal repercussions. Fewer still would do the work on a fixed price contract.

The Architect's Approval

As a basic protection to the contractor, the AIA General Conditions does not allow the owner to make a unilateral determination of the contractor's default. The owner must obtain the architect's written certification that sufficient cause exists to justify the owner's proposed action.

The architect who executes such a certificate should make a positive independent determination that one or more of the causes specified in Subparagraph 14.2.1, are in

Termination of the Construction Contract

actual existence and can be proved by documentary, photographic, or other convincing evidence. In practically every case, the basis of the certificate will be vigorously contested and will have to be proved to the satisfaction of an arbitrator or judge. It will be minutely dissected by the contractor's technical and legal advisors.

It will not be very impressive to arbitrators or judges if the architect's certificate is dated after the owner's termination notice to the contractor or if the certificate had not been appended to the notice.

The architect must find that one or more of these reasons exists and must certify, in writing, that there is sufficient cause to justify termination of the contract. The owner must give the contractor and the surety, if any, 7 days' written notice. Then the owner, subject to any prior rights of the surety, may

1. Take possession of the site and of all materials, equipment, tools, and construction equipment and machinery thereon owned by the contractor
2. Accept assignment of subcontractors, as provided in the contract
3. Finish the work by whatever reasonable method the owner may deem expedient (A201, 14.2.2)

Owner's Rights After Termination

The owner may undertake to finish the work by whatever reasonable means the owner deems expedient. This means, of course, that the owner must be very diligent not to waste funds nor incur any upgrading expense. Any project betterment costs must be strictly segregated because they cannot later be charged to the defaulting contractor. Competent record keeping is imperative because all expenditures will be carefully scrutinized later by the contractor's accountants.

All of these owner rights are subject to any prior rights of the surety if there is one. As a practical matter, the owner has more to gain than lose by cooperating with the contractor's surety. The surety's objective will be to gain prompt and economical completion of the contract. The surety will probably have more familiarity with the contractor's financial condition and sources of funds than the owner could possibly have. The surety will also have better rapport and communication and considerably more influence with the contractor than would the owner under the circumstances.

The surety, in furtherance of its own interests will often provide financial advice and assistance to the contractor. The surety is entitled to use its own judgment in deciding how to finish the work of the contract. It may use the same contractor, hire a new contractor and subcontractors, continue with the existing subcontractors, or pay money to the owner.

Balance of the Contract Sum

After termination and during the completion of the contract, the owner is not obligated to pay any further funds to the contractor. (A201, 14.2.3.)

If the cost of completion exceeds the unpaid balance of the contract sum, the excess will be charged to the contractor. In the usually unlikely event that the completion costs are less than the unpaid balance, the remaining funds are to be paid to the contractor.

The architect is required to issue a certificate determining the final payment from or to the contractor. The obligation for payment pursuant to the architect's certificate survives the termination of the contract. (A201, 14.2.4)

In either case, the architect's additional services and expenses made necessary by the termination are paid by the owner and charged to the contractor.

In most cases it would be extremely difficult for the owner to finish the work for the funds remaining in the contract. Furthermore, the owner should realize that any cost overruns would be nearly impossible to collect from a bankrupt or impecunious contractor. If there is a surety bond, prospects might not be so bleak. However, both the owner's and the architect's actions will be strictly monitored by the surety in their efforts to keep the completion costs under control and limit their potential losses.

Considering the charged and contentious climate in which all these termination procedures will be taking place, the architect and owner must be exceptionally diligent in following the stipulated procedures and in their stewardship of the construction funds.

A Practical Alternative. Apparently the owner has the clear right to terminate if the grounds exist. However, if the contractor makes a successful effort to rehabilitate its position during the notice period, the owner could decide to set the termination aside. Economically, this is often the best solution to the problem for both parties, although the owner might have lost confidence in the contractor. The practicality of this outcome often rests more heavily on the temperaments and personalities of the parties than on considerations of economics and legal rights.

Suspension by the Owner for Convenience

Owners often require the flexibility of suspending a construction contract in order to accommodate other unpredictable factors or events such as changes in cash availability or business requirements. The contract allows the owner at any time, without stating any cause, to order the contractor to suspend, delay, or interrupt the

Termination of the Construction Contract

work of the contract for any length of time suitable to the owner's requirements. If this occurs, the owner must reimburse the contractor for the extra costs of performing the contract plus an agreed fixed or percentage fee. The owner's order must be in writing. (A201, 14.3)

Position of the Surety

If the contract is bonded using the AIA standard bond form (Performance Bond and Payment Bond, Document A312, December 1984 Edition), the surety and contractor must be given a chance to cure the deficiencies and reinstate the contractor. The bond agreement requires the owner to give the surety and contractor written notice of a conference, to be held within 15 days of the notice, to discuss methods of the contractor's performing of the contract. If the owner, contractor, and surety can agree, the contractor will be given a reasonable time to perform the contract. In the event of no such agreement or if the contractor fails to properly resume production, the owner's declaration of termination cannot be made sooner than 20 days after the conference notice. (A312, 3.1 and 3.2)

One of the principal advantages of the conference will be to enlist the aid of the surety in motivating the contractor to rehabilitate its organization and finances and to reestablish its workmanship standard, coordination, and scheduling. The surety can exercise considerable financial pressure on the contractor and will prove to be very helpful to both parties. In some cases, to protect its own investment, the surety will provide financial advice and assistance to the contractor.

If the original contractor cannot by itself satisfactorily proceed with the work of the contract, the surety may use its own judgment and exercise its own choice as to method of completing the contract. It may hire a new contractor, use the original contractor, contract directly with subcontractors and suppliers, or pay monetary damages to the owner. (A312, Paragraph 4) The owner should insist that a completion and payment bond equivalent to the original bond be in effect for the completion of the contract. The owner must continue paying the balance of the contract price to the bonding company or to its designated contractor. (See Chapter 11, Construction Administration When the Contract Is Bonded.)

Position of the Architect

The periods leading up to and following termination are generally very stressful for all involved parties including the architect. Opportunities for maladministration by the architect are profuse, and the professional liability implications are limited only by the creative imaginations of lawyers hired by the owner, contractor, and surety.

All professional acts of the architect must be performed with deliberation and in strict accordance with the owner-architect and construction agreements and the standard of care for architects in the community. The architect should try to keep the lines of communication open among the parties, insist that all actions be in writing, and carefully maintain and preserve all documentation. The architect should avoid being a bottleneck, impeding rapid and open communication.

An architect's certificate of sufficient cause upon which an owner will base its right of termination must be an independent opinion founded on solid provable evidence. If later a court or arbitrator deems that sufficient cause is lacking, the owner may be assessed damages in favor of the contractor.

The Architect's Certificate

The architect's certificate should carefully correlate each of the contractor's alleged transgressions with the corresponding reason in the General Conditions, Subparagraph 14.2.1. The architect should have appropriate backup documentation or other proof of each of the allegations set forth in the certificate. It is inevitable that the certificate will be minutely dissected, thoroughly discussed, and painstakingly analyzed by the contractor's team of technical advisors and legal counsel.

When an owner in a moment of panic or intemperance fires the contractor and then seeks legal advice, counsel should review the contract's termination provisions and remind the owner of the required architect's certificate. A certificate dated after the expulsion date is not very impressive to construction arbitrators, while a fraudulently pre-dated certificate can produce obvious problems. A prudent owner would have served a copy of the architect's certificate on the contractor simultaneously with the termination notice.

There is always the risk that an ejected contractor will be able to convince arbitrators that the termination was unjustified. Then, the owner's burden will be doubly onerous: not only would there be the excessive completion costs and delayed completion with a new contractor, but also the payment of lost profits and damages to the discharged contractor. All this would be further compounded by the necessarily extensive legal costs.

Practical Considerations in the Decision to Terminate

Considering the far-reaching legal and economic consequences of ejecting the contractor, both the owner and architect must be meticulous in applying the dismissal procedure delineated in the contract. There is little likelihood that an ousted contractor will voluntarily concede to an expulsion without rigorous contention.

Architect's Independent Evaluation. The dissatisfied owner should be advised to seek the architect's considered evaluation of the contractor's supposed shortcomings. If they can be honestly sustained, in the architect's independent judgment, then a written certificate should promptly issue. The architect's opinion should be based upon current firsthand knowledge of field conditions and personal examination of the relevant documents and not simply an unconfirmed reiteration of the owner's position.

Protecting and Completing the Work. When the contractor is gone from the jobsite, whether by voluntary abandonment or owner's ejection, the owner is faced with the serious problem of protecting the work in process, the site, materials, and equipment until such time as a replacement contractor has been engaged. It is usually very difficult to find a new contractor who will complete the work for the remaining contract sum or, for that matter, for any fixed sum. It is almost always necessary to contract on a cost-plus-fee basis without a guaranteed maximum price. Time will always be lost, as it will be necessary for the new contractor to become intimately acquainted with the contract requirements as well as with the status of all aspects of the work in the field. All the subcontractors and suppliers must be contacted and decisions made as to who will be retained and who will be replaced. The architect must decide the methods and extent of correcting all defective work. If any defective work is to remain uncorrected, the owner's concurrence must be obtained for any compromise of the original contract standards. Budgets and time schedules must be reestablished. The subcontractors, employees, and materials must be remobilized and bonds and insurance must be in place before the new operation can commence.

Late Completion. The inevitable late completion will create additional expenses and duties for the owner: deferral of furniture, stock and equipment, and machinery deliveries, or if it is too late for postponement, provisions for receiving, protecting, and temporary storage; deferral of new or transferred employees; and notification of governmental authority and customers. Alternate financial arrangements may have to be provided.

Economic Conservation. It is only a remote possibility that the work can be completed on time or for the funds remaining in the contract. In those situations where the contractor has been involuntarily removed and thus financially responsible for the construction cost overrun, there will almost always be an economic shortfall. There is little likelihood that a contractor in those circumstances will willingly pay the difference. Even though the owner may not be sympathetically disposed toward the contractor, there remains a serious duty to the contractor to eliminate waste, to exercise conservative administration of the funds, and to mitigate damages. Any contract changes which amount to economic upgrading or betterment of the project cannot be charged to the discharged contractor.

From the contractor's standpoint, the decision to abandon the contract is potentially costly, as the losses could be considerably more than the mere loss of a job. Usually the reason the owner has defaulted is insurmountable financial or legal obstacles, and the owner will not be able or willing to pay its obligations to the contractor.

Whether the owner or contractor has initiated the termination procedure, both sides will be exposed to financial losses, legal claims and counterclaims, legal fees, attacks on reputations, and considerable loss of time and effort. No termination should be initiated before complete analysis and advice from legal counsel. If after thorough consideration of the consequences and alternatives termination is deemed to be the least onerous option, it should be cautiously undertaken. Termination of a construction contract is a technical legal procedure and must be pursued strictly in accordance with the contract. Obviously, all aspects of the termination procedure should be in writing.

Bankrupt Contractor or Owner. If the defaulting owner or contractor happens to be in or on the verge of bankruptcy proceedings, this will have a profound impact on the termination procedure. Permission of the bankruptcy court will be required for any moves against the financial interests of the bankrupt party. It is not possible to predict whether the trustee acting for the bankrupt will approve continuation of the contract or will favor the termination procedure. The decision will depend on which course would, in the trustee's judgment, yield the maximum conservation of the bankrupt's assets and the impartial treatment of all the creditors. Legal advice necessary for dealing with the bankrupt contractor should be furnished by the owner. In the event that the architect's client (the owner) is the bankrupt, the architect might have to confer with its own legal counsel.

Warranty Responsibility

The changing of contractors before completion of the contract raises a knotty problem relating to responsibility for warranties required by the contract. Confusion will arise as to which contractor is responsible, particularly for latent defects, those which are not known at the time of the change of contractors. The first contractor will always claim that the second contractor damaged or destroyed the original work while the second contractor will claim that the work of the first contractor was unsuitable or defective. Written and photographic records which describe the precise condition of the project at the time of changeover will be invaluable in determining the validity of buck-passing arguments.

In the event that the first contractor is gone on account of bankruptcy or insolvency, the warranties for its work are worthless unless the project is bonded.

21

Architect's Decisions Based on Design Concept, Aesthetic Effect, and Intent of the Contract Documents

Imprecise Standards for Architect's Decisions

Construction contracts include an impressive assemblage of documents drafted with all of the precision and technology available to informed architects and engineers and experienced construction industry lawyers. The construction drawings precisely delineate complex physical structures and operations while the specifications accurately describe particular procedures and specific materials. The agreement exactly specifies the contract time to the day and the contract sum to the dollar. The supplementary conditions customize the unique contract conditions for the specific project at hand. The AIA General Conditions is a collection of highly coordinated contract conditions based on the principles that form the very essence of most construction contracts in the United States. It clearly reflects the standard and customary procedures of the U.S. construction industry.

With all this graphic and verbal technological and legal precision it would no doubt be surprising to the uninitiated to find in the AIA General Conditions that owners and contractors routinely agree to abide by the architect's decisions in some extremely abstract, nebulous, and undefined areas. It is particularly surprising that contractors, noted for their practicality, would agree to be bound by such indefinite provisions.

Intent of the Contract Documents

The cryptic phrase "intent of the contract documents" is defined in an uncertain manner in Subparagraph 1.2.3 of the AIA General Conditions: "The *intent* of the Contract Documents is to include all items necessary for the proper execution and completion of the Work by the Contractor." Further, in the same paragraph, "performance by the Contractor shall be required only to the extent consistent with the Contract Documents and *reasonably inferable* from them as being necessary to produce the *intended results*." (All italics have been supplied by me for emphasis. The AIA General Conditions has no italics.)

Only the architect can supply a definitive answer as to what physical results were actually intended. It would be far better if architects could always express their design intentions by specific contract requirements that would produce the desired results. Then the contractor would know exactly what was expected and could provide for it in the construction scheduling and the cost breakdown. As to what precisely is "reasonably inferable", there can be many an honest disagreement between well-intentioned competent contractors and fair-minded owners. If anything of substantial economic consequence is left off the drawings or omitted from the specifications, it is not proper to charge the contractor for it merely by the owner's or architect's declaration that it was reasonably inferable and should be included in the contract price. An architect, making a ruling to decide this disagreement (AIA Document A201, Subparagraph 2.6.15), must often rule in favor of the contractor. The term "reasonably inferable from the contract documents" means that the architect's opinion or conclusion must be well-balanced, rational, logically reasoned, and based on the existence of some reasonable evidence. The architect is justifiable in relying on this catchall phrase only in such situations where all of the pieces and parts customarily included in a specified installation or assembly are normally expected to be supplied without naming each part. For example the contractor should include all fastenings, washers, gaskets, adhesives, lubricants, tools and equipment normally incidental to the installation of the object or assembly specified.

Items are not reasonably inferable if the specification is written in such a way that competent experienced bidders would not normally include them in their cost breakdown. If they have been truly omitted from the contractor's cost breakdown, then the owner has not paid for them.

The architect is authorized to order additional testing or inspection of the work "Whenever the Architect considers it necessary or advisable for implementation of the *intent of the Contract Documents*...." (A201, 4.2.6) Clearly, any testing ordered under this provision will be chargeable to the owner unless the tests disclose defective or nonconforming work, in which case the testing would be charged to the contractor. (A201, 13.5.1, 13.5.2 and 13.5.3)

Design Concept

Another vague expression used in the AIA General Conditions is the term "design concept" which appears in Subparagraphs 3.12.4 and 4.2.7, in both cases relating to the contractor's submittals. The purpose of submitting shop drawings, product data and samples "is to demonstrate...the way the Contractor proposes to conform to...the *design concept* expressed in the Contract Documents"; and "The Architect will review and approve...the Contractor's submittals...but only for the limited purpose of checking for conformance with...the *design concept* expressed in the Contract Documents." In each of these subparagraphs, the architect is limited by what is actually expressed in the contract documents. Regardless of the architect's mental image of the "design concept", no more can be required of the contractor than is actually explicitly stated in the contract documents.

If the architect wishes to extend the contract requirements by means of the shop drawing process, a change order should be initiated to cover the cost of any variations or elaborations. By use of the term "reasonably inferable," these two subparagraphs do not extend the meaning of design concept beyond what is specifically shown in the contract documents.

Architect's Decisions

During the construction period, when differences of opinion arise between owner and contractor, the architect is invested with the duty of interpreting the documents and rendering a decision. The decision will be final and binding upon the parties if not appealed to the arbitration process within 30 days. (A201, 4.3 and 4.4) (See Chapter 22, Resolution of Construction Disputes.)

The AIA General Conditions, Subparagraph 4.2.12, requires, "Interpretations and decisions of the Architect will be consistent with the *intent* of and *reasonably inferable* from the Contract Documents....". This places the burden of impartiality and intellectual honesty squarely on the architect's shoulders. This is particularly true if the disagreement between the owner and contractor is related to shortcomings in the contract documents prepared by the architect. If the architect is not scrupulously fair in making such decisions, an impartial arbitrator will later find no difficulty in reversing or revising the decision.

Aesthetic Effect Decisions

A unique characteristic of construction contracts which include the AIA General Conditions is the power given to the architect to make quasi-judicial determinations on the claims of the contracting parties. This is a practical procedure and a necessity to keep the work of the contract moving expeditiously in the field. However it relies heavily on the competence and professional integrity of the architect. The architect must also possess the ability to resist pressure from either the client or the contractor. The contractual procedure must also be capable of functioning in the rare event that an architect is unable to render objective decisions acceptable to the parties or to summon the necessary courage to rule against the intimidations of an outraged client or a threatening contractor.

If either party is dissatisfied with the architect's determination, it can be appealed to arbitration. If the arbitrators deem that the architect's decision was ill-founded or erroneous in any way, they have the power to revise or reverse it when promulgating their own award. The arbitrators' award will be final and binding if not appealed within 100 days (in California) to the court having jurisdiction in the matter. The grounds for setting aside an arbitration award are few and are limited to matters relating to fraud, corruption, and fairness of the procedure. Judges will not hear appeals based on the merits of the controversy.

All appeals from the architect's decisions as well as other controversies not submitted to the architect will ultimately be referred to arbitration as set out in Subparagraph 4.5.1 of the General Conditions, which states in part that "Any controversy or Claim arising out of or related to the Contract, or the breach thereof, shall be settled by arbitration...., except controversies or claims relating to *aesthetic effect*...."

Closely related to this is Subparagraph 4.2.13, which provides that "The Architect's decisions on matters relating to *aesthetic effect* will be final if consistent with the intent expressed in the Contract Documents." Aesthetic effect is not defined in the AIA General Conditions, so we must rely on the dictionary, which defines aesthetic as "relating to the beautiful as distinguished from the merely pleasing...the useful and utilitarian." It is synonymous with artistic.

Architect's Decisions Based on Design Concept

The purpose of these seemingly loose aesthetic effect provisions is to protect the integrity of the architectural design and the design judgment and decisions of the architect. Someone has to be the final authority in everything and it would be inappropriate for an arbitrator, even if an architect, to have final design prerogative superior to that of the design architect.

Practical application of the relevant contract provisions provide ample protection to the owner and contractor in the unlikely but possible event that an architect would get outrageously out of hand. If the owner and contractor are in unanimous opposition to the architect's aesthetic effect decisions, the architect will simply be bypassed. They will do whatever they mutually agree to do without regard to the architect's wishes.

Aesthetic effect decisions are final, not subject to arbitration, but with two notable exceptions: first, when the decision is ruled to be not consistent with the intent expressed in the contract drawings, and second, when the arbitrators rule that the determination was not strictly an aesthetic effect decision, but was artfully so classified by the architect to place it beyond the reach of arbitrators.

Consequently, architects who wish to make aesthetic effect decisions which will remain in effect must adhere strictly to the terms of the contract. The decision must be in writing. (A201, 4.4.4) The decision must be consistent with the intent expressed in the contract documents. (A201, 4.2.13) It must be expressed, not merely reasonably inferable. The decision must be actually an aesthetic effect matter, not just so classified for strategic reasons or for convenience.

The newly issued General Conditions of the Contract for Construction, Fourteenth Edition, AIA Document A201, although substantially revised in 1987, is not materially different from the previous edition (1976) in respect to the matters covered in this chapter. Various provisions of the new edition have been incompletely cited or paraphrased for brevity and should be reviewed in their entirety and in proper context. (See Appendix D.)

Architect's Authority to Reject Defective Work

On those rare occasions when workmanship or materials must be condemned, the architect always hopes that the contractor will accede willingly and without argument or rancor. However, when the rejection of work imposes a significant financial burden and loss of contract time, some contractors will not give in easily and will instead resist by all available means. Contractors will defend rejected work even more strenuously when time delay would result in liability for payment of liquidated damages or loss of an early completion bonus.

The architect's authority to reject unsatisfactory work must be founded and defined in the construction contract. Without contractual authority, the architect lacks the power to pass judgment on the contractor's workmanship or materials.

The AIA General Conditions confers upon the architect the power to reject work that does not conform to the requirements of the contract documents:

> Subparagraph 3.5.1 states that work not conforming to the contract documents and substitutions not properly approved and authorized may be considered defective.

> Subparagraph 4.2.6 gives the architect the authority to reject work which does not conform to the contract documents.

Considering the substantial possibility that the contractor will vigorously oppose the architect's rejection of work, the architect should be very careful in judging the work and meticulous in documenting physical conditions, judgment standards, and all administrative actions taken.

Standards for Judging

The primary criteria to be applied in weighing the acceptability of the contractor's work are the requirements specified in the contract documents. In judging the suitability of workmanship or materials, the specified standards such as ASTM specifications, industry standards, building codes, and other published standards should be applied to determine what is acceptable.

The architect cannot insist on a higher or different standard than that actually specified. If an architect expects more than the minimum acceptable under a certain standard, then it should have been so specified so the contractor could have priced it accordingly.

In condemning work, the architect should be prepared to make reference to the specific manner in which the specified standards have been violated. In some cases the basis of rejection will be failure to comply with previously submitted and approved samples, product data, or shop drawings.

The architect's rejection of unacceptable work should always be prompt. It is not fair to the contractor to allow unsatisfactory work to stand for some time after it is available for examination before it is rejected.

The rejection does not have to be in any particular format and could be as a comment in a field observation report or in the form of a memorandum or letter. It should always be in writing and dated, with a copy to the owner.

Architect's Decisions Based on Design Concept 253

In some circumstances, the architect cannot make a final determination of the acceptability without testing for strength or other physical characteristics. The General Conditions, Subparagraph 4.2.6, authorizes the architect to require additional testing or inspection of the work if it is considered necessary or advisable. Such additional testing must be paid for by the owner if the work proves to be in conformity with the specifications and by the contractor if it does not. (A201, 13.5.2 and 13.5.3)

Owner Can Accept Defective Work

The contract allows the owner (not the architect) to accept nonconforming or defective work if that is preferable rather than requiring its removal and replacement or correction. The contract price would then be reduced as appropriate and equitable. In the event that the owner and contractor cannot agree on a price adjustment, the architect must make a decision for them. (A201, 12.3.1) If the architect's determination is not acceptable to either or both parties, the matter may be appealed to arbitration within 30 days. Otherwise, the architect's decision becomes final. (A201, 4.5)

If the matter is one involving aesthetic effect, the architect's decision is final without possibility of appeal to arbitration. (A201, 4.2.13 and 4.5.1)

The AIA General Conditions does not authorize the architect to unilaterally accept defective or non-conforming work. To do so without the owner's knowledge and concurrence could be considered a breach of the architect's fiduciary duty to the owner.

Substantial Performance

If a contractor in good faith (an honest mistake) uses a wrong material that is as good as or almost as good as what was specified, judges and arbitrators would be reluctant to order removal and replacement, because this would seem harsh and wasteful. The contractor would have been considered to have *substantially performed* the contract. They would be more inclined to allow a credit to the owner in an amount that would compensate for any reduction in cost or market value. For example, using ash cabinets when birch was specified, or using brand X hardware when brand Y was specified would probably not reduce the market value, even though the cost might be less.

If the contractor had not acted in good faith, however, the doctrine of substantial performance would not apply and in that case judges and arbitrators would have no trouble in charging the contractor for the value of removal and replacement or correction of the non-conforming work. It would be considered bad faith if a contractor submits a request for substitution that is denied by the architect and the contractor goes ahead and uses the rejected material. This would not qualify as substantial performance.

If an inadvertent error, no matter how innocent, would result in conditions not suitable for the use or function of the building, it would have to be properly rectified by the contractor.

Consequential Damages

When a construction error results in damage to other parts of the building or to its contents, the architect's rejection decision should include the cost of repairing any resulting consequential damage. For example, if roofing or flashing is improperly installed and allows rainwater to get in, any resulting damage to interior surfaces would be considered consequential damage. The cost of repairing such consequential damage should be charged to the contractor. If rectification of the damage involves additional architectural fees, such as for redesign or excessive administration, they should be billed to the owner and the construction contract sum reduced accordingly. (A201, 4.3.9, 10.2.1.2, and 10.2.5)

If the owner has occupancy or control of the building, as after completion or in remodeling work, the owner must do all that is possible to limit the amount of damage caused by the contractor's defective or nonconforming work. For example, when there is a roof leak, the owner should move, remove, or cover valuables which might get wet and place protective plastic sheeting or vessels where needed. This is in satisfaction of the implied contractual duty to mitigate damages. If the owner fails to mitigate damages when possible, it would be difficult to hold the contractor responsible for consequential damages. All of the owner's extra costs for necessary mitigation efforts should be charged to the contractor.

Architect's Decision to Reject Work

The architect must make the final decision as to whether or not to reject the contractor's defective or nonconforming work. Although the architect does not have the power to accept any deviations from the contract requirements, the architect in some cases would advise the owner to accept certain deviating work. The architect's duty would include a fair and impartial analysis of the advantages and disadvantages of retaining the contractor's work. Factors to be considered in the architect's recommendation would include safety, longevity, maintenance costs, appearance, and suitability for the intended purpose.

The architect's final decision to reject should be an independent determination, without consideration of pressures imposed by the contractor or owner. The fact that a building inspector, lender's representative, lessee, or other third party may be willing to accept a deviation from the contract requirements is irrelevant.

Architect's Decisions Based on Design Concept

As with all other final determinations, the architect should always examine the contract documents, do the necessary technical research, order testing if appropriate, and obtain the full position of the owner and contractor before finalizing the decision. The rejection should be dated, in writing, should make reference to the specified standards, and should clearly state the basis and conditions of rejection.

Architect's Minor Changes

Contractors generally accept that minor changes in the work will have to be made during construction even with the best of drawings and specifications. Some refining adjustments must always be made to smooth out imperfections in the documents or to reflect realistic field conditions.

Concerns of the Parties. Architects are concerned that the contract documents be faithfully executed and that unconsidered deviations initiated by others do not get out of the architect's purview. For this reason, only the architect usually has the authority to order minor deviations from the contract documents. Contractors are concerned that the architect's authority to order minor changes does not become a blank check or lead to chaotic project management. Owners are rightfully apprehensive about the possibility of the architect and construction people "wheeling and dealing" with the contract.

Clarifying the Contract Documents. Contractors, subcontractors, and suppliers cannot always perceive the exact meaning or application of the construction drawings and specifications that have been furnished for the project. This is not necessarily because they are inexperienced or incapable of reading technical drawings or specifications. Often the field condition is not exactly as the architect's designer or drafter had visualized and thus a field adaptation or adjustment must be devised. Sometimes, ambiguity or conflict in the documents must be clarified or resolved and erroneous or unrealistic information or omission must be explained. Occasionally the architect will see where minor adjustments can improve the utility or refine the design effect. Standard construction contracts must allow for the practical necessity of minor adjustments while respecting the rights and concerns of the parties involved. (A201, 4.2.8 and 7.4.1)

Architect's Interpretation of the Contract Documents

It is among the architect's contracted duties to interpret the meaning and intent of the documents when necessary. In most cases, the contractor could probably devise a satisfactory solution to the anomaly and avoid disruption of construction progress. But this course of action carries with it the risk of later disapproval if the contractor's solution does not match the architect's thinking.

When the contractor encounters a situation that requires interpretation, the architect should be notified, preferably in writing. The architect's interpretations should also be in writing or in the form of drawings. Informal oral notifications should be later confirmed in writing.

An architect's interpretation or clarification cannot be used as a device to impose a higher quality or standard of workmanship or materials than was originally specified. If the interpretation involves additional construction cost or time, a change order should be prepared for the owner's and contractor's concurrence.

Limitations on Architect's Authority to Order Minor Changes

In many situations the architect will find it necessary or advantageous to make minor changes in the work. The authority to order minor changes during construction is granted to the architect in the General Conditions, Subparagraphs 4.2.8 and 7.4.1. However, this power is not without qualification. Three significant limitations are specified:

1. The change must cause no cost adjustment
2. It must not affect the construction time
3. It must not be inconsistent with the intent of the contract documents

The architect's minor change must be effected by written order. (A201, 7.4.1)

Although the architect's changes are specified to be binding on the owner and contractor, this may not always be practical to enforce. If the owner and contractor both object to the change, it would be very difficult for the architect to require their consent. They would just tacitly by-pass the architect.

If the contractor opposed the change, the mere quotation of a justifiable cost or time consequence would place it outside the ambit of a minor change. Then it would require a change order and the owner's approval, not only for the change itself, but also for the added cost or construction time.

If the contractor accepts the minor change but the owner objects, then the architect must decide whether imposing the change is worth the resulting alienation of the client.

Abuse of the Minor Change Process

The minor change power cannot be used to compel the contractor to remove or change any work which has already been done in conformance with the contract documents. For example, moving a door or window in wood construction already framed in accordance with the drawings. As a minor change, the architect could have ordered them to be moved before they were framed. The same would be true in the case of a door swing which could be properly ordered before the door is hung and hardware installed, but not afterwards. To change the color or finish of a purchased item such as a lighting fixture would be a minor change if ordered before the fixtures were purchased, but might entail considerable time, cost, and trouble to change if the fixtures were already purchased and on hand in the electrical subcontractor's warehouse. In each of these examples, a change order agreeable to owner and contractor must be used to authorize additional costs or time.

Clearly, the architect lacks the authority to order changes involving additional cost or time unless previously empowered by the owner, preferably in writing. (A201, 7.4.1)

Obtaining Contractor's and Owner's Viewpoints

When an architect is making what is intended to be a minor change, it would be advisable to confirm with the contractor beforehand that there is in actuality no adjustment to be made in the contract price or time. It would also be advisable to determine that the owner is not opposed. As subparagraph 7.4.1 requires architect's orders for minor changes to be in writing, it could be simply a minute in the architect's field observation report or in the form of a letter or memorandum.

Another practical possibility would be to process it as a "no cost-no time" change order so that it will bear the signed acquiescence of owner and contractor.

The minor change authority when properly exercised can be very convenient in making subtle adjustments to account for differing field conditions and to effect advantageous design refinements.

22

Resolution of Construction Disputes

Misunderstandings During Construction

Even under the most favorable of circumstances, contractors and owners will not always be able to agree immediately when confronted with unexpected claims or demands from each other during the construction period. When the owner is disappointed in workmanship standards or other contract performance, the contractor might not be sympathetic with the owner's expectations, criticisms, and demands. The contractor might feel that the owner is unrealistic, unjust, or over-reaching. When contract conditions vary from the contractor's expectations, the owner might be shocked and appalled upon receiving the contractor's claim for additional compensation or contract time. Honest, conscientious contractors and high-principled, fair-minded owners sometimes do not see things in the same way.

Construction contracts are lengthy and complex written agreements, including numerous detailed drawings and voluminous technical specifications, describing the expected conduct of owner and contractor and their relationship during the period of construction and for a time thereafter. The simplest of projects require contract documents detailing all of the contractor's and owner's respective responsibilities and their duties to each other. No matter how carefully drawn the contract documents may be, situations can and usually do arise which were not exactly as visualized by the parties, even though they might have been contemplated in principle by the contract documents, for example, the encountering of unexpected subsurface conditions not revealed by foundation investigations or soil tests. The contractor will rightfully expect to be reimbursed for all extra labor and materials as well as overhead and profit, and will want an appropriate contract time extension to make up for any time lost in the operation. The owner will often be stunned or even outraged over these unanticipated and unbudgeted costs which will adversely impact the building's economics, possibly for years to come. The owner could be somewhat resentful toward the contractor for apparent profiting from the owner's bad luck. And the time schedule is placed in arrears at the very beginning of the project. Under these circumstances, many owners will find it hard to readily and cheerfully accede to the contractor's claim for additional compensation and time. The contractor would be adamant in its demand for compensation because the costs and time were incurred in good faith, were unavoidable, and were not in the cost or time estimates. If the owner refuses to pay this legitimate bill and to grant the necessary time extension, the owner's ill fortune will become the contractor's loss. The contractor is not likely to give in. Many other realistic examples could be cited to illustrate the areas in which construction disputes can arise.

General Conditions of the Contract

The AIA General Conditions of the Contract for Construction, Fourteenth Edition, Document A201, published in 1987, continues to provide for a system of dispute resolution mechanisms in which the architect is the central element. The system is based on the concept that the agreement between the owner and contractor will be administered by the architect. (Subparagraph 4.2.1) The owner and contractor agree to route all communications to each other through the architect. (4.2.4) Furthermore, the architect is authorized to "interpret and decide matters concerning performance under and requirements of the Contract Documents on written request of either the Owner or Contractor." (4.2.11) Thus all claims or demands of the parties against each other will first be presented to the architect for consideration and a decision. Under certain circumstances, the architect's decision will be final and binding and not subject to appeal.

This decision-making process embraces all manner of claims including those based on alleged errors or omissions by the architect. (4.3.2) When performing the quasi-judicial function of interpreting and deciding all matters submitted, the architect is under the pressure of the owner's and contractor's critical and sometimes skeptical scrutiny. They have the right to expect the ultimate in fairness and impartiality from the architect. These duties can be extremely difficult for the architect when the proper decision has to be against the owner, who after all is paying the architect's fees. The weight of responsibility is particularly oppressive when the architect must render a decision in which defective drawings or specifications or other imperfections of the architect's own actions must be acknowledged. Extra construction costs caused by errors in the architect's work, in most cases, must be charged to the owner, not the contractor. The architect will then have to deal separately with the owner to apportion the liability for the error. (See Chapter 24, Professional Standard of Care.)

In many situations, both owner and contractor will be dissatisfied to some degree with the decision. The contract provides that the architect will not be liable for results of interpretations or decisions rendered in good faith.(4.2.12) The architect must have confidence in the righteousness of decisions and remain resolute against any clandestine lobbying by contractor or owner. The administrative burden on the architect imposed by the General Conditions implies that the architect should keep all of the claim processes moving and to watch the time limits so the parties will not lose important or valuable rights through inadvertence.

Claims and Disputes

In general, there is a 21-day maximum time limit for the parties to make a claim against one another. The time is measured from the date of the first occurrence or discovery of the event, whichever is later. Each claim must be in writing and be submitted to the architect within the 21-day limitation, or presumably the right to the claim is lost. (A201, 4.3.3, 4.3.6 through 4.3.9, and 4.4.1) All time limitations are expressed in calendar days. (A201 8.1.4)

For the dispute resolution process to function efficiently, the architect must act diligently upon receipt of a written claim from either party. The architect must render the decision in writing with "reasonable promptness" or within a time limit mutually agreed upon between the parties and the architect. If no time limit has been agreed upon then the architect has an indefinite period but is still required to act with reasonable promptness. Neither owner nor contractor can create pressure on the architect for interpretation or decision by claiming delay until 15 days after the written claim is made. (A201, 4.2.11) The architect can, however, within 10 days after the claim is made, take one or more of five preliminary actions:

1. Request additional supporting data from the claimant
2. Submit a schedule to the parties indicating when the architect expects to take action
3. Reject the claim in whole or in part, stating reasons for rejection
4. Recommend approval of the claim by the other party
5. Suggest a compromise

The architect may also, but is not obligated to, notify the surety, if any, of the nature and amount of the claim. (A201, 4.4.1)

At any time that a claim has been satisfactorily concluded, the architect is required to prepare or obtain appropriate documentation for execution by the parties. (A201, 4.4.2) If a claim has not been resolved after the architect has taken the preliminary actions, the claimant, within 10 days thereafter, is required to take one or more of three possible further actions:

1. Submit additional supporting data requested by the architect
2. Modify the initial claim
3. Notify the architect that the initial claim stands (A201, 4.4.3)

After an unspecified time, but presumably with reasonable promptness, the architect, after reviewing any further responses from either of the parties will give written notice that the decision will be forthcoming within 7 days. The notice should state that the decision will be final and binding on the parties but is subject to arbitration. (A201, 4.4.4) The architect in preparation for rendering the decision should be extremely careful to obtain each party's full position and should carefully and completely review the contract documents so that the decision is technically competent as well as fair. The decision should be dispositive of all elements of the claim to avoid the necessity of amending the decision after review by the parties. The decision should include all appropriate monetary and time adjustments to the contract.

The architect's final decision should be properly drafted to comply with the specific requirements of Subparagraph 4.5.4.1 of the General Conditions. The decision should specifically state that the decision is final but subject to arbitration and that a demand for arbitration of the matter included in the decision must be made within 30 days after the date on which the arbitration claimant received the final written decision. If an arbitration demand is not made by either party within the 30 day period, then the architect's decision becomes final and binding upon the parties. (A201, 4.5.4.1) The mere written or oral objection of owner or contractor to the decision is not sufficient to prevent the final decision from becoming binding; it is necessary that arbitration proceedings be initiated by the filing of a demand.

Architect's Aesthetic Effect Decisions

Decisions made by the architect on matters relating to aesthetic effect will be final and not subject to any further review by arbitrators if they are consistent with the intent expressed in the contract documents. (A201, 4.2.13) This places a burden of intellectual honesty on the architect to refrain from classifying questionable decisions as aesthetic when they are not. Although arbitrators are not entitled to review and possibly redecide aesthetic effect decisions, they might be called upon to rule on whether the architect has properly classified a decision as an aesthetic effect matter. The architect's mere classification of a decision as aesthetic does not necessarily place it beyond the purview of arbitrators or judges.

Resolution of Construction Disputes

Should an arbitrator or judge decide that an architect's aesthetic effect decision was not consistent with the intent expressed in the contract documents or that it was not in fact a matter of aesthetic effect, the decision would no longer be final and might be reversed or modified. (See Chapter 21, Architect's Decisions Based on Design Concept, Aesthetic Effect, and Intent of the Contract Documents.)

Architect's Minor Changes

The architect has authority to order minor changes in the work not involving any change in the contract sum or time and not inconsistent with the intent of the contract documents. Such orders must be in writing and are binding on the owner and contractor. (A201, 7.4.1) The owner or contractor or both may disagree with the architect's exercise of this prerogative, in which case the only appeal is to arbitration unless the architect is willing to amend or withdraw the order.

Assistance from Surety

If the contractor has furnished a performance bond and a labor and material payment bond, the parties will find the surety a helpful ally in the search for a solution of certain types of controversies. If a claim has been ruled upon by the architect and acceptance by the parties is not forthcoming, and there appears to be a possibility of a contractor's default, the architect may, but is not obligated to, notify the surety to request assistance in resolving the controversy. (A201, 4.4.4) The surety will have some economic leverage upon the contractor. It is in the surety's financial interest to prevent the default of the contractor. The surety will often provide their contractors with helpful economic advice and assistance.

Architect's Failure to Render a Final Decision

The architect has the role, assigned by the General Conditions, of quasi-judicial interpreter and decision maker for the owner and contractor during the construction period. This duty must be promptly and competently carried out to satisfy the expectations of the parties to the contract. In most cases, the architect will discharge this contracted professional responsibility with proficiency and fairness. On those rare occasions where the post of architect becomes vacant or if the architect, for other reasons, has not rendered the required written interpretation or final decision within 45 days after the initial written claim to the architect, then either party may initiate arbitration proceedings, but not before. (A201, 4.5.1)

Alternative Dispute Resolution Methods

Negotiation. Upon the emergence of an apparently irresolvable difference, the first resort of a typical business-oriented person will be negotiation. This is the process of rational discussion, compromise, and, ultimately, agreement. This is in the recognition that sums conceded in negotiation will more than offset the legal and administrative costs of proceeding to a more formal forum for dispute resolution. If concurrence can be reached by normal business negotiation, time, effort, and money will usually be saved. Negotiation will not yield satisfactory results when both sides persist in clinging tenaciously to firm or extreme positions and refuse to enter into the spirit of mutual concessions. It is the way most practical business people prefer to resolve their day-to-day problems and differences. It is quick and effective and does not require the services of uninvolved outsiders such as lawyers, technical experts, and judges.

Mediation. Should negotiation fail to resolve the matter, the parties could agree to mediation. This is a procedure in which a mutually acceptable impartial intermediary talks to both sides, together and separately, and assists them in their negotiations. The mediator will privately advise each side of the strengths and weaknesses of their respective positions and attempt to get the parties to see their positions more objectively. The mediator does not impose any decisions on the parties but instead helps them to arrive at their own voluntary resolution.

Mediators can be lawyers, accountants, contractors, architects, engineers, or business people, in short, anyone who understands the problem and who has the necessary mediation skills. The mediator has to be acceptable to both sides. In some cases there will be more than one mediator if the parties feel that additional expertise is needed. The mediator's fees and expenses are usually split evenly between or among the parties, but could be agreed otherwise.

In some cases, the project architect, being intimately acquainted with the background information vital to understanding the dispute, will prove to be the ideal mediator. This is assuming that the architect is respected by and acceptable to both sides. The parties must also consider the architect to be trustworthy, technically competent, unprejudiced, and of the temperament to counsel both sides in a negotiation likely to result in an amicable accord.

If the mediator succeeds in getting the parties to agree, the time and expense of more formal procedures can be avoided. A mediated settlement will leave the parties in a friendlier state so they can continue doing business with each other. This would be extremely valuable when the project has still to be completed.

Like negotiation, mediation is voluntary and not provided for in most construction agreements. Negotiation and mediation can take place before the arbitration demand is filed or while waiting for the first arbitration hearing to take place. If the dispute can be resolved in this way, the time, trouble, and expense of the arbitration can be avoided. Otherwise, the arbitration will proceed in accordance with the construction agreement.

Conciliation. This is very similar to and could be considered a form of mediation. The principal difference is that the conciliator will try to convince the parties to adopt a solution devised by and considered fair by the conciliator. The parties are under no obligation to accept the conciliator's opinion. If the conciliation fails, the dispute can proceed to arbitration or court, depending on provisions of the contract.

Mediators and conciliators must be acceptable to the parties and neither have the power to enforce their opinions or awards. The conclusions can be rejected by either side and will be effective only when voluntarily accepted by the parties. Thus, the parties cannot lose their cases on purely legal technicality as they must agree on the resolution.

Arbitration. When the parties find that they cannot voluntarily agree through negotiation, mediation, or conciliation, then someone will have to make the decision for them. This can be done in the court system or in arbitration.

The two main drawbacks to court litigation, lack of judicial construction expertise and slowness of the system, can be overcome by the arbitration process. The arbitrators are chosen by the parties for their known qualities of fairness and appropriate expertise. Panels of three arbitrators can include three different areas of expertise, such as, a construction lawyer or accountant, a contractor or subcontractor, and an architect or engineer. The hearings can be organized as quickly as the parties wish and in the desired locale.

Arbitrations are conducted in the form of hearings where the parties can present their claims, counterclaims, and defenses by means of witnesses, documentary evidence, and argument. The hearings are as formal or informal as required by the presiding arbitrators.

Arbitrators have the power to make final and binding decisions on all matters that are submitted to them. Their award is legally equivalent to a judgment of the court that would have had jurisdiction and the court will help enforce it.

Arbitration is a voluntary process that takes the place of court litigation. No one is required to sign an agreement to arbitrate. However, if a signed agreement contains an arbitration clause then arbitration must be used unless both parties agree otherwise. If one party tries to block or stall an arbitration, the Construction Industry Rules allows the AAA to proceed with setting up the arbitration and it will go ahead.

Filing An Arbitration Demand

Arbitration is a method of dispute resolution which is voluntarily selected by the parties and agreed to in a contract as an alternative to using the court system. Most

AIA form agreements, including the General Conditions, contain arbitration clauses. Where both parties would prefer using the court system, the arbitration clause can be eliminated at any time either before or after the contract is signed. However, after the agreement is signed, the arbitration clause will prevail if only one party wishes to eliminate it. If the construction agreement does not contain an arbitration clause, and the parties later wish to submit their dispute to arbitration, this is accomplished by execution of a submission agreement signed by both owner and contractor.

American Arbitration Association (AAA). Arbitration required by the AIA form documents will be in accordance with the Construction Industry Arbitration Rules of the American Arbitration Association. The decision of the arbitrator or arbitrators, called an award, may be entered in any court having jurisdiction. (A201, 4.5.1) A copy of the Construction Industry Arbitration Rules may be obtained at no cost from any AAA regional office. All architectural construction administrators should have a copy of the rules at hand for reference.

To initiate an arbitration, notice must be given by the claimant to the other party, called the respondent. The notice can be given on a form printed by the AAA called a Demand for Arbitration or simply by writing a letter addressed to the respondent. The demand must include a statement setting forth the nature of the dispute, the amount involved and the remedy sought. This should be sent to the other party, and 3 copies should be sent to the regional office of the AAA together with 3 copies of the arbitration provisions of the contract and the appropriate filing fee.

AAA's Construction Industry Arbitration Rules provide for administration of arbitrations under one of three "tracks" depending on the size of the claim involved:

>Fast Track, for claims involving up to $50,000
>Regular Track, for claims falling between $50,000 and $1,000,000
>Large, Complex Case Track, for claims involving over $1,000,000

A non-refundable filing fee is payable in full to the AAA by a filing party when a claim, counterclaim, or additional claim is filed.

AAA Fees. Filing fees are based on the amount of the claim. In addition, fees are charged for hearings, postponements, and hearing room rental. The arbitrators' fees are in addition to AAA's administrative charges. Arbitrators do not charge for their services for the first day of hearing where the claim is under $10,000.

The AAA has 37 regional offices spread across the continental United States, Hawaii, and Puerto Rico. Hearings may be held at locations convenient to the parties and are not limited to cities with AAA offices.

The AAA, in addition to administering arbitrations also administers mediation proceedings and will assist in the selection of a suitable experienced mediator.

Upon receipt of the demand, copies of the arbitration provisions of the contract, and the filing fee, the AAA will start the administration process leading to the selection of arbitrators and setting of the first hearing date. In the event that the architect renders the final decision after arbitration proceedings have been initiated, the decision may be used as evidence in the arbitration but the arbitrator's decision will take precedence unless the architect's decision is acceptable to all parties concerned. (A201, 4.5.4.1)

Any demand for arbitration must include all arbitrable claims then known to the party making the demand. The arbitrator or arbitrators may allow later amendment of a demand to include any claims omitted through inadvertence, oversight, or excusable neglect. Claims which arise subsequently may be added by amendment. (A201, 4.5.6) The arbitration process allows for the submission of counterclaims by the respondent.

Use of Lawyers

In mediations and conciliations as well as arbitrations, the parties may be represented by legal counsel if they wish. Legal representation is allowed but not required. If a party knows its case, is articulate, is reasonably persuasive, and has the necessary time to prepare its presentation, then it could risk presenting its own position without professional counsel. However, important cases involving large sums or complex issues will usually have a more satisfactory outcome when presented by a competent construction lawyer.

Participants' Control Over the Outcome

In negotiation, mediation, or conciliation, the disagreeing parties will have the most influence in the outcome of their dispute because their voluntary concurrence is required at every step in the process. No outside entity will impose a decision on the parties. The parties involved will always know their cases more intimately than any arbitrator or judge possibly can.

In arbitration or court litigations, arbitrators or judges will impose their views on the parties.

Continuing Contract Performance

While awaiting the outcome of an architect's decision or an arbitrator's award, the contractor is required to continue with performance of the construction and the owner is obligated to continue making payments, unless the parties agree in writing to suspend construction and/or payments. (A201, 4.3.4)

Arbitration Award

The arbitrator's award, in accordance with the Construction Industry Arbitration Rules, must be in writing and should be made promptly and no later than 30 days after close of hearings or, if oral hearings have been waived, from the date of submission of final written statements and proofs to the arbitrator.

The award will be final and binding on the parties with no appeal available on the merits of the arguments. The only legal grounds for overturning of an award are based on fraud or corruption of the arbitrator, the arbitrator's refusal to accept relevant evidence, an award which does not dispose of all issues or which makes rulings on matters not submitted. When heard by AAA arbitrators and administered by the AAA, an arbitration award would seldom be deficient in form or technicality.

Conclusion

The foregoing analysis of the dispute resolution system provided in the new AIA General Conditions is presented from the vantage point of a practicing architect. For convenience, many of the provisions of the General Conditions have been partially quoted or paraphrased. The reader is advised to refer to the complete texts of the AIA General Conditions and Construction Industry Arbitration Rules of the AAA to obtain the complete language in its proper context.

The architect engaged in construction contract administration would be well advised to recommend the owner and contractor to consult with their respective legal counsel whenever project conditions and business relationships have deteriorated to the point that the architect's interpretations and final decisions are no longer acceptable to the parties. Although it is not absolutely necessary or legally required for the parties to be represented by counsel in an arbitration proceeding, there is no question that experienced construction industry lawyers are well equipped to represent owners and contractors in arbitrations under the Construction Industry Rules.

23

Preventing Time and Delay Disputes in Construction Contracts

Changed Conditions

Most day-to-day construction industry disputes are readily resolved by the owner and contractor or acceptably decided by the architect. Those that are not easily settled are usually based on conditions that were not contemplated in the drafting of the original contract. Unforeseen or changed conditions generally have an adverse impact on the time needed to accomplish the work of the contract. Loss of time will almost always result in economic loss accompanied by disagreement as to which party should be required to bear the unexpected burden.

AIA General Conditions

The AIA General Conditions of the Contract for Construction, provide contractual requirements for the owner's and the contractor's conduct and practical procedures for the architect's administration of the construction contract. Compliance with the specified procedures will not guarantee freedom from disputes but should reduce the frequency and magnitude of contention. Methodical, timely, and even-handed architectural administration should lessen owner and contractor apprehension of the risks and vagaries of the construction process. Meticulous following of the contractually mandated procedures, particularly when completely and properly documented, concurrent with construction, will serve the parties well in the event of subsequent claims to higher authority, such as submission to arbitration. All procedures hereinafter described will be those of the AIA General Conditions.

Liquidated Damages

Most owners need or want their building projects available for use at a particular time. Owners, as a basic premise, expect the contractor to start the construction promptly when ordered, to pursue the work diligently, competently, and purposefully and to complete the work on or before the promised due date. Usually they will have made major financial commitments based on reasonable expectation of completion on the scheduled date. Failure to have use of the building when anticipated will result in losses of rentals and profits and incurred costs such as interest on invested capital, rents on duplicate facilities, and other costs. Thus construction contracts usually have provisions relating to when the construction should start, how long the construction process should take, and when the work should be completed.

Owners expect to suffer economic loss if the building is not available for use when contractually promised, so contracts often include payments to the owner from the contractor to compensate for monetary losses caused by the contractor's failure to complete on time. Payments, called "liquidated damages", imposed on the contractor for late completion of construction, are not to be construed as a penalty, as the courts and arbitrators are reluctant to enforce penalties. Liquidated damages, estimated and established as realistically as possible, represent the best ascertainment of the actual losses which the owner will suffer if the completed construction is not delivered on time. Liquidated damages may be expressed as lump sum amounts and/or as a certain amount per day, with or without a maximum cumulative limitation. When a contract contains liquidated damages provisions, the architect is under serious obligation to administer the time provisions of the contract fairly and equitably to both owner and contractor.

Working Days Versus Calendar Days

Time provisions expressed anywhere in the contract documents should be well thought out and written as clearly as possible. Expressing time lapse for construction in terms of working days is ambiguous and usually causes arguments in calculation as well as concept. It is more certain to express all time periods in terms of calendar days. It is easy enough to make the necessary adjustments to account for Saturdays, Sundays, and holidays when determining the number of calendar days to allow for a certain operation. When possible, it is advisable to express starting and completion times as exact calendar dates, using month, day, and year. Calendar days are used in the AIA General Conditions. (Subparagraph 8.1.4)

Construction contracts must contain adequate provision to accommodate the inevitable changes in conditions from those upon which the contract was originally based. Any change order procedure should have provision for the number of calendar days to be added to or deducted from the contract in addition to any change in the contract price. If a change affects price only and there is to be no change in contract time, the change order should state zero days (or no time adjustment) in writing. A change order silent on time will merely provoke contention because the owner will assume that silence connotes no time extension will be allowed, while the contractor will assume that silence means the time extension is still open to discussion. A change order based on a condition or event which affects time only should be prepared and executed even though there is no change in contract price.

Unforeseen or Differing Conditions

Unforeseen conditions, such as those encountered during subterranean excavation, or differing conditions, such as are frequently found in alteration of existing buildings, will usually result in justifiable contract time extensions. Time extensions should always be formalized by written change orders. Occasionally, unforeseen conditions will prove to be easier or less complex than expected and, if it shortens the critical path, should result in a time credit change order in fairness to the owner. In a lump sum contract the cost advantage of performing such simpler conditions will accrue to the contractor if the subterranean risk of higher cost is assumed by the contractor. The owner will benefit by cost savings in unexpectedly easier conditions only when contractually exposed to the extra cost of unexpectedly difficult conditions.

Delay Caused by Owner or Architect

Actions of the owner or architect which create delay should be promptly quantified and an appropriate time change order should be executed. Materials, services, or separate contractors to be supplied or coordinated by the owner, if not provided in time to suit the prime contractor, will adversely impact the overall time schedule. Sometimes delay is caused by the owner's inability to make available the complete site and all rights of way and easements. The owner's failure to obtain all necessary governmental approvals can also cause delay. Architects can be and sometimes are the cause of construction delay by failure to provide timely reviews of submittals, failure to make required determinations or decisions in a reasonable length of time, and by furnishing of defective contract documents.

Construction Schedule

In liquidated damages contracts, the construction time schedule is an extremely important document. The General Conditions (3.10) requires the contractor, prior to starting work, to prepare a construction schedule to be submitted for the owner's and architect's information. The architect is not asked to approve or disapprove it; however it would be good practice, in the very least, to review it and be acquainted with its provisions. Without a time schedule it would be practically impossible to objectively measure whether certain events are on time, early, or late. The contractor is required to revise the schedule periodically to conform with the actual progress and conditions of the project.

The contractor is also required to submit at the same time, for the architect's approval, a schedule of submittals which has been coordinated with the construction schedule. (A201, 3.10.2) All duties of the owner or architect upon which the contractor will claim reliance should be entered in the time schedules. For example, the dates on which each of the various shop drawings are supposed to be submitted to the architect should be scheduled as well as the dates on which the architect is expected to send them back to the contractor. The architect's review of the submittal schedule provides opportunity for comment in advance of unrealistic time allowances for submittal review. The schedule might later have to be adjusted by the contractor to accommodate redrafting and resubmittal of rejected shop drawings. The submittal schedule will also enable the architect to know in advance when to schedule personnel and review time in the architect's own office. Often, contractors will unexpectedly submit shop drawings at the last minute, informing the architect that the shop drawing review is holding up the job. The process of scheduling shop drawing submittals and reviews should be helpful in clarifying the time responsibilities of both contractor and architect. Delays caused by outside agencies, such as government, beyond the control of owner, contractor, or architect, will give rise to justifiable extensions of contract time.

Weather Delays

Delays caused by adverse weather are a major source of dispute in claims for construction time extension. Inclement weather conditions including rain, snow, wind, and excessively high or low temperatures are clearly beyond anyone's control, although they could be reasonably expected to occur. Some construction contracts allow time extension for any and all weather delays, whereas others allow only for abnormal weather or adverse weather conditions not reasonably anticipated. The AIA General Conditions provides in Clause 4.3.8.2 that

> "If adverse weather conditions are the basis for a Claim for additional time, such Claim shall be documented by data substantiating that weather conditions were abnormal for the period of time and could not have been reasonably anticipated,

and that weather conditions had an adverse effect on the scheduled construction".

In every day practice, this provision could cause as many disputes as it was intended to resolve. Moreover, it creates administrative headaches for architects and contractors alike. This clause, in its attempt to provide certainty, fairness, and equity, creates a procedure which is too unwieldy for all but the largest, most costly projects. This procedure would be more appropriate for the courtroom or arbitration hearing rather than routine construction administration. In any event, compliance with this clause makes it difficult to comply with Subparagraph 4.3.3, which requires written notice within 21 days after occurrence of the event giving rise to such claim, and Clause 4.3.8.1 which requires the claim to include an estimate of cost and probable effect of delay on progress of the work.

In the circumstances prevailing in more mundane projects, it would seem preferable to employ a simpler procedure. One possibility would be for the contractor to state the number of days of adverse weather that had been anticipated and included in its base bid and time schedule. The architect would then authorize time extensions for all adverse weather which impacts construction time after allowing for the number of days initially stated by the contractor. Another possibility is to specify that the construction time schedule be prepared to reflect only actual estimated construction time without allowance for any adverse weather conditions. In this situation, the contractor would be entitled to time extension for all adverse weather which impacts construction. These circumstances would then reflect actual weather conditions, would not give the contractor a windfall profit in case of exceptionally fine weather, and would be fair to owner and contractor. Ideally, the contractor should not profit from fine weather at the owner's expense or suffer a loss from adverse weather. In effect, weather should be the owner's risk. There is no logical reason why a contractor should be considered to be an accurate long range weather prophet. Why should a contractor be expected to pay for the inability to accurately predict the weather months or even years in advance? Naturally, the parties can reverse the risk situation by contract provision.

Whatever method of measuring time is used, an additional source of weather-based dispute is the muddy, impassable site which prevails for several days following each rainfall. When site paving and landscape work are completed and the main work is under cover, time extensions for rain and mud would be inappropriate and should not be allowed. Rain which occurs on weekends or holidays or during other concurrent delays (such as labor union stoppages) does not extend the contract time except for the resultant muddy site conditions affecting normal working days. Windy days will cause postponement of certain operations such as roofing or exterior painting and will thus affect the contractor's schedule accordingly. Adjustment for adverse weather which has no impact on the critical path cannot be allowed.

Critical Path

Analysis of the effects of adverse weather on the critical path from the opposing viewpoints of owner and contractor will reveal yet another source of dispute: To which party does the "float" in the critical path belong? There appears to be no industry wide settled answer as to who has the right to the float. Therefore, this type of question will have to be resolved by the architect on the basis of specific project circumstances and application of common sense, fairness, and equity to both parties to the contract.

Should the contractor receive additional time on the entire project in order to perform an operation added by change order which can be achieved simultaneously with previously scheduled activities? If the whole added operation can be performed in a simultaneous manner, obviously the contract time should not be extended. However, if a portion of the work impacts the critical path, then the change order should provide accordingly for an appropriate time extension.

The contractor, according to the AIA General Conditions, is entitled to a time extension for delay in progress of the work caused by the owner's or architect's act or neglect, labor disputes, and other causes beyond control of the contractor. (8.3.1)

Substantial Completion

In construction contracts governed by the AIA General Conditions, the ending date of the construction time schedule is the date of Substantial Completion. (8.1.1) The AIA General Conditions defines substantial completion as "the stage in the progress of the Work when the Work or designated portion thereof is sufficiently complete in accordance with the Contract Documents so the Owner can occupy or utilize the Work for its intended use." (9.8.1) This generalized definition provides a good conceptual impression of what constitutes substantial completion but is not sufficiently complete and certain to eliminate disputes in specific situations between owners and contractors. The AIA General Conditions leaves it to the judgment of the architect for determination of the date of substantial completion. (8.1.3) The architect should look to the definition for guidance in making the determination and in issuing the certificate. The determination of specific starting and ending dates as well as the number of authorized days of time extension is crucial to the avoidance of controversy in the calculation of liquidated damages. The architect should treat these obligations very seriously and should make all required determinations fairly, promptly, and in writing.

Delay Damages

When the total elapsed construction time period is extended appreciably beyond the time initially allotted, the contractor's overhead costs will undoubtedly be increased.

Among those direct and indirect costs are the wages of supervisorial and administrative personnel, rental value of facilities and equipment on the job site, interest on invested capital, and extended involvement of the contractor's organization, thereby causing possible loss of alternative business opportunities. The contractor may also have incurred similar obligations to subcontractors or suppliers for their similar costs during the same delays.

Whether change orders for changed conditions can include any element of delay cost, in addition to the mere time extension, is a matter for contract interpretation and analysis of the specific circumstances prevailing on the project. The AIA General Conditions " does not preclude recovery of damages for delay by either party under other provisions of the Contract Documents." (8.3.3) The contractor should not recover for any delay costs when the delay was caused by or could have been prevented by the contractor.

Time delay caused by the contractor can expose the owner to additional losses and expenses such as loss of interest, increased rental expense or loss of rental income, and increased personnel and other costs. However, these costs all merge into and are limited by the agreed sum expressed in the liquidated damages clause. In contracts without provision for liquidated damages, the owner could claim without limit the actual amount of all proved damages caused by the contractor.

In the event of a delay claim which cannot be amicably resolved by the owner and contractor, the architect is obligated by the AIA General Conditions to make a determination. The architect's decision will become final and binding on the parties if not contested by the timely filing of a demand for arbitration in accordance with the procedures set out in Article 4 of the General Conditions. Some contracts contain carefully drafted "no damages for delay" clauses in an attempt by owners to eliminate the possibility of damage claims by contractors for certain delay causes such as acts of the owner or architect. The AIA General Conditions does not include such a clause.

The Moment of Truth

In all situations in which the architect is obligated to make determinations or render decisions, the architect should obtain the positions of both of the parties, preferably in writing, before deciding such matters. All decisions must be fair, just, and equitable and be arrived at with meticulous impartiality. This is often an unpleasant and difficult task, particularly when the decision must run counter to the interest of the architect's client.

All extra costs, time, or delay damages caused by the architect's shortcomings must be charged to the owner, not the contractor, and later resolved between the owner and architect. Such determinations are the ultimate test of an architect's professionalism, fairness, and objectivity.

24

Professional Standard of Care

Professional Responsibility

Architects and engineers do their best to carry out the terms and conditions of the professional service agreements they enter into. They all strive to do their very best professional work, but occasionally the physical or administrative results do not measure up to the client's expectations.

Owners may feel that the designers have not been sufficiently diligent or have performed some part of their duties at a substandard level. The project may have exceeded the client's budgetary or completion time expectations. There may be defects in the finished product for which the client would seek to hold the designer financially responsible. As a matter of fact, architects and engineers can be held legally responsible for a client's financial losses if proved to have been negligent.

Professional Negligence

Negligence is a technical legal term, so how can one know what it means? Design professionals can be held negligent if they are not reasonably competent or have failed to exercise due care under the circumstances. Designers, specifiers, and contract administrators must possess the requisite skills for projects undertaken and should conduct their professional activities with due diligence and reasonable care. Architects are required to conduct themselves in at least the same manner as the average of other architects practicing in the same community under the same circumstances.

However, this does not mean that they can do any less than is required by professional service contracts, even if the standard in the community would have allowed for less.

Conversely, they are not expected to do any more than the contracts require as long as the quality of services contracted for meets the standard of care.

Design professionals must not undertake projects which are clearly beyond their own technical abilities or those of the personnel available to work on the assignment. Only experienced, competent, and qualified staff should be assigned to each task. Junior and inexperienced personnel must be carefully supervised by fully qualified architects. Whenever necessary, the firm should retain outside engineering and other consultants to supplement its own capabilities.

Regional Variations in Practice

Design professionals are not required to be the best practitioner in the community nor even to be above average. But they must meet the minimum acceptable standard in the community.

What is meant by community? In some regional localities, standard architectural practices vary in some respects from other areas. So it is necessary to know of any local peculiarities or variations in practice. This is of particular importance when the architect has projects which are to be constructed in a new and unfamiliar jurisdiction. However, most aspects of architectural practice and usage are very similar all over the U.S. This is becoming more prevalent as common use of AIA standard contract and practice documents increases nationally.

The extensive national architectural press, being widely read by architects, tends to encourage standardization. Regional differences are becoming less significant as national transportation and technical communication is improved. So, in some specific respects, the community might represent a local district or state, while in most aspects the architectural and construction community comprise the entire nation.

Determining Professional Standards

How can an architect know what the professional standard of care is? Becoming aware of the standard of care is closely interrelated with continuing education. The burden is on each architect to keep informed on a current continuing basis and on each firm to assure that its personnel maintain and improve their professional skills.

Professional Standard of Care

There are numerous ways in which to build awareness of the standard of care simultaneously with enhancing professional skills:

> Subscribe to and read the professional press including national, regional, and local architectural, engineering, and construction magazines and publications. Compile and maintain, or have access to, a library of professional literature, textbooks, and reference works as necessary to properly research all assignments undertaken.
>
> Review the latest manufacturers' literature. Often this will be found in daily "junk mail".
>
> Associate with other professionals. This is most easily accomplished by joining and becoming active in professional societies such as the American Institute of Architects or the Construction Specifications Institute and is most effective when one participates in the activities of interesting committees.
>
> Attend seminars, workshops, and other continuing education offerings.
>
> Examine the work of others and review their construction documents whenever possible.
>
> Participate in peer review programs, both as reviewer and reviewee, wherever they are offered.
>
> Teach courses, give talks, and write articles whenever the opportunity arises. Preparation and study for giving a talk will be far more educational for the teacher than it will be for the audience.

All of these diverse activities will provide a general overview of the ways in which other architects think, approach their assignments, and conduct their practices. It will also present opportunities consult with fellow professionals on practice problems and to discuss subjects of mutual interest.

Younger participants should make friends with more experienced practitioners so they can pick up the telephone at any time for an informal consultation. All professionals, regardless of age, should have friends or associates they can use as telephone resources as the need arises or to obtain a candid second opinion.

Failure of Materials or Procedures

It is extremely important that architects and engineers keep themselves aware of new developments in construction technology as they occur. This is particularly true of recent discoveries of unsatisfactory materials or processes. Architects will usually be found negligent if they continue specifying materials or procedures that have been proven harmful or unsuccessful. This was the case with asbestos products and polychlorinated biphenyl (PCB) which were widely used until their hazardous properties became generally known. A number of other materials in common use were later discredited for certain uses, as more extensive experience revealed their shortcomings.

Some newly developed materials and procedures have not been in use long enough for any to know how effectively they will perform over an extended period of time.

Architects and engineers are not expected to conduct physical testing programs in the laboratory or field. However they are required to be aware of the physical and engineering properties of the materials and processes which they specify and must properly apply the generally accepted principles of proper construction design. They are also required to keep informed of new technological developments when the information becomes generally available in the technical publications.

The Hazards of Innovation

The possibility of error is raised appreciably when design projects or parts of projects are experimental or innovative. All creative people aspire to be innovators, to be on the cutting edge of their professions. New methods, creatively conceived forms, novel ways of doing old things, and imaginative but untested systems bring with them the possibility of a miscalculation or an unexpected result. Such costly and embarrassing failures will run the gamut from inconspicuous imperfections to utter and embarrassing fiascoes. Innovation mistakes are difficult to anticipate and eliminate. Even extensive checking by additional experts will not completely uncover and eliminate innovative shortcomings.

When venturing into imaginative new forms, structures, and procedures, it is difficult and sometimes nearly impossible to accurately visualize and anticipate all of the unique conditions and configurations. These could produce such unexpected adverse side effects as dangerous stress concentrations, adverse chemical reactions, or the unforeseen need for sealing or flashing.

Innovation cannot be entirely circumvented, however. It is what makes some new projects better than the old ones. It is the primary means by which progress can be achieved. Without imagination and responsible experimentation, forward development would cease. There would be no advancement in technology nor improvement in efficiency and effectiveness. There would be no new life or progress in architectural design. Creativity would degenerate and stagnate. Innovation is the essence of creativity and the stepping stone into the future.

But if the new material, new use of an old material, or a new process fails to perform as anticipated, the specifying architect could be found negligent. To limit this

Professional Standard of Care

possibility, it is necessary to undertake appropriate investigation and obtain the informed consent of the client when pioneering in new areas.

Innovation should always be disclosed to the client, advantages, downside risks, and possible side effects discussed, and informed permission obtained before proceeding.

The same principles apply to use of inexperienced or unrecommended contractors, suppliers, or engineering consultants.

Another form of innovation is in the area of contract document presentation or contract administration procedures. For example, if a new method or format of notation or dimensioning is used, any losses attributed to failure of communication could be charged to the author. Novel, experimental, or unique procedures which fail in some way could be held to be a deviation from the architectural standard of care.

Several years ago an architect reasoned that an effective way of eliminating liability for erroneous or inadvertent approval of defective shop drawings would be to specify that no shop drawings need be submitted. This would certainly be successful in achieving the stated objective but could prove to be disastrous if an adverse situation arose on the project that might have been avoided if shop drawings had been reviewed by the architect. If architects in a community generally review shop drawings of certain trades, then it is potentially perilous to eliminate or deviate from this procedure.

Unsuccessful Innovation

Unfortunately, some innovations will not be successful. New and unusual ideas do not always come out as expected. They may finally succeed but only after the time and expense of redesign, redrawing, and reconstruction, often on a costly trial and error basis. Other cases are destined to fail abysmally even after investment of inordinate amounts of time, trouble and expense. The typical project owner will not be willing to pay the unbudgeted costs of repairing new construction that did not work out as expected. Furthermore, it is not fair to expect contractors to shoulder the costs of experimentation when they are required to use untested and unfamiliar materials and systems.

Considering that the client will ultimately have to pay the bill for all repairing and reworking of building defects caused by faulty design, the architect should make a complete disclosure of all proposed innovations. The client should be made aware of the architect's intentions to specify any materials or processes that have not been in use sufficiently long to have a proven track record.

If the architect feels that the untested procedure has a definite cost/benefit advantage and a reasonably good chance of succeeding, then there may be some justification for its further consideration.

Responsible Innovation

Clients are entitled to a clear explanation of the innovative proposal and a rational evaluation of the risks as well as the prospective advantages. For example, a newly-formulated paint coating of favorable cost may be offered that can be applied in a single application rather than in the usual two or three coats. The economic and time advantages could be substantial if the material is free from disadvantageous characteristics, such as limited color range, excessive application cost, or short service life. A careful architect would properly research and analyze the new material before forming the judgment to use it. Then, a complete disclosure must be made to the client. The explanation should be in plain language and sufficiently comprehensive to enable the client to make an informed decision. The realistic prospects of success or failure should be openly disclosed along with the likely economic consequences of a negative result. Even with the technical explanation, the client will be relying on the architect to provide a reasonable recommendation based on a balanced, mature judgment.

The client's approval is often more an acknowledgment of faith in the architect's recommendation rather than a concurrence in the technical reasoning. The main thrust of the architect's disclosure should be to explain the real risks and to determine if the client is willing to accept the inconvenience, time loss, and monetary cost of a possible failure. If the client is not willing to accept all of the inherent risks and costs, then the architect is bound to seek safer, more conventional solutions.

Innovation by Engineering Consultants

Architects are also at risk to their client for undisclosed innovation when their engineering consultants experiment without prior discussion and approval. The relationship between architects and their engineering consultants is usually close enough that they can candidly discuss all aspects of the design solution. If the consultant considers it advisable to employ some innovative element, the architect should have the opportunity to participate in the decision and to decide whether it needs to be taken up with the client.

Other Forms of Innovation

Architects can inadvertently mislead their clients by proposing the use of materials or firms that are not completely familiar to them. Architects should not include materials, processes, or equipment in their specifications if they are not reasonably aware of all relevant characteristics. Inclusion of a brand name in the specifications is tantamount to a recommendation. The client is justified in assuming that the specifier has properly researched the specified item and concluded that it is appropriate in all material respects for the intended use.

Similarly, when compiling a list of proposed bidders for a project, architects should realize that the client will rightfully assume that each contractor on the list has been deemed suitable. The client's reasonable assumption is that the architect has properly investigated each contractor's experience, qualifications, financial capacity, and reputation, or has had previous satisfactory experience with them. When including unfamiliar or unknown contractors on a list of prospective bidders, the architect should disclose this important information. The client can then decide whether or not to retain the names on the list or to authorize more investigation.

Obtaining the Client's Concurrence

When proposing the employment of experimental, unfamiliar, or untried materials, methods, or firms, the architect should take the precaution of documenting the circumstances of the decision. All research materials should be retained in the project file as evidence of appropriate consideration of the procedure. The disclosure to the client can be in the form of a letter to be signed and returned as an indication of the client's concurrence in the decision and acceptance of the risks. Alternatively, the disclosure could be made in a discussion in which the client orally expresses acceptance of the risks. In this case, the architect should take the precaution of memorializing this exchange by writing a confirming letter or memorandum to the client. It should summarize the discussion and conclusions. It should also confirm that the client fully understands the situation and its inherent problems, concurs in the decision, and accepts the attendant risks.

If a client refuses to sign a disclosure letter or protests upon receiving a confirming letter or memorandum, this will give a forewarning of the client's likely attitude in the regrettable consequence of an innovative failure.

Reviewing One's Own Work

It is common practice for architects to check their own work and the work of their engineering consultants. They should review the drawings, specifications, and bidding documents for compliance with clients' instructions and program, the building code, proper use of materials, and correct application of construction techniques and processes. They should also check for coordination within the documents and for carrying out the recommendations of expert consultants, such as soils engineers and land surveyors. Architects who do not utilize normal checking procedures are not meeting the standard of care when it is a general practice among architects to do so.

Research by Architects

In analyzing new materials or processes, architects are not obligated to maintain testing laboratories or to conduct any kind of physical or chemical testing programs.

In assessing the advisability of using new materials or processes, the architect should require the vendor to submit samples, specifications, details, test reports, engineering studies, photographs, addresses and dates of installations, and the names of previous specifiers and users. The offerer should be asked to reveal the advantages, disadvantages, limitations, recommended uses, and recognized misuses of the material or process. The architect should critically review the supplier's submissions and, where deemed appropriate, examine previous installations and talk to the concerned owners, contractors, engineers, and architects.

In some cases the architect should ask for more testing or additional technical information. It is usually helpful to review the manufacturer's written warranty, if any, as it will provide information about known limitations and the extent of the supplier's willingness to back its product. If there is to be specific reliance on the terms of the warranty, the architect should ask the client to obtain a legal opinion on the sufficiency of the warranty language. The architect should also discuss the proposed use of the product with its supplier to avoid inadvertent misapplication.

Finally, the architect should apply careful reasoning in analyzing the phenomenon or proposed procedure to make sure that no known general principles are being violated. For example, no innovation, no matter how clever, can overcome the law of gravity, the shingle principle, capillarity, galvanic corrosion, heat loss, thermal expansion and contraction, or any other of our immutable physical laws and principles. If a new material or process seems too good to be true, it probably is.

Others' Reliance on the Architect's Skill

Not only the owner, but contractors, subcontractors, suppliers, and sureties also rely on the architect's special skill and adherence to the professional standard of care. If any of these entities perceive that their major financial losses, injuries, and inconveniences are in any way caused by the architect's negligence then they will undoubtedly be advised by their counsel to pursue a legal claim against the architect.

Expert Witnesses

The architect's or engineer's main legal defense to a negligence claim is proof of compliance with the professional standard of care. This is frequently demonstrated through the medium of testimony of expert witnesses skilled in the same discipline. Judges and juries will not normally presume to judge compliance with the standard of professional care but will rely on the expert testimony of qualified architects or engineers. However, if experts on both sides of a controversy disagree, the judge, jury, or arbitrator will then have to decide who to believe.

Is Perfection Possible?

Architects and engineers all know there is no such thing as a perfect set of construction documents. Contractors and subcontractors have always known this to be true. But some owners have not yet been informed of the fact and that architects are only human and not perfect.

The repercussions of the design professional's shortcomings vary from project to project. Sometimes imperfections are easily rectified and the embarrassment passes. Other times the inadequacies are more substantial and if they are not corrected in time, considerable harm could be done and money wasted. Even more rarely, the blunder is of catastrophic proportions and the price is high in time, money, loss of confidence and reputation, and possibly personal injury or death. Occasionally, the mistake is sufficiently concealed to remain undiscovered for years before the latent misfortune materializes.

Selling Architectural Services

In the process of attracting clients and marketing professional services, one accents the positive aspects of one's abilities. This is only natural, as most would not want to risk alienating prospective clients. Professionals point out their educational and experience backgrounds and extol the virtues of their associates and consultants. Thus, the professional carefully conditions the prospective clients to expect nothing less than excellence and perfection. This may set the stage for unrealistic expectations by clients.

Upon mature reflection, a perceptive and intelligent client would probably realize that designers may not be as good as they say they are. However, professionals generally fail to inform the client of the realistic prospect of making errors of one type or another, regardless of impressive pedigrees, an unusually high degree of care, and extraordinary design capabilities.

The client's unrealistically high expectations will accrue to the disadvantage of the professional when the inevitable error or omission becomes evident. One would have been better off if the prospective client had been amiably preconditioned with the gentle forewarning that perfection is not likely. It would appear less threatening to the client to receive such a disclosure at a time when no actual errors are yet under consideration.

When a professional discloses the possibility of imperfections, it should be explained realistically what happens if a mistake in the construction documents appears during construction.

If the error or omission does not directly effect construction cost or contract time, the architect will voluntarily correct the documents, will properly instruct the contractor, and will not charge the client for any additional architectural services made necessary by the mistake.

If the mistake in the documents results in added construction cost or time, it would be manifestly unfair to hold the contractor financially responsible for it. The owner, rather than the contractor, will have to pay any added construction cost attributable to an architectural error or omission. Moreover, if additional construction time is required on account of the mistake, the contractor should also be granted an appropriate time extension. Again, the architect should furnish all related architectural services without additional charge to the owner.

When Should the Architect Pay for Errors?

This significant question remains to be answered. When should the architect be held financially responsible for the added construction costs caused by anomalies in the documents? The generic answer is fairly simple: the architect can be held financially responsible if the mistake has resulted from the negligence or intentional misconduct of the architect, architect's employees, or engineering consultants.

An architect or engineer would be considered negligent when the professional standard of care has not been met. Without a doubt, it would be considered negligent practice when an error in the documents has resulted from lack of proper research, absence of checking procedures, assignment of unqualified personnel, improper supervision of junior personnel, deliberate disregard of the client's program or instructions, or failure to retain appropriate consultants when needed.

If the architect's mistake does not arise from negligence or intentional misconduct, the extra construction cost and time will have to be absorbed by the owner.

Many architectural firms, as a matter of policy, and to better serve their clients, will voluntarily pay for nominal added construction costs caused by their errors, even in the absence of negligence. However, if the added cost is beyond the reasonable financial capacity of the architectural firm, the owner will be primarily responsible to meet the contractor's additional charges. Voluntarily paying for affordable errors will also regrettably forestall a complete discussion with the client which would have revealed the client's obligation to pay for the architect's non-negligent errors.

Negotiated Compromise. Sometimes in the spirit of compromise, the architect will negotiate an equitably shared assumption of the economic burden by the contractor, owner, and architect. This is possible, however, only when there has been a history of cooperation and rapport among the parties. In those situations where the three parties are constantly in a state of mutual distrust, hostility, and lack of respect, there is little likelihood of their ever reaching a cost sharing accord. It will also be unlikely in those situations where the costs are unaffordably high.

Measuring the Architect's Liability

When the architect has been negligent, the amount for which the architect can be found liable to the owner will be the additional construction cost over and above what it would have cost had there been no error or omission in the documents or professional services.

When necessary work is erroneously specified in or omitted from the architect's documents, it must be paid for by the owner when it is added back into the construction contract. However, the architect could be liable for the extra costs due to the work having to be done out of normal sequence.

Often, change order work will cost 10 to 15 percent higher because of the lack of competitive pressure which would have been present during the initial bidding process.

The architect could also be responsible for the consequential costs of errors, such as rain water damage and loss of use of the building or liquidated damages. The costs of related removal and reconstruction, restocking charges, and wasted materials and labor could also be chargeable to the architect.

Architects Do Not Guarantee their Work

At first glance, it may seen unfair to charge the owner with mistakes made by the architect. However, architects cannot and do not guarantee that their judgment will be flawless and that their designs and documents will be completely free from error or omission.

The courts agree that architects and engineers sell service not insurance and they will not be liable in the absence of negligence or intentional misconduct. An architect is required only to act with reasonable diligence and will not be liable for damages even if the results of the work are unsatisfactory.

Considering the enormous volume of information that is normally included in a typical set of construction documents, usually hundreds of thousands of bits of data, it would be humanly impossible to avoid error completely. Typically, architectural fees are quoted at a level to account only for the time expended in producing the service plus a profit. Architects do not charge for, and owners do not pay, an additional fee designated and set aside to provide a pool of funds to cover the cost of architectural errors.

Standard of Care

A California supreme court case is frequently cited with respect to the standard of care expected of professionals.

> The court in that case said that services of experts are sought because of their special skill. They have a duty to exercise the ordinary skill and competence of members of their profession, and a failure to discharge that duty will subject them to liability for negligence. On the other hand, those who hire such persons are not justified in expecting infallibility, but can expect only reasonable care and competence. They purchased service, not insurance. Gagne v. Bertran (1954) 43 Cal 2d 481, 275 P2d 15.

Numerous other courts have agreed with this reasoning and have reached the same conclusions. Thus we are fairly certain that architects and engineers who do not violate the standard of professional care will not be held financially responsible in the absence of negligence, recklessness, or intentional misconduct.

> The law provides that absent a special agreement, an architect does not imply or guarantee a perfect plan.... This court has held that the responsibility of an architect does not differ from that of a lawyer or a physician. When he possesses the requisite skill and knowledge, and in the exercise thereof has used his best judgment, he has done all the law requires. The architect is not a warrantor of his plans and specifications. The result may show a mistake or defect, although he may have exercised the reasonable skill required. Lukowski v. Vecta Educ Corp. (1980) 401 Ne 2d 781 (Ind. Ct. App.).

All of the following legal cases stand for the proposition that architects and engineers are not liable in the absence of negligence:

> United States v. Peachy (1888) 36 F 160 (Sd Ohio)
>
> White v. Pallay (1926) 119 Or 97, 247 p 316
>
> Gagne v. Bertran (1954) 43 Cal 2d 481, 275 P2d 15
>
> Bonadiman-McCann, Inc. v. Snow (1960) 183 Cal. App.2d 58, 6 Cal. Rptr. 52
>
> Allied Properties v. John A. Blume & Associates, Engineers (1972) 25 Cal. App. 3d 848, 102 Cal.Rptr. 259
>
> Swett v. Gribaldo Jones & Assoc. (1974) 40 Cal. App.3d 573, 115 Cal. Rptr. 99

Lukowski v. Vecta Educ Corp. (1980) 401 Ne 2d 781 (Ind. Ct. App.)

Del Mar Beach Club Owners Assn. v. Imperial Contracting Co. (1981) 123 Cal. App. 3d 898, 176 Cal. Reptr. 886

Additional reading on this subject and discussion of these and other relevant cases is available in these two recent books by James Acret, a prominent construction attorney:

Architects & Engineers, Third Edition, Shepard's/McGraw-Hill, 1993

California Construction Law Manual, Building News Books,
BNi Publications, Inc.

Insuring a Guarantee of Perfection

Professional liability insurance commonly carried by architects and engineers will provide funds for legal defense and liability for negligence. The insurance would not cover any guarantee of error-free work product should an architect ever unwisely include such a guarantee in pre-contract discussions or in the architectural services agreement.

Disclosure to Owner

Architects, painfully aware that they are far from perfect, and knowing that their clients will have to pay the costs of non-negligent architectural error, should at the very least, discuss the entire matter with all prospective clients before signing architectural service agreements. This is not legally required but would clear away some of the mystery for the client and lessen the probability of legal attack when there is an unfortunate lapse in professional judgment or service.

Practical Recommendations

Realistic owners are mindful of the possibility of having to fund unexpected demands for increased construction costs due to errors or omissions in the contract documents. An appropriate contingency reserve should be carried in the owner's overall project budget. Experienced owners employ a factor of from two to seven percent of the construction contract, depending on the scope, complexity, or uniqueness of the project.

When establishing the completion date of the project, the owner should recognize that rectification of some document errors could also effect the time of construction, delaying completion and occupancy accordingly.

Fortunately, most projects proceed smoothly to satisfactory completion with little or no additional construction cost or contract time charged to the owner. However, as a matter of simple business prudence, the owner should think ahead and anticipate how it would live with the extra costs and possible delayed completion resulting from imperfect construction documents.

Handling Mistakes Realistically

It is a universally accepted belief in the construction industry that a perfect set of construction documents has never been produced. Not that it is impossible, but because it is a practical improbability. Most design professionals have never actually produced (or even seen) a complete set of documents entirely free from any kind of error.

Once it is accepted that imperfection is the norm, then it is obvious that contract administration procedures should be revised to recognize and accommodate realistic conditions. At present, attitudes and procedures are based on the idealistic concept that errors will not occur.

Conventional attitudes of owners, contractors, and architects must be adjusted to accept the inevitability of anomalies and ambiguities in the contract documents. New attitudes must reflect the current state of the law but above all must be fair to all parties involved.

Most of the construction documents are prepared in the offices of architects or engineers. The sheer volume of technical material that is ordinarily included renders it statistically improbable that all elements and interrelationships could be perfect. The principal unknowns in any set of documents will be the magnitude and identity of the errors and omissions.

Accordingly, it should no longer be a question of whether there will be error, but rather, how many errors? When will they become apparent? How serious will they be?

Architects and engineers do not like to talk about their errors, particularly those that may be discovered in the future. They know that they are there but do not want to speculate about them. They cannot avoid talking about and pondering errors discovered in the past. Those who are affected by a design professional's mistakes rarely let them slide quietly by.

Errors arise from various sources and reasons well known to construction designers. They are concerned on a continual basis with the elimination of conditions that could eventually lead to embarrassing and costly professional lapses. Errors, whether of commission or omission, are usually caused either by ignorance or inadvertence.

Errors Stemming From Ignorance

Ignorance, or lack of knowledge, can be controlled in the first instance by an architectural firm's policy of declining work that is beyond the capability of the firm and its personnel.

The most insidious element of ignorance is that one never knows the breadth and depth of one's lack of knowledge. Therefore when design firms take on projects for which they are currently inexperienced, unequipped, and improperly staffed, they can unintentionally wander into situations that more appropriately experienced practitioners would know enough to avoid. They might also be unaware of specialized problems that must be taken into consideration. Firms that take on unfamiliar work must voluntarily assume the burden of fresh research. They will have to be their own judge of how much restaffing and outside consultation they must incur to compensate for their unfamiliarity with the new work type.

Even with competent and experienced personnel, some errors will creep into the work because of novel and unusual configurations and peculiar situations never before encountered.

The principal defense against errors caused by ignorance is the careful screening of all personnel at the time of hiring. All technical personnel should be appropriately educated and experienced for the work to which they will be assigned. Staff designers and technicians should always be supervised by seasoned architects and engineers capable of doing the work. Junior members who are performing technical assignments for the first time should be closely supervised by qualified people who know how it should be done.

Errors of Inadvertence

The enormous volume of information included in a typical set of construction documents easily leads to the possibility of clerical or bookkeeping error. These are simply errors in presentation and format that include transpositions, misspellings, mathematical miscalculations, improper references to other documents, improper numbers of code sections, wrong stock numbers, and miscellaneous omissions of information. These errors would usually be recognized by their perpetrators if it were not for oversight, inattention, or inadvertence, all human failings. The use of typewriters and computers does not lessen the probability of this type of error in the absence of careful proofreading. Teams of people working on the same drawings and specifications will compound the problem if they do not work to the same standard methods. Simple errors are constantly being noticed and corrected as the documents pass through various hands, but some inevitably are overlooked and remain to surface unexpectedly later.

Budget Problems

Undershooting. One class of error that does not necessarily cause a building defect does affect the budget. Excessively conservative design might cause the building cost to fall considerably short of the allowable budget. This would ordinarily be good news if the architect had not cut the program back severely, thereby depriving the client of needed elements or quality that could have been afforded.

Overshooting. The reverse of this situation, exceeding the budget, is the more common. If the architect and owner had followed the procedures outlined in the AIA architectural services agreement, (B141, 2.2.2, 2.3.2, 2.4.3, and 2.5.1) the owner would have to decide whether to raise the budgetary limit or reduce the program (or a combination of both) and authorize the architect to revise the documents at the owner's expense. However, if the owner and architect had agreed in writing to a fixed limit of construction cost, then the architect would have to assume the expense of adjusting the contract documents as necessary to meet the budget. (B141, 5.2.2 through 5.2.5) (See Chapter 28, Designing to a Program and a Budget.)

Finding and Eliminating Errors

All competent architectural and engineering offices have their own methods and procedures for examining their work. Over time they will have developed comprehensive checklists and policy manuals.

Checking the office work product is a demanding task and should not be relegated to junior personnel. It should be performed by experienced people who are highly conversant with the design criteria and the client's program. They should also be intimately acquainted with the relevant building regulations and industry standards for the building type. They should have the ability to visualize and form a valid judgment in relation to the buildability of the construction depicted in the documents. The documents also should be examined to determine that all engineering and other specialist advice has been properly incorporated. The checker should perform a coordination check of the architectural, structural, civil, electrical, mechanical, landscape, and any other specialized documents.

Any and all errors, anomalies, ambiguities, discrepancies, omissions, and mistakes will be caught if the checking is perfect, but in reality that never happens. Unfortunately, the checking will be done by humans and that is the root of the problem. Mistakes will lie undiscovered to surface later at a crucial moment.

Professional Standard of Care

Offices that fail to check their work, give internal review little or no attention, or that assign it to unqualified personnel increase the possibility of overlooking serious error in their work product. They also eliminate one important defense argument in the event that a significant error slips through that could have been discovered during a comprehensive checking procedure.

Repercussions From Errors

When errors in the documents are discovered during the bidding process, it is a simple matter to issue a clarifying bulletin to all document holders.

The errors that cause most of the trouble are those that are discovered during the construction process, either before, during, or after carrying out the affected work. Of these, some will be innocuous and can be simply clarified in an addendum to the contract documents. There will be no economic consequence to anyone.

Some errors are such that their timely correction in the field will cost no more than the original cost would have been had there been no error or omission. This would be the usual case where a specification is missing or a size or spacing has to be corrected. The owner is not put to any additional expense but is faced with a disappointingly unexpected and unbudgeted expense.

Errors that cause defects to be built into the construction that must then be removed or corrected, will cause the owner extra expense. Similar errors that cannot be physically or economically removed or rectified will cause permanent defects to be built into the construction. The project would then forever carry a defect that may or may not affect the value or utility but could be anomalous.

A latent building defect may not be discovered for years after construction is completed. Depending on its nature, it may have minor or significant financial impact on the then owner of the project. Serious latent defects could involve extensive repair costs or irreparable depreciation in building value.

A final type of error in the documents is one that is never discovered and apparently had no noticeable effect on either the construction or the project.

Owners' Attitudes

Architects have conditioned their clients to expect perfection in their professional work product. All of the sales efforts in attracting new clients concentrate on their education, abilities, experience, and past accomplishments. Naturally, weaknesses are not emphasized nor past mistakes mentioned. Architects do not tell the owner that there is a distinct possibility, even a probability that the design and documentation may be imperfect. Clients have not considered in advance what would happen if their architects and engineers made errors on their projects.

Neither have contractors spent any time explaining to owners their firm's policy in respect to errors in the contract documents. When an error comes to light, usually by discovery of the contractor, the owner is not at all prepared for the consequences. If the error causes an extra construction expense, the owner naturally assumes that the cost will be borne by the contractor or the architect. This is not a realistic assumption but normally the owner has not been informed otherwise.

Contractors' Attitudes

Experienced contractors are not at all surprised when they encounter an error or ambiguity in the contract documents. It happens to a greater or lesser degree on every project. Even architectural firms with a fine reputation for excellent design and documentation have difficulty in producing an absolutely error-free product. An otherwise perfect set of documents might have minor imperfections that do not affect the construction or cost. Imperfections per se are not unusual.

When the contractor realizes that a documentation error has created misunderstanding or might have some cost or time impact, the usual reaction is to call it to the architect's attention. The architect must then analyze the situation and decide how to rectify it. If the contractor deems the architect's clarification to affect the contract cost or time, a change order will ensue.

The contractor has no obligation to rectify the problem without compensation. Most contractors are aware that if they go ahead and rectify the problem without consultation they will often face difficulty in getting paid for the extra work. There is also the added risk that the owner or architect might not agree with the contractor's method of solving the problem.

Architects' Attitudes

Although architects and engineers know that their work cannot be perfect, they are normally not aware of any specific undiscovered errors in their construction documents. If they did, they would immediately rectify them. They usually have spent considerable time, effort, and resources in checking their work. All known errors in the documents have been ferreted out and corrected.

Therefore, it comes as a shock and a disappointment when unknown errors become apparent during construction. Fortunately, after the documents have been carefully prepared by qualified people, and properly checked, the majority of errors and omissions remaining to be discovered are not often of high significance. Nevertheless, errors of any sort are a source of irritation and embarrassment to serious minded architects and engineers. A good part of their distress stems from the necessity of having to explain to the client why the contractor is entitled to be paid for an extra work change order. Some architects are so disconcerted about discovery of an error that they will voluntarily pay the extra costs, sometimes directly to the contractor, rather than to face up to explaining the complete situation to the client.

How much more open and satisfactory it would be to educate the client in the early stages of consultation. It would be prudent to disclose the obvious fact that none of us is perfect. It should be openly and frankly discussed. The client must be warned that it is highly unlikely that all of the documentation will be entirely free from error. It should also be explained that the contractor cannot be expected to pay the extra costs resulting from errors in the contract documents.

Finally, it should be explained to the client that such extra costs are customarily added to the contract sum and if extra time is also involved, the contract completion date must be extended. Thus the owner pays both in money and in time.

Purpose of Professional Liability Insurance

The client might well ask at this point, what is the purpose of the architect's professional liability insurance? The architect should explain that the insurance is to cover extra costs caused by the architect's negligent errors. This would be error caused by improper office procedures such as assigning unqualified personnel, specifying known discredited materials or processes, or failing to check one's work product. The insurance will not cover non-negligent error. Neither the architect nor the liability insurance carrier is a guarantor of perfection.

An Open Discussion Could Be Revealing

Construction professionals should seriously consider the possibility of mitigating in advance the shock that is usually generated when a mistake is discovered in the documents. Although the consternation to the client is often severe, the impact on the architect or engineer is also damaging.

The architect and client will both benefit from a complete and candid discussion of what each party realistically expects of the other should an error be discovered.

They must openly discuss both the architect's and the owner's attitudes toward liability for error and who is expected to pay for errors in the documents. If the discussion is held before any known errors are in existence, it can be kept on an objective, academic, and hypothetical level.

It is logical to include in all construction budgets a reserve allowance to cover the owner's exposure to extra costs caused by errors discovered after the contract price is established. This would be an owner's contingency, not included in the construction contract.

If the client cannot accept the architect's viewpoint and this line of reasoning when no claims are on the table, the designer should know what to expect if any errors are discovered during construction.

Architects' Decisions

All during the construction period, it is the architect's duty to make rulings on claims of the owner or contractor against each other. (A201, 4.3.2) The architect as the contract administrator also must render interpretations of the meaning of the contract documents. (A201, 4.2.11 and 4.2.12) Such rulings will be binding on the owner and contractor if not appealed to arbitration within 30 days after the architect's written decision. (A201, 4.5.4.1) The architect must be scrupulously fair and equitable when passing judgment on these matters and cannot show favoritism toward either party. (A201, 4.2.12)

Sometimes the architect's decisions or interpretations will be to clarify the meaning of documents prepared by the architect's own office or by that architect personally. The temptation and opportunity would be present to misrepresent the true state and reasonable meaning of the documents. It is not enough for an architect to explain what was intended to be included in erroneous, or incomplete documents. What was intended must be reasonably inferable from the actual contents of the contract documents. The architect must apply common sense, technical expertise, and intellectual honesty. If after a proper analysis the just ruling must require the architect's admission of error in the documents, then the decision will have to go against the owner, the architect's client. (See Chapter 21, Architect's Decisions Based on Design Concept, Aesthetic Effect, and Intent of the Documents.)

25

Owner's and Contractor's Legal Claims Against Architects

When The Architect Gets Sued

Notwithstanding the noblest efforts to conduct one's architectural practice to the highest attainable standards, occasionally an owner or contractor will perceive, rightly or wrongly, that the architect is the cause of their problems. They might feel that the architect improperly administered the contract or furnished defective construction documents. Consultation with their legal counsel and other advisors will often confirm their perceptions and the architect is then faced with a formalized claim in the form of an arbitration demand or a lawsuit.

It is seldom that the formal claim arrives without some form of prior warning. Except in the case of a personal injury or a construction defect that has suddenly manifested itself, there was most likely a period of gradual deterioration of the relationships among the architect, owner, and contractor. A lapse in communication or a declining tolerance of each other's singularities can lead to reciprocal loss of respect and mutual hardening of attitudes. Flexible dispositions, reasonable discussion, and friendly negotiation will no longer be possible. One or more of the parties will then confer with counsel thereby transferring the controversy to the arena of legal compulsion.

After the initial shock has worn off and the architect has sufficiently dealt with such mental images as the ungrateful client or the grasping contractor, the feelings of recrimination toward colleagues and associates, the self-pity, the unfairness of it all, and "why me?", something positive must be done. The architect must organize a response and establish a legal position

Arbitration Demand or Lawsuit

If the claim against the architect is being made by someone with whom there is a written agreement, such as a client or consulting engineer, and if the agreement contained an arbitration clause, then there would be a demand for arbitration. In an arbitration the parties are called claimant and respondent. All standard form contracts of the American Institute of Architects contain arbitration clauses naming the American Arbitration Association as administrator and incorporating the AAA Construction Industry Arbitration Rules.

If the claim is being made by someone not contracted with or by someone with whom the agreement did not have an arbitration clause, a lawsuit will be served. In a lawsuit the parties are called plaintiff and defendant.

Responding to the Complaint

In either event, the architect will have to answer the demand or complaint in a legally sufficient manner and in proper time in order to be protected. If the suit is in the public court system, legal counsel must be engaged, and it should be done promptly. Time will be needed to confer with the lawyer and for preparation of the legal response to the lawsuit or arbitration demand within the time constraints of the court or arbitration system.

An architect who has received an arbitration demand is not required to be represented by a lawyer. Legally, one is entitled to represent oneself in an arbitration.

In some situations where the issues are straightforward and the stakes not excessive, an articulate architect with an organized mind, a businesslike practice, and enough free time can self-represent and can often prevail. This of course also presupposes a detached objectivity and freedom from emotional involvement. However, this would be the exceptional case. In the event that an arbitration is lost, for all practical purposes there is no appeal. Most architects will find more comfort and security in hiring an experienced construction industry lawyer for sound and objective legal advice, to answer the arbitration claim, to organize the case, and to present it to the arbitrators.

Professional Liability Insurance

If the architect has a professional liability insurance policy, the insurer should be notified immediately upon receipt of an arbitration demand or lawsuit. The insurance company will promptly assign the case to an attorney, usually a local one in the vicinity. The assigned lawyer will need to confer with the architect to become acquainted with the subject matter and to gather sufficient information to analyze the legal issues and answer the arbitration demand or lawsuit.

Even if the architect does not carry professional liability insurance, the legal or insurance counsel should carefully examine any general liability policy that may have been in effect at the time of the alleged negligence and at the time of the alleged loss. In rare cases, they will find some insurance coverage for cost of defense or payment of loss.

Legal Counsel

Insurance companies in the professional liability field generally hire competent lawyers or law firms with considerable experience in architectural, engineering, and construction industry matters. The main problem associated with the insurance company's selection of the attorney is the matter of possible conflict of interest. Sometimes the interests of insurer and insured are not in all ways identical, in which case the architect would be better served with independent counsel. An example of a conflict of interest is where an architect's client has made a claim alleging professional negligence and has refused to pay the architectural fee. The insurance company would consider bargaining away the architectural fee to settle the negligence claim. Independent counsel might be more diligent in asserting the fee claim as well as defending against the negligence claim. Also, in situations where the insurer is trying to deny coverage, the assigned lawyer might find something in the architect's files which would aid the insurer in this regard. Business secrets would be handled with confidentiality by one's own lawyer but not necessarily by the lawyer shared with the insurance company. It is customary for the insurer-assigned lawyer to make periodic confidential progress reports to the insurance company. In cases where there is a conflict of interest between the architect and the insurance carrier, most states require the insurer to pay the fees of independent counsel selected by the insured.

Selecting a Lawyer

In the quest for independent legal counsel, it is best to seek a lawyer or law firm with prior experience in the representation of architects and engineers as well as other segments of the building industry. Lawyers who express an ingrained prejudice against arbitration should be avoided, because the architect will not receive a balanced judgment when there is a choice between arbitration and litigation in the court system. Many lawyers specialize in construction industry litigation and professional liability matters and will not need extensive familiarization with background information in preparation for the representation.

After legal counsel has been retained, or assigned by the insurer, it is important that one conducts oneself in accordance with the attorney's advice. In particular, the case should not be discussed with anyone without the lawyer's prior knowledge and concurrence. If the architect is being sued by the client or the contractor while the construction progresses, the contracted professional services and advice to which the client is entitled must continue to be rendered, although the relationship might be strained.

Preparations to Assist the Lawyer

If the case is to be heard by an arbitration panel, the attorney must be prepared and ready for the first hearing in as little as 30 days up to seldom longer than six months. Lawsuits tried in the courts do not usually come up for trial any sooner than 3 to 5 years, depending on the trial backlog of the jurisdiction.

As time passes it becomes more difficult and time-consuming to gather all of the relevant documents and to preserve the testimony of the people involved in the circumstances. If the claim is made during the construction period or shortly thereafter, all of the project records would still be in use and most of the concerned personnel still in the vicinity. The records which will be helpful in proving one's position must be collected and presented for the lawyer's examination, evaluation, case preparation and, ultimately, trial of the matter. The materials which should be collected will depend to a great extent on the type of claim which is being made.

Claims made against architects are of many varieties and occur at various times. The list in recent years has become longer, broader, and more innovative. Construction defects based on alleged design error are sometimes discovered while the construction is still in progress while others go unnoticed for years. Claims based on maladministration of the construction contract such as overcertifying contractor's payments, are usually made promptly either during or directly upon completion of the construction. Claims for construction-related personal injuries will be made during or shortly after the construction period. Some personal injury claims will surface after many years of use of the project.

When a claim is made years later, as would be the case when a building defect suddenly becomes apparent or if someone is injured on the premises, it will be more difficult if not impossible to find all of the relevant records and the persons who

Owner's and Contractor's Legal Claims Against Architects

have useful information, helpful to one's defense, regarding the events which occurred many years previously. The following checklists are offered as a general guide to the types of records and documentation which would be of interest to an attorney. Some of these items will be more relevant to one type of claim than to another.

If the controversy relates to events occurring prior to start of construction:

Owner-architect agreement and all amendments and early drafts

Maps, photographs, drawings, logs, and reports received from owner describing the site

Photographs of existing site conditions before construction

Minutes and informal notes of all meetings with owner, material and equipment manufacturers, and consultants

Owner's design program and criteria

All design criteria developed by architect or consultants

Copies of all disclaimers and disclosures sent to owner by architect or consultants

Boundary and topographic surveys and all reports and calculations received from consultants in all disciplines including but not limited to geotechnical, civil, structural, electrical, mechanical, and acoustical engineers

Reports received from building cost consultants and construction schedulers

Correspondence to and from owner and all owner's advisers such as lawyers, accountants, financial, real estate lenders, marketing, and advertising

Environmental impact statements

Memoranda to file

Insurance policies

Applicable building codes, zoning ordinances, and other governmental regulations

Building department plan correction lists

Industry standards and manufacturer's literature for relevant materials

Representations made by material and equipment manufacturers in respect to sizes, capacities, models, variations, recommended uses, and limitations

Approvals from building department and other governmental agencies

Correspondence to and from utility companies serving the site

Appointment books, telephone logs, and itemized telephone bills

Building cost estimates

Design sketches, presentation drawings, and presentation writings

Construction drawings and specifications and all early drafts and check prints

Building code analyses and research notes

Time cards of all involved personnel

(See also other lists for relevant items)

If the controversy relates to events occurring during the construction period:

Construction contract documents including owner-contractor agreement, general, supplementary, and special conditions, construction drawings, specifications, and all addenda; early drafts of all contract documents

Modifications to the construction contract documents such as contract amendments, change orders, construction change directives, and architect's written orders for minor changes

Advertisement for bids, bidding instructions, and sample forms

Contractor's proposal and proposals received from unsuccessful bidders. Resumes and promotional literature submitted by contractors

Contractor's surety bond and bid bond

Certificates of insurance submitted by contractor or owner

Minutes and informal notes of all meetings with contractors, subcontractors, and suppliers

Shop drawings, both rejected and approved; product data and samples submitted by the contractor

Correspondence and memoranda to and from contractors, subcontractors, and suppliers

Reports of all materials sampling, testing, and inspection

Engineering calculations and certificates submitted by materials suppliers

Architect's site observation reports

Contractor's daily log

Requests for information, memoranda of architect's decisions, and interpretations

Reports, notices, and memoranda from building inspectors and all other governmental agents

Construction progress and submittal schedules, outdated and interim, as well as final

Contractor's payment requests and architect's certificates for payment

Contractor's schedule of values

Lists of subcontractors and suppliers submitted by contractor

Architect's certificates of substantial completion and final completion

Reconciliation of cash allowances

Computation of liquidated damages

Architect's inspection lists, Contractor's inspection lists

Photographs and videos taken during construction

(See also other lists for relevant items)

If the controversy relates to events occurring after completion of the construction:

Contractor's written guarantees and mechanic's lien releases

Correspondence relating to owner's requests for warranty service

Operating instructions and equipment lists submitted by contractor

Record drawings and marked-up prints and specifications submitted by contractor

Photographs of completed project, photographs illustrating maintenance standards

Articles or advertisements from periodicals offering or describing the project

Certificate of occupancy issued by building department or other governmental agency

Notice of Completion

Bids or contracts for maintenance services

(See also other lists for relevant items)

If the controversy involves owner's advisors or architect's consultants:

Agreements with advisors and consultants and all amendments and early drafts, accounting for all fee charges and payments

Certificates of insurance submitted by consultants

Minutes and informal notes of all meetings with advisors and consultants, correspondence and memoranda to and from all advisors and consultants

Proposals and professional resumes submitted by consultants

Reports, calculations, sketches, drawings, specifications, and other work product of consultants

All representations and disclosures received from consultants relating to materials, systems, and equipment to be used and to design criteria

Building department plan check correction lists relating to work of consultants

Applicable building codes, governmental regulations, industry standards and design criteria relating to work designed by consultants

Reports of site observation examinations made by consultants

Photographs and videos taken by consultants

(See also other lists for relevant items)

If the controversy involves general business matters of the architectural firm:

Insurance policies

Financial statements

Partnership agreements, articles of incorporation, and corporation minutes

Purchase and rental agreements

Check records, paid bills, accounts with clients, consultants, suppliers, lessor, accountant, lawyer, insurance agent, and other advisors; correspondence to and from all of the preceding

Employee records including applications for employment, resumes, time cards, and payroll records; records of vacations, holidays, sick leaves, and leaves of absence; reprimands and probation periods; statements of personnel policy; policies for affirmation action, anti-discrimination, and equal employment opportunities.

(See also other lists for relevant items)

In addition to gathering the relevant documents for the lawyer, it is also helpful to prepare a chronology of events as soon as possible before the subject matter becomes too stale. This will assist the attorney to become quickly oriented and to assimilate the circumstances which comprise the background setting of the problem.

Another useful tool which can be prepared for the lawyer is a list of all of the persons, firms, and agencies and their roles involved in the situation. Names, addresses, and telephone numbers should be included.

When the claim against the architect concerns a premises which has not been seen for several months or years, it would be beneficial to visit the building or site to observe and record present conditions and to obtain current photographs. However this will require permission from the owner and/or occupant as well as prior approval from the lawyer.

How Long Should Records be Kept?

It is very burdensome and space-consuming to preserve the vast volume of original drawings, prints, papers, and records which seem to clutter offices; and the volume continues to proliferate. There are various and sundry governmental laws and regulations requiring the maintenance of records for periods of time differing from 4 to 7 years. However, if certain records in the architect's possession would be useful (or even crucial) for a defense or in the assertion of a claim, it would be in the architect's interest to have the record or document available whenever the need arises. This merely emphasizes the advisability of preserving all records and documents for a period considerably longer than required by law. Needless to say, a professional should have not only an efficient and convenient filing system but one which is not allowed to fall into a state of ill maintenance. There is no point in saving reams, rolls, bundles, and piles of documentation if it is not properly categorized, marked, indexed, and carefully preserved, so it may be efficiently retrieved. Improperly organized materials will waste time.

Statutes of Limitations

So how long should records really be kept? One hears of legal claims being made against architects, engineers, and contractors for damages caused by alleged design or construction defects or for personal injuries as long as 10 or 15 years after design of the project. The problem lies with the statutes of limitations for commencing of legal action which apply in the various states. A statute of limitation defines a time period within which a particular type of lawsuit must be filed. There is not complete uniformity in application of the laws and the time periods vary from state to state. In some states, as in California, there is a four year limit on patent construction defects. These are defects which would be apparent to a prudent observer. The time limit in California is increased to 10 years in the case of latent defects, that is, defects which are hidden or at least not apparent upon reasonable inspection. There are other time limits for initiating lawsuits, depending upon the specific type of legal action.

One of the problems associated with the time limits imposed by these statutes is the difficulty in determining the time from which the right of legal action is measured. For example, in an alleged design defect, the starting time could be when the mistake was made on the drawings, or when the project was completed, or when the damage was discovered.

Architects who use the AIA Standard Form of Agreement Between Owner and Architect, Fourteenth Edition, Document B141, 1987, will find that the starting time for statutes of limitations is defined in Paragraph 9.3. Construction agreements which include the AIA General Conditions of the contract for Construction, Fourteenth Edition, Document A201, 1987, have the starting time defined in Paragraph 13.7. Also included in both form contracts is a stipulation that arbitration demands must be made no later than when a lawsuit on the same matter could have been filed. (Document B141, Paragraph 7.2, and Document A201, Subparagraph 4.5.4.2) These contract stipulations will be very helpful in limiting the time in which legal action may be initiated if the dispute is among contracting parties. They will not be of much use when the claim is initiated by third parties such as lessees, later owners, occupants, and passersby.

An additional serious problem with statutes of limitation is that some judges and juries look at them as a way of depriving innocent and deserving plaintiffs of their recourse at law. Some also feel that the defendant's risk is very likely covered by insurance and that this is a desirable method of spreading the risk of bad luck. Skillful plaintiff's attorneys have kept the risk of liability open for periods longer than would seem possible under the statutes of limitation.

Regardless of statutes of limitations, any insurance policy or certificate of insurance that covers property damage, builder's risk, professional liability, or general liability should *never* be discarded.

Settlement

See Chapter 22, Resolution of Construction Disputes, for a discussion of alternative dispute resolution methods. Often a lawsuit or arbitration can be avoided by a mutually acceptable settlement arrived at through negotiation or mediation. These alternatives should be carefully investigated and seriously considered by the professional and the lawyer. After the formality of an arbitration demand or a lawsuit filing, the lawyer will take the lead in initiating all compromise discussions and settlement procedures. If a negotiated or mediated settlement is reached, it is important that its terms are properly documented and formally agreed to. Any pending legal proceedings must then be terminated in a legally effective manner. All of these legal formalities will be administered by the lawyer.

Appearing as a Witness

If a legal claim has been made against the architect, it is usually necessary for the architect to testify in the ensuing legal proceeding, either in an arbitration hearing or a court trial. Before appearing to testify, there should have been a complete discussion between architect and attorney in order to be properly prepared. Attorneys will explain the purpose and value of testimony and the types of questions which will be asked in direct testimony. The attorney will also be able to predict the types of questions which are likely to be asked by opposing counsel during the cross-examination.

Arbitration hearings and court trials are similar in that testimony of witnesses is usually elicited in question and answer format. Trials are more formal than arbitrations and strictly follow rules of evidence. Arbitrators may relax or eliminate the rules and may conduct their hearings in whatever manner they find most efficient.

During testimony, the architect should listen very attentively to the questions which are asked and should not answer unless the question is thoroughly understood. If the question is not completely unambiguous and understandable, the architect should not answer it but should ask that it be rephrased or explained. It is not a good idea to guess what the attorney is trying to get at. If the question is compound, that is, requiring two answers, ask that it be separated before answering.

All questions should be answered honestly, openly, and completely. Answers should be addressed to the judge or arbitrators. Answers should be succinct and directly to the point of the questions without rambling on beyond the actual question asked.

If an answer is not known or cannot remembered, the architect should reply, "I do not know" or "I do not remember". It is not required to know and remember everything.

When a question is asked, it is wise to hesitate slightly before answering to make sure the question is properly digested and to give the lawyers a chance to object if it is deemed necessary. If an objection is made, it is better not to answer until after the judge or arbitrator has ruled on it.

Files and notes should be taken along if necessary to refresh the memory. But if testifying from a file or notes, opposing counsel may ask to see the entire file.

If a question is asked that involves documents, such as drawings, specifications, change orders, or contracts, the architect should ask to see them before answering. Sometimes a question will be asked with a demand for a yes or no answer which in fact needs to be explained. The best way to answer is to state that the answer must be explained, then answer yes or no, then explain the answer. If the cross-examining lawyer will not allow an explanation, the architect's own attorney will later ask for the explanation during redirect examination, if it is considered relevant and necessary.

In the event that an owner and contractor of a project resort to arbitration or litigation against each other, the architect will undoubtedly be called upon to testify as a witness. The advice for testifying is basically the same as above. However, it should be always be kept in mind that the lawyers retained by the client and the contractor are not the architect's lawyers and if it is found that taking their advice would be injurious to the best interests of the architect, the counsel of the architect's lawyer should be sought for unbiased opinions and recommendations.

Third Party Legal Claims Against Architects.

Pitfalls for the architect who prepares plans and specifications for owner-builder-developers. When a contractor-developer undertakes to build a project for its own account, the architect is usually engaged on a minimal-service basis. This is normally the case in the construction of speculative projects such as tract housing, neighborhood shopping centers, apartment houses, condominiums, and light industrial, office, or warehouse complexes.

Complete Architectural Services. When an architect is hired by a conventional owner, the plans and specifications and other documents produced by the architect will be used to form the contract between the owner and general contractor. Such contract documents are generally expected to be complete and will comprehensively define all of the contractual obligations of the owner and contractor. Completeness of these documents is imperative because they will be the basis of all charges and credits to the contract sum. The general contractor will use them to delegate many of the responsibilities to subcontractors and suppliers. During the course of construction, the architect will render the usual construction services including checking of shop drawings, visiting the project to observe the progress of the work, administering periodic payments to the contractor, making rulings on the claims of owner and contractor, processing change orders and time extensions, and making construction progress reports to the owner.

Upon completion of construction, the architect will perform various other duties such as issuing the Certificate of Substantial Completion, collecting contractor's warranties, as-built drawings and operating instructions for the owner and certifying the final payment to the contractor. All of these services are necessary and valuable to the conventional owner.

Architectural Services for Owner-Builders. A typical contractor-owner-builder-developer, (hereinafter for brevity called developer), does not usually need the full spectrum of services normally offered by architects. The developer, being the owner of the project, at least before and during the construction period, will not usually look to the architect for any more than the meager necessities: only basic design concepts and minimal construction drawings. The drawings will be considered to serve only two principal criteria: to satisfy applicable governmental agencies sufficiently in order to obtain all necessary approvals and permits, and to construct the building. The developer will generally not require any more than the barest of outline specifications.

Owner's and Contractor's Legal Claims Against Architects

For additional economy of architectural fee, the architect is commonly expected by the developer to eliminate from the drawings all of the commonplace standard details and instructions. All supplementary clarifying details, sections and elevations are to be omitted. This genre of documents will be referred to as owner-builder documents, as distinguished from complete contract documents.

It is the developer's expectation that all of the omitted specifications, details, notes and drawings will be accounted for in two ways: by the inclusion of this information in the subcontract agreements and purchase orders which the developer will write and enter into with the subcontractors and suppliers, and by relying on the contractor's own experience and expertise and that of the subcontractors and suppliers who will be coordinating and constructing the project.

Some developers are very skillful in this process and have been extremely successful in completing construction projects by this system. Many architects who are accustomed to these procedures produce competent and appropriate documents which, in the hands of capable developers, result in creditable projects.

So where is the problem? If developers are obtaining good quality architectural services tailored to their needs and if architects freely enter into these contracts, everyone should be reasonably happy. If developers and architects are dissatisfied with the situation in any way and need to make changes in their methods of dealing with each other, they can easily resolve these matters in negotiating their professional service agreements.

Third Party Claims. The major problems for architects which arise out of this type of professional activity are presented by third parties, that is from people who were not the architect's client. The developer-client is not the usual primary claimant.

Then, who are the third-party claimants? During the construction period, it could be any of the subcontractors, sub-subcontractors, suppliers, their employees, lenders, insurers, sureties, passers-by or trespassers. After the project is completed, third-party claimants could be lessees, their employees, customers, invitees, repairmen or cleaning personnel. Often, the building is sold by the developer upon completion, so the new owner is added to the list of possible claimants.

In the case of a condominium residential development, there will be a multiplicity of owners plus a homeowners' association. Some of these third-party claimant entities are quite militant and well-organized for litigation. Homeowner associations are capable of pursuing extensive class action litigation, financed by nominal monthly contributions from each participant.

The Legal Claim. Usually the third-party claimant first comes to the architect's attention in an indirect way. The aggrieved party will have filed a lawsuit against the developer, among others, alleging a design defect in the drawings or the building, and the developer will react by filing a cross-complaint against the architect for indemnity. Even though the architect and the developer-client may have enjoyed a satisfactory, long-standing, and continuing business relationship and a sensitive working rapport, the architect will certainly be sued because this will be the advice of counsel for the developer or insurer.

Developer's Autonomy. Aside from the abbreviated features of owner-builder documents, there are two additional characteristics of their use that create potential problems for the architect: First, the developer is under no obligation to the architect to follow the drawings, although there is an obligation to comply with the drawings insofar as compliance with governmental regulation is depicted. There may also be, in some situations, a contractual obligation to a lender or a lessee to follow some aspects of the drawings. But, essentially the developer, being the owner of the land and building, in its sole discretion, can comply with the drawings or not. And, second, the architect's services which should be performed during the construction period, are commonly not contracted for.

Consequently, misconceptions in interpretations of the documents are frequently carried into construction. Often features which the architect would have regarded as important to design, safety or function will be changed or left out, either through inadvertence, misunderstanding, lack of coordination, or by the developer's considered decision. Had the architect been retained for field observation of construction and checking of shop drawings, such alterations or omissions could have been at least discussed at the appropriate time for informed decisions or to avert untimely extra costs.

When third-party claimants and their lawyers and other expert consultants are reviewing the owner-builder documents, commonly they are judging them, erroneously, on the same scale as contract documents produced for conventional owners. This error in understanding of the quality of the architectural work product has proven to be a very costly burden to the architectural profession.

The developer's completed building may not have been constructed in conformance with the documents produced by the architect and frequently an attempt will be made to hold the architect legally responsible for the differences, even in the absence of any architectural control over the construction process. If the owner-builder has substituted its pragmatic personal judgment for the architect's by exercising the right of not complying with the construction documents, then this is an assumption of design liability. To the extent that a design defect has been perpetrated, the owner-builder must be found liable.

Abbreviated Documents Another source of complaint against the architect springs from the abbreviated form of documents. The claimant will attempt to show that the normal architectural work product would have included such determinations that are usually omitted from owner-builder documents. Logically, the owner-builder should

Owner's and Contractor's Legal Claims Against Architects

be legally liable for the portions of the design determinations for which the owner-builder has assumed responsibility by dispensing with the architect's services.

Non-architects have difficulty in discriminating between conventional contract documents and owner-builder documents.

Expensive Legal Process Usually such claims against architects will not ultimately be perfected in the court room because there will be adequate opportunity for architect's counsel to present well-researched briefs, expert testimony, and compelling legal arguments. This is a very time-consuming and expensive process, even if the merit of the architect's position is sustained.

On many occasions, the expense of trial and the risk of losing can be forestalled by a pre-trial voluntary settlement. This is often made after considerable legal expense and research has already been incurred and the settlement contribution is heaping insult on injury. It is tantamount in the architect's mind to extortion. This is particularly offensive to the architect's sense of fair play and justice when the professional work product is within the normal standard of care for owner-builder documents.

Indemnity: A possible solution. Attempts to prevent this type of problem by the mere addition of caveats and exculpatory notes on the drawings is not highly effective, as many of the possible third-party claimants will not have ever seen the documents.

So what practical measures can an architect take to minimize or eliminate this exposure? Considering that the real underlying cause of this problem is the developer's desire to obtain low cost professional services, it would seem that the owner-builder should, as a quid pro quo, indemnify the architect against third-party suits in which the plaintiff's damages are not proven to be caused by the architect's negligence. An indemnity clause to this effect could be negotiated for and included in the architect's professional service agreement with the developer-client. Advice of legal counsel should be sought to obtain contract language which will be effective in limiting this area of professional liability.

AIA's documents committee has recently devised a new architectural services agreement for privately funded housing projects where the owner is a builder-developer. It provides for limited architectural services and answers most of the questions raised above, including an indemnity provision holding the owner responsible for damages resulting from the owner's authorization of deviations from the architect's drawings without the architect's written agreement.

> Standard Form of Agreement Between Owner and Architect for Limited Architectural Services for Housing Projects, AIA Document B188

This agreement does not coordinate with other existing AIA documents such as A201, General Conditions of the Contract. It should not be confused with B181, Standard Form of Agreement Between owner and Architect for Housing Services, which is designed for the use primarily in publicly funded single- and multifamily housing projects.

How to Attract a Professional Liability Lawsuit.

Sources of Professional Liability Claims. The last thing an architect wants is a professional liability claim. Fortunately, most claims against architects and engineers do not rise to the level of a lawsuit. Of those that do, there are very few that finally result in a judgment against the design professional. Nevertheless, it is costly, time consuming, and unpleasant to have to deal with dissatisfaction, controversy, and potential legal claims.

Professional liability claims can originate from many quarters. Claims against architects are made by all segments of the building industry that are affected by their professional activity and by owners, users, and occupants of the projects designed. By far, the most prolific source of claims is the client. These are the ones that the architect is closest to and presumably would have the most influence over. Design professionals should be able to reduce the incidence of claims from clients if they treated them better than they apparently do.

At the beginning of every client relationship, the architect's popularity is at its zenith. The client has just selected this architect from among the competition and awarded the commission. The client is impressed with the architect's design abilities and administrative talents, and possibly as a valued friend, advisor, and confidant. From that point on, if the relationship deteriorates, it could be caused by the architect. Clients are not always ideal, and sometimes because of insensitivity architects bring out the owner's most negative and unendearing personality traits.

This is a compendium, not at all exhaustive, of the most common types of thoughtlessness and imprudent actions that could provide motivation to clients for assertion of claims against their architects.

Overselling: The Empty Promise of Perfection. Sometimes an architect creates unrealistically high expectations in the minds of prospective clients. Clients should be reminded that an architect is only human and incapable of producing perfection. Primary sales efforts are generally directed to displaying a fine education, superior design capabilities, and comprehensive experience background.

It is not surprising that the client would be disappointed, and perhaps resentful, when the architect commits an error that results in additional construction costs or extra maintenance costs that last forever. It would have been better to have had a frank discussion about the possibility of imperfection early in the relationship so the client would have had a realistic view of what to expect.

Exhibiting Professional Liability Insurance. Architects should not lead their clients to regard the existence of professional liability insurance as a guarantee of perfection. The insurance policy should not be flaunted as a sales tool to convince reluctant prospective clients of faultlessness. When liability insurance is discussed, it should be to explain its realistic purpose, that is, to protect the client if the architect is negligent and thereby becomes liable for damages. The architect will be liable only upon failure to meet the standard of professional care. It could be pointed out to prospective clients that architects usually conduct their practices carefully and diligently and are seldom found guilty of a violation of professional standards. Failure of judgment and lack of perfection is not necessarily negligence and is not necessarily covered by insurance. The architect's insurance policy does not cover the client's risk of paying for the architect's non-negligent errors.

Lack of Communication with the Client. Keeping the client informed is not only a contractual requirement but a practical necessity as well. Architects sometimes get so engrossed in the details of design and the technical requirements of the assignment, that they neglect to inform the client of the current status of the work. The client may not even be aware that the architect is performing many services on the client's behalf. Good examples of these are the processing of shop drawings, an important and time-consuming procedure, and representing the client's interests with building and planning agencies and other regulatory bodies.

In the interest of advancing understanding and good will with the client, regular communications are of prime importance. Written minutes should be issued to report all meetings with the client, contractors, and others, and to confirm understandings reached and decisions made. During the construction period, a written report should be issued for each site visit. Copies of all correspondence should be sent to the client as a means of promoting awareness of all the behind-the-scenes activities.

Whenever discussions take place and understandings are reached or issues decided, they should be promptly confirmed in writing. This should also be done whenever the architect's recommendations are declined by the client. The letter should briefly explain the reasoning behind the recommendation in the event of a later claim against the architect based on rejected advice. It is the client's prerogative to accept or reject professional advice, provided it does not involve matters of health, safety, or building regulation.

Failure to Answer Letters and Return Telephone Calls. Potential trouble, dissatisfaction, and deteriorating conditions should be detected at the earliest possible moment and dealt with promptly. Letters should always be answered immediately. If a proper and complete answer requires research or redesign, then a quick letter should be sent out explaining the delay and the likely date of the complete reply. An unexplained delay will cause the client to question the architect's ability, efficiency, and intentions. Failure to answer phone calls in a timely fashion raises similar doubts in the client's mind. Unresolved small problems, questions, and misunderstandings, if allowed to fester, can grow needlessly into major concerns.

Sitting on Complaints. A related matter is that of putting off decision-making to a later time. Complaints should be dealt with quickly and fairly. All involved should be kept informed and the dispute resolution procedures in the contract should be followed. Procrastination only allows complaints to accumulate and proliferate. Complaints should be properly researched, and a decision should be made. This will allow complainants to get on with their appeal procedures if they are not willing to accept the architect's decision. When decisions are not forthcoming or not made at all, the uncertainty tends to stimulate some complainants to escalation of their demands.

Not Checking the Work Product. Architects and engineers normally check all of their work product before it leaves the office. Failure to do so would definitely constitute negligence. Document production should not be considered complete until all have been thoroughly scrutinized by a competent person acquainted with the project requirements. Even a cursory check would usually disclose obvious errors that could be corrected before anyone is caused any inconvenience or economic loss. A complete check would ascertain that the client's program has been complied with in all respects and that recommendations of all engineering consultants have been incorporated into the contract documents. The documents should also be examined to find errors in dimensions, improper use of materials, and building code violations. There should be a complete coordination and consistency check among all of the architectural drawings and specifications and those prepared by the engineering consultants. An overall constructibility check should be made before the documents are issued for bidding or construction.

Previously undetected errors, omissions, and ambiguities in the documents that are discovered during construction are a major source of client complaints against architects and engineers. A thorough checking procedure will go a long way in reducing this cause of client dissatisfaction.

Violation of Sound Business Principles. Professional liability claims against architects are often triggered by the financial insolvency of the client. In a typical scenario, the owner falls in arrears in paying the architect's bills for fees and reimbursements, and finally quits paying altogether. The architect responds by trying to collect the past due amounts by repeated billing and gentle persuasion and, failing in this effort, eventually consults with legal counsel. An arbitration demand or lawsuit is then filed.

The client's response after conferring with legal counsel is to file a counterclaim alleging imperfections in the architect's services and claiming extravagant damages far in excess of the unpaid architect's fees. The client's legal strategy will be to settle the claims and counterclaims by offsetting one against the other, thereby wiping out the architect's fees. This tactic often succeeds as the architect's counsel will advise that such a settlement could be less costly than protracted hearings or trial and there is no guarantee that the architect will prevail.

How could this eventuality have been circumvented? The basic cause of the problem was the architect's unwise extension of credit to an uncreditworthy person. When dealing with an old client of known credit delinquency or a new client of unknown credit rating, an appropriate retainer should be required, to be credited to the final payment. The architectural services agreement should provide for monthly payments, and should give the architect the right to stop work whenever any payment is 5 or 10 days overdue. The agreement should also provide for remobilization costs if the architect has to stop work and then resume work later.

Frequent Payments. Professional services agreements should always provide for frequent payments, no less than monthly. This will obviate lack of the realization that the client has stopped paying. It prevents the bill from continuing to mount when realistically there is no hope for payment.

All accounts in the architect's office should be billed every month, even when there has been no activity. Each monthly statement should show the previous month's balance, payments made, new charges and credits, interest added, and the new balance. The process of monthly billing will remind the architect as well as the client of the current status of each account. When a payment has not been made within the time limit agreed in the contract, the architect should immediately communicate with the client to ascertain the reason the payment has not been made and when it will be made. If the contract allows the work to stop when awaiting past due amounts, then it should be stopped.

Establishing an Appropriate Fee. When undertaking new work, it is important to establish a proper fee. If the agreed fee is insufficient to perform the necessary services, there would be economic pressure to reduce the quantity or quality of the service. This could easily result in defective work and a legitimate negligence claim. Shortage of available fee is not a legally or ethically acceptable excuse for cutting quality or stopping work. If the architect cannot negotiate a fee commensurate with the required scope of services, then it would be prudent to decline the assignment.

Maintaining High Quality Office Standards. Sometimes claims against architects are based on errors that could have been prevented by appropriate supervision and the assignment of competent personnel. Work assignments should not be undertaken if they are beyond the realistic capabilities of the firm's principals and available personnel.

The office should be properly equipped including all necessary research materials. The principals and all technical staff should attend to their own continuing education. They should read the professional literature on an ongoing basis and should participate in the relevant seminars, educational programs, and technical committees provided by the professional society.

In Summary. The common thread obvious in all of these paragraphs is the absolute necessity of maintaining continuous amicable realistic communication with the client. Many times the professional liability claim is simply a striking back at the architect for not treating the client with the proper degree of consideration and respect. When the architect neglects to follow the client's program, fails to recognize legitimate time constraints, or allows the budget to be breached, the client might retaliate by refusing to pay the architect's bill or by filing a legal claim.

26

Analyzing Liability for Construction Defects

Types of Defects

While the vast majority of construction defects are not dramatic ones leading to total collapse or loss of human life, considerable sums of money are often involved, necessitating appropriate action by the owner. Building and site development defects appear in a multitude of forms and variations, including:

>Inadequate strength or stiffness, structural instability, settling of foundations, or distressed structural members as evidenced by cracking, movement, and excessive deflection

>Unsightly or dangerous cracking, crazing, scaling, peeling, discoloring, swelling, blisters, or excessive smoothness or roughness

>Weather and moisture intrusion caused by failure of roofing, exterior walls, floors, or openings

>Premature depreciation such as abnormal wear, decay, erosion, corrosion, or disintegration, beyond the effects of normal wear and tear

Inadequate capacity or function of electrical, mechanical, vertical transportation, or environmental systems including wiring, piping, ducts, devices, and equipment

Some of these defects will result in buildings or parts of buildings that are unsuitable or unusable. Others will result in excessive operating and maintenance costs, premature replacement costs, or depreciated or unacceptable appearance -- and sometimes property damage, personal injury, or death.

Analyzing the Situation

It is usually difficult for an owner to know how to go about rectifying the situation and to whom to look for redress. Someone has to analyze the situation and provide answers to the key questions:

> What is the phenomenon? How can it be described and explained? What is its exact cause?

> How can it be remedied? By removal and replacement? By repair? By recoating? By strengthening? By recognizing shortened life and the necessity of premature replacement? By acceptance of the defect and a monetary adjustment? By some equivalent to the originally specified product or procedure? Or by some creative or innovative procedure?

> What are the costs of remedial work? Of analysis and recommendation? Of redesign? Of inspection and testing? Of labor, materials, tools, equipment, coordination, and supervision?

> What are the costs of consequential damages? Interior damage caused by roofing and enclosure leaks? Value of loss of use of the building, in whole or in part? Damages for personal injuries?

This analysis can be performed by a qualified and knowledgeable architect or engineer experienced in forensic analyses of construction problems. If there is a combination of problems, the advisor-analyst will usually consult with other appropriate technical experts. Sampling, testing, examining, and research might have to be performed in identifying the problem and devising a solution. If arbitration or litigation is possible, the analysis and the standards upon which it is based should be presented in writing including all supporting materials such as descriptions, measurements, photographs, cost estimates, specifications, and remedial recommendations.

Analyzing Liability for Construction Defects

Identifying the Source of Responsibility

The final question which must be answered is, Who is responsible for the failure? There are only three possibilities in determining the basic responsibility:

>Is it a *design* problem? Was the construction project properly designed, devised, selected or specified in the first place? Was it engineered properly? Were the contract documents faulty?

>Is it a *construction* problem? Was the project built in accordance with the contract requirements? Was it properly constructed? Were the workmanship practices at fault? Were the materials and equipment furnished as specified and in accordance with industry standards?

>Is it a *maintenance* or *usage* problem? Has the project and all its systems been properly cared for? Has the project been abused or vandalized by its users? Has the project been used improperly?

Design problems are the responsibility of the architects and engineers of record. Design adequacy must be gauged by the opinions of other professionals in the same discipline. If the design flaw is founded on a deviation from accepted standards and practices of design, care, and diligence for that profession, the designer could be held liable. However, if the basis of the defect is faulty judgment, there might not be liability and the owner's recourse will be severely limited or nonexistent. Each case must be considered on its own facts and merits.

Construction problems are the prime responsibility of the general contractor. The actual party at fault might be an employee, a subcontractor, a supplier, or a manufacturer of building materials or equipment. When a contractor fails to follow the requirements of the contract documents, or unilaterally decides to deviate therefrom or make substitutions, any unsuitable or defective result is the responsibility of the contractor. If the contractor faithfully carries out the dictates of erroneous construction drawings or specifications, the contractor should not be held liable but could be held partly liable if the error is of a type that could have been recognized by a competent contractor.

Maintenance or usage problems are the owner's responsibility. Sometimes owners of a new facility are lulled into inaction by the newness of materials and systems and do not commence their maintenance programs soon enough. Some maintenance personnel are not properly trained and cause damage by improper or negligent actions. The occupants and users of buildings often abuse them by rough or inappropriate usage or vandalism. Normal wear and tear is not a construction defect. Sometimes the answer springs from a combination of causative factors in which more than one party is proportionately responsible.

Mediation, a Possible Solution

Identifying the responsible party can be difficult for an owner when the explanations and excuses offered by the involved contractors and architects are expressed in plausible sounding terminology. To bring all parties into litigation or arbitration in a "shotgun type" approach, hoping to snare the responsible one, could be inefficient, unjust, and time-consuming and would be counterproductive should the owner be found responsible for the defects. A more satisfactory course would be to persuade the parties to submit the issue to an experienced neutral mediator, familiar with standard construction industry customs and practices and the standard of care of architects and engineers. The mediator can examine the contract documentation, the construction project, and the alleged defects and will evaluate the viewpoints of the owner, contractor, and architect. The mediator can then offer an impartial, informed assessment of proportionate liability for consideration by the parties. In the event that a voluntary, mutually acceptable settlement cannot be then effected, the owner will at least be in a position to proceed with confidence and certainty to negotiate, arbitrate, or litigate with those liable without any further inconvenience to the blameless party.

27

Obtaining The Owner's Instructions

An Efficient Way of Obtaining the Owner's Instructions

It is not often that architectural practice procedures can be simplified, save time, and simultaneously reduce the professional liability exposure to architects. But these are the realistic rewards for using AIA's standard document with the lengthy title:

> Owner's Instructions Regarding the Construction Contract, Insurance and Bonds, and Bidding Procedures, AIA Document G612, 1987 (Appendix F)

This form provides a convenient medium by which the architect may extract and record important facts, decisions, and preferences from the owner. This information is needed for preparing the contract documents and the bidding instructions. Most of these decisions could not be made by the architect without the owner's input and the contract documents cannot be completed without them.

The document is in the form of a questionnaire to be filled in by the owner. It is divided into three parts covering the owner's requirements regarding the construction contract, insurance and bonds, and the bidding procedure. All three parts should be submitted to the owner simultaneously so that the entire scope of information needed will be known. However the parts are separable and may be returned to the architect individually as the owner gathers the information and finalizes the decisions. The three parts are arranged chronologically as the information is needed by the architect.

Part A, Owner's Instructions Regarding the Construction Contract

The information transmitted in this part will be needed first, soon after the construction documents are commenced.

In it, the owner will state the project title to be used, the owner's legal name and address, type of legal entity, and the legal description of the property. The owner will name its construction phase representative and field representative, if any.

The owner will have to decide such matters as the construction contracting system to be used, method of selecting the contractor, whether by bidding or negotiation, the form of construction contract to be used, and the system and timing of contract payments.

The owner's wishes regarding retainage and liquidated damages will also be conveyed.

Finally, owner preferences in a variety of issues may be expressed, including equal opportunity requirements, extensions of time, lien waivers, phased occupancy, special time periods during which the contractor cannot perform construction, and work by the owner's own forces, to name a few.

Part B, Owner's Instructions for Insurance and Bonds

Although most experienced specifiers are generally conversant with the usual construction insurance, the owner should always rely on its own insurance advisor to establish all insurance requirements for the project.

There are sections in the form to be filled in to specify the coverage and monetary limits for all insurance required of the contractor for workers' compensation, general liability, property damage, contractual liability, personal injury, umbrella excess liability, and automobile, aircraft and watercraft liability.

Coverage and limits must be stated for the owner's liability insurance, property insurance, and loss of use insurance. The owner should also clarify any additional insurance requirements that must be specified by the architect in the construction documents.

The owner must decide whether or not a performance bond and payment bond will be required and may state the amount and form to be used.

Part C, Owner's Instructions Regarding Bidding Procedures

In this section the owner may instruct the architect on the forms to be used for instructions to bidders and bid proposals and whether bids will be solicited by public advertisement or private invitation. The owner may state desired bidding conditions such as bid security required, use of plan rooms, bid date, who prepares bid tabulation forms, how bids are to be opened and announced, when construction may commence, and the desired completion date. Any additional or unusual bidding requirements may be included. Most of Part C will not be needed when the contract is to be negotiated with a single previously selected contractor.

The form will serve as an efficient discussion agenda and comprehensive check list and will help the owner to assemble all of the information at one time instead of instructing the architect in a piecemeal manner.

As each section of the questionnaire is separated from the form for submission to the architect, the owner should sign and date it. This will enable the owner to know that the architect has properly received the correct information and will provide the architect a written record of the owner's specific instructions.

The average owner will not ordinarily be able to fill in the questionnaire completely without some degree of consultation with legal, accounting, and insurance advisors. Although experienced architects generously furnish practical advice to their clients of the advantages and disadvantages of the options available, owners should be encouraged to consult with their own specialized consultants who are recognized experts in their respective fields.

28

Designing To a Program And a Budget

Designing Within the Budget, A Shared Responsibility

It is a traumatic experience for the owner and architect alike when the project is priced and the lowest responsible bid exceeds the available funds. The project will have to be redesigned to a lower quality or reduced size. Sometimes it will result in a canceled project and much disappointment, ill-will, and legal repercussions. Commonly, the owner will hold the architect entirely responsible for the whole unfortunate situation.

However, it is very difficult for an architect to design within a stated budget without the active cooperation of the owner. This important concept is recognized in the AIA Owner-Architect Agreement, Fourteenth Edition, Document B141, 1987 (Appendix A).

Figure 28.1 Owner's Overall Budget

Project: _____ **Date of Estimate** _____

A. Land Acquisition, Feasibility, and Financing Costs
1. Land cost $ _____
2. Rights-of-ways _____
3. Title report _____
4. Real estate appraisal _____
5. Financing costs and loan fees _____
6. Bonds and assessments _____
7. Community development fees _____
8. Legal and accounting fees _____
9. Topographic, boundary, and aerial surveys _____
10. Geophysical investigation and report _____
11. Environmental impact report _____
12. Feasibility studies _____
13. Sales, leasing, and advertising costs _____

                                                   A. Total .... $_____

B. Design Costs
1. Architectural and engineering fees $ _____
2. Landscape architectural fees _____
3. Interior design, graphics, and color consultation _____
4. Special engineering (solar, acoustical, etc.) _____

                                                   B. Total .... $_____

C. General Building Construction Cost
1. _____ square feet @ $ ____ per square foot = $ _____
2. _____ square feet @ $ ____ per square foot = _____
3. _____ square feet @ $ ____ per square foot = _____

                                                   C. Total .... $_____

D. Other Construction Costs
1. Off-site development (utilities, streets, curbs, gutters, sidewalks, fire hydrants, street trees, etc.) $ _____
2. On-site development (grading, retaining walls, fences, walks, paving, etc.) _____
3. Landscaping, planting, and irrigation _____
4. Recreational Features (swimming pool, tennis court, etc.) _____

                                                   D. Total .... $_____

E. Construction-related Costs
1. Cost estimating $ _____
2. Permit fees _____
3. Construction taxes required by various government agencies _____
4. Insurance and bonds _____
5. Materials testing and inspections _____
6. Property taxes during construction _____
7. Utility costs during construction _____
8. Construction funds disbursement service _____
9. Construction Management _____
10. Auditing of construction costs _____

                                                   E. Total .... $_____

F. Furnishings
1. Interior finishes, flooring, blinds, and draperies $ _____
2. Furnishings, fixtures, appliances, and equipment _____
3. Graphics _____

                                                   F. Total .... $_____

                                       Total of A, B, C, D, E, and F ...... $_____

G. Contingency for estimating errors, design errors, and unforeseen conditions .... $_____

                                                   New Total .... $_____

H. Adjustment for inflation (to date: _____) x _____ % .... $_____

                                                 Total Overall Budget .... $_____

The Overall Budget

The owner is required to "establish and update an *overall budget* for the project, including the construction cost, the owner's other costs, and reasonable contingencies related to all of these costs." (B141, 4.2)

Architects should insist that owners comply in full with this provision. It will help minimize misunderstandings as to which elements are included in the architect's design budget and which are not. Often owners do not realize the magnitude of related and incidental costs and their impact on the total budget. The actual building cost can be and often is less than 50 percent of the overall budget including land costs.

A suggested format for an Owner's Overall Budget is shown in Figure 28.1. This can be used as a guide for owners who need help in compiling their comprehensive project cost figures. Some of their costs will be known or already have been incurred, such as land acquisition costs. All other anticipated costs will have to be estimated. If the owner does not have the means to estimate the rest of the costs, consultation should be sought with a knowledgeable building contractor, real estate appraiser, or building cost estimator. Some of the information will be obtainable by consultation with the owner's insurance advisor, banker, accountant, and lawyer. Some architects maintain a file of current construction costs such as those published by Building News to assist their clients in preparation of their overall budgets.

After reviewing the owner's overall budget form, it is obvious that architectural design should not be commenced before the overall budget is fairly well developed, the total determined, and it is acceptable to the owner.

If the overall budget is comprehensively considered before any funds are committed it would still be possible to make beneficial adjustments to the individual components as a value engineering technique and for controlling the total cost. As time passes, and more funds have been committed to specific items, it becomes less possible to make effective and acceptable adjustments.

The Owner's Program. The budget depends on and is an expression of the owner's program. Many of the budget items cannot be realistically estimated before the program is finalized. A number of assumptions will have to be made in estimating the quantity, quality, and current prices of the various budget line items. The form should be checked for completeness, adding supplementary items as appropriate for a specific project.

Every line item should be analyzed as carefully as possible, multiplying quantities by unit costs where applicable. When estimating, it is better to make errors on the high side, but excessive errors will yield a distorted picture and possibly a discouraged owner and an aborted project. An erroneously high estimate will demand cutbacks in scope and other economies which will later be regretted when the true costs are then known and it is too late to revise the project.

The Overall Time Schedule. The owner should also prepare a provisional time schedule compatible with the program and budget. This is different from the contractor's construction schedule. It should state the key dates for starting and ending each of the major events such as financing, design, bidding, construction, and completion. Reasonable time allowances should be included at all stages for reviews and approvals by financiers, government, and owner.

The program, budget, and time schedule, all prepared by the owner, should be presented to the architect for evaluation each in terms of the other, as provided in B141, 2.2.2. It is only after the balancing of these three documents to the mutual satisfaction of owner and architect that attention should be turned to commencement of architectural design.

Complying with the Client's Program

The responsible design process starts with the owner's program. A program is a detailed word description of needs, wants, and aspirations as perceived by the owner.

The program can be communicated to the architect in writing or orally in face to face discussions. Usually, the client's preliminary concepts will have to be interpreted by the architect and the program reorganized to make it understandable and workable. The architect will commit the program to writing and will add in the necessary realistic adjuncts such as spaces for structural elements, circulation, and environmental systems, always with due regard to convenient arrangement. The owner can then review and confirm the architect's reiteration and suggest further program adjustments, if necessary.

After further refinement and the owner's final approval, the search for a specific design solution can then proceed. To have proceeded with design studies before confirmation of the program would have been wasteful and premature. The skilled architect will soon know whether or not the program is susceptible of being developed into a practical design solution. If it cannot, the program must be adjusted accordingly.

The resulting project design would not be acceptable to most clients if it falls short of program compliance in any material respect. In the real world, designs that do not substantially comply with the program never get built.

Designing to a Budget

The program is always subservient to the budget. All building projects have budgetary restraints, some extremely stringent, while some few others may be more liberal. In either case, the design proposal must reflect a project of appropriate size and quality to remain within the imposed budgetary limitations.

Some budgets are arbitrarily established based on available funds. Often the budget is insufficient to produce the desired program. It seems to be human nature to want more than one can afford. Budgets for business or revenue producing buildings are limited to a definite economic relationship with projected income and to the market conditions prevailing at the time in the same geographical vicinity.

Designing an awe inspiring building that is beyond the financial means of the client is a futile exercise, disappointing to designer and owner alike. Violation of the budget would not knowingly be tolerated by most clients. They would simply find a more suitable architect. Apparent inability to design within the budget would mark a design professional as incompetent.

Relating the Program to the Budget Price

The program requirements must be summarized and reduced to a definite size or *quantity* of construction. Obviously, if the building area is larger, it will cost more, and conversely, will cost less when the building area is reduced.

The other co-equal factor in the price equation is the *quality* of the construction. It is self-evident that higher quality will cost more than lower quality. The measure of quality is affected by various characteristics of the proposed construction.

The relative richness of materials, such as using marble instead of laminated plastic or rare hardwoods instead of common softwoods, will raise the quality and the price.

A higher intensity of use of materials and equipment is another measure of quality. Two drinking fountains will cost more than one. Kitchens, laboratories, and toilet rooms, abundant in wiring, piping, and equipment, cost more than storage areas, offices, and meeting rooms. Highly compartmented plans will always cost more than the same area subdivided into fewer spaces.

Building complexity is another measure of quality. More complicated structural and environmental systems will cost more than simpler systems and equipment. Unusual building shapes, extravagant volumes, and innovative concepts are usually priced higher by the building industry which achieves higher production rates and lower costs with conventional construction.

A realistic program then must always take into account quantity and quality, the two conclusive factors which are determinative of the price.

Measuring Quantity and Quality

Quantity is an expression of the size of the building and is usually expressed in square feet.

Quality is determined by richness, intensity of application, and complexity and is normally expressed in terms of unit cost, reduced to dollars per square foot.

The Building Portion of the Budget

The budget price, therefor, is a function of the program. Clearly, the quantity (square feet) multiplied by the quality (dollars per square foot) will be the cost of construction (in dollars). The client decides the program and determines the budget. It is up to the designer to monitor these decisions so that an appropriate design may be conceived. The objective is to produce a design that, when evaluated by competent contractors, will cost no more nor less than the budget price, give or take a reasonable tolerance for estimating error and unexpected variations in conditions.

Schematic Design Phase

It is probable in many cases that the architect will have completed the schematic design, at least provisionally, in order to validate the program and thereby allow that portion of the overall budget to be completed.

The AIA architectural services agreement provides that during the Schematic Design Phase the architect will prepare "a preliminary estimate of construction cost based on current area, volume, or other unit costs." (B141, 2.2.5) At this stage the architect must have firm knowledge that the owner's designated program will fit in the building area allotted in the budget and that the quality corresponds with the assumed unit prices. The assumed quantities must not exceed the budget description of the elements being designed by the architect.

The assumed unit cost must be compatible with the quality of construction as determined and controlled by complexity of structure and systems, expense of materials, and refinement of detailing and workmanship. If the quality designated by the architect is too high, then the size of the project will have to be reduced to compensate. If the designed building area is excessive, then the quality must be reduced accordingly.

Sensitive design judgment must be exercised in determining the appropriate balance between quantity and quality consistent with the owner's program and budget. It is

Designing To a Program And a Budget

essential that the owner and architect are in complete agreement at each point along the way as to contents and meaning of the program, schedule, and budget. Only effective and open communication will make this possible.

Design Development Phase

As the project design is developed in an increasingly detailed manner, the architect must be very careful not to increase the size or number of any element of the project. And if the complexity and quality remain unchanged, the program, schedule, and budget will remain intact. The architect must be ever alert to inform the owner of the economic and time impact of any requests for deviation from the previously approved schematic design. Moreover, the architect must resolutely fight the urge to elaborate on or enlarge the project without the owner's informed approval and appropriate adjustments to the program, budget, and schedule.

The owner-architect agreement at this point requires the architect to "advise the owner of any adjustments to the preliminary estimate of construction cost." (B141, 2.3.2) If the owner and architect are still in accord with the design and are in agreement with the program, budget, and schedule, they are getting closer to a successful project.

Construction Documents Phase

If the design development documents were realistic and complete, it should not be difficult to prepare the construction and bidding documents without violating the sanctity of the program, budget, or schedule. Although the owner has the unquestionable authority to make (and pay for) deletions, changes, and additions to the scope and quality of the project at any time, the architect must constantly alert the owner whenever these new decisions produce some effect on previously approved programs, budgets, and schedules.

If the architect does not immediately raise the alarm when appropriate, the owner might not realize the consequences of what might have seemed harmless adjustments. Needless to say, the architect is not authorized to increase costs by enlarging or embellishing the project. The owner's informed approval must be obtained for all changes in the previously approved program, budget, or schedule.

Bidding or Negotiation Phase

At the end of the Construction Documents Phase, the owner and architect should once again be in complete harmony and both secure in the knowledge that the program has not been amended except with the owner's knowledge and consent. The portion of the overall owner's budget which is comprised of components designed by the architect should still be valid if all changes in quantity and quality have been ratified by the owner and the budget adjusted accordingly. The estimated time schedule will remain intact if all events up to this point have taken place on time and the schedule adjusted to reality where necessary.

At the Bidding or Negotiation Phase the construction documents will be tested in the marketplace. The budget for the portion being bid should be adjusted if necessary to account for inflation from the date of the estimate to the date of tendering. Receiving of bids 5 or 10 percent over or under the estimate would not be unusual. An additional variation above the low bid among competent bidders is to be expected. When bidding documents are skillfully prepared, variations among bidders will diminish.

When variations among the main body of bids are small, exceptionally high or low bids are usually erroneous.

Cost Estimates Not Guaranteed

The architect's estimates of cost submitted to the owner in the Schematic, Design Development, and Construction Document Phases should be based on realistic current costs which are available in Building News Costbooks and other publications, to which reference can be made. These estimating factors should be carefully applied, using reasonable skill and judgment. Avoid the temptation of wishful thinking or unrealistic expectations that a favored design will somehow materialize.

Architects cannot and do not guarantee such estimates and B141 makes this clear in Subparagraph 5.2.1 which states in part, that the architect's estimates "represent the Architect's best judgment as a design professional familiar with the construction industry. It is recognized, however, that neither the Architect nor the Owner has control over the cost of labor, materials or equipment, over the Contractor's methods of determining bid prices, or over competitive bidding, market or negotiating conditions. Accordingly, the Architect cannot and does not warrant or represent that bids or negotiated prices will not vary from the Owner's Project budget or from any estimate of Construction Cost or evaluation prepared or agreed to by the Architect."

Fixed Limit of Construction Cost

The architectural agreement clarifies that there is no fixed limit of construction cost inherent in the contract unless the owner and architect specifically agree otherwise in writing. If a fixed limit is so established, the architect is allowed to include contingencies for design, bidding, and price escalations. The architect is also permitted to determine what systems and construction materials are to be used and to adjust the scope of the project as necessary. The architect is also allowed to include

Designing To a Program And a Budget

alternate bids for adjusting the price to meet the fixed limit. Fixed limits are adjusted to account for inflation. (B141, 5.2.2)

If the fixed limit, after all allowable adjustments, is exceeded by the lowest bona fide bid or negotiated proposal, the owner is given four options for proceeding:

1. Approve an increase in the fixed limit,
2. Approve rebidding or renegotiating the project within a reasonable time,
3. Abandon the project, or
4. Cooperate in revising the project scope and quality as required to meet the fixed limit of cost. (B141, 5.2.4)

If option 4 is chosen, the architect is obligated to revise the contract documents at no additional cost to the owner. This is the maximum extent of the architect's responsibility arising out of the establishment of a fixed limit of construction cost. (B141, 5.2.5)

Elimination of Cost Estimating from Architectural Services

Some architectural offices would prefer to eliminate all cost estimating services from their contracts along with all of the attendant responsibilities. This can be easily accomplished by amending the AIA Owner-Architect Agreement, Document B141:

> Eliminate all references to budgets or cost estimates in Subparagraphs 2.2.2, 2.2.4, 2.3.1, 2.4.1, and 2.5.1.

> Eliminate paragraphs 2.2.5, 2.3.2, 2.4.3, 4.2, and 5.2 in their entirety.

This will be acceptable to many clients who are interested primarily in their chosen architect's design ability and the production of the contract documents. Such a client would rely on other experts for controlling the budgetary aspects of the project.

29

Written Communication

What Must be in Writing

The AIA General Conditions provides that various administrative functions must be performed in writing. They may be conveniently classified in five general categories:

1. Certificates given by architect or contractor

2. Notices to and from owner, contractor, architect, surety, and others

3. Submittals to and from owner and contractor

4. Additional agreements needed to facilitate progress or the work of the original contract

5. Orders, authorizations, approvals, and objections and miscellaneous communications among the owner, contractor, and architect

In most cases, there are no special forms which must be followed. There are a few standard AIA forms which may be used, and some construction industry textbooks and legal formbooks provide additional formats and guidelines. In writing or devising documents, it is important to use plain, unambiguous, direct English. Relevant contract provisions should be checked to make certain that all of the requirements are met.

Architect's Certifications

In the case of architect's certifications, it is important to make sure that the wording is realistic and does not assert facts beyond the architect's ability to know, or promise results beyond a capacity to deliver. For example, some lenders and owners ask architects to sign a certificate similar to the following: "I hereby certify that the work was completed in strict compliance with the contract documents and in conformance with all applicable zoning laws and building codes."

This is much too broad and very likely would be construed as a warranty which would be excluded from the coverage of most professional liability insurance policies. There is no way an architect could possibly know all applicable zoning laws and building codes, let alone know whether all aspects of the work are in conformance. A more realistic and acceptable wording would be the following:

> "Based on my on-site observations, I hereby certify that to the best of my knowledge, information, and belief, the work was completed in compliance with the contract documents and applicable zoning laws and building codes except for (list exceptions)."

Certifications should always have all necessary modifying or limiting conditions completely described and should be dated. An architect could be found liable for losses incurred by anyone who places reliance on an architect's certificate which later proves to be erroneous or misleading. This is particularly true with respect to lenders, sureties, and owners when the architect has overcertified payments to contractors. If the architect is not absolutely certain of the suitability of the wording of a certification of any sort, the safest course is to review it with legal counsel before signing.

Certificates Required by the AIA General Conditions

By its very nature a certificate would have to be in writing. A certificate is always a document. The purpose of a certificate is to attest to the genuineness or truthfulness of the information that is the subject of the certificate. Therefore, the certifier must be prepared to stand behind the facts upon which the certificate is based.

Contractor's Certificate for Payment

The Application and Certificate for Payment, AIA Document G702, submitted by the contractor for each payment requested, contains the following certificate to be signed by the contractor and sworn to before a notary public:

> "The undersigned Contractor certifies that to the best of the Contractor's knowledge, information and belief the Work covered by this Application for

Written Communication

Payment has been completed in accordance with the Contract Documents, that all amounts have been paid by the Contractor for Work for which previous Certificates for Payment were issued and payments received from the Owner and that current payment shown herein is now due."

The contractor's application and certificate are in conformance with the payment procedure set out in paragraph 9.3 of A201.

Architect's Certificate for Contractor's Payment

The same application form (G702) contains an architect's certificate for payment which states:

"In accordance with the Contract Documents, based on on-site observations and the data comprising the above application, the Architect certifies to the Owner that to the best of the Architect's knowledge, information and belief the Work has progressed as indicated, and the Contractor is entitled to payment of the AMOUNT CERTIFIED."

The architect's payment certificate is in conformance with the certification procedure set out in paragraph 9.4 of A201.

The architect should not issue the certificate without having recently visited the site to ascertain that the work has progressed at least to the point being claimed by the contractor. The certificate is also an assurance to the owner that the work meets the standard of quality specified in the contract documents.

The certificate is a representation of the architect's opinion that the amount certified is due from the owner to the contractor.

The certificate is not unqualified, however. In the last part of A201, 9.4.2, are four important qualifications, limiting the breadth of the certificate:

However, the issuance of a Certificate for payment will not be a representation that the Architect has:

1. Made exhaustive or continuous on-site inspections to check the quality or quantity of the Work
2. Reviewed construction means, methods, techniques, sequences or procedures
3. Reviewed copies of requisitions received from subcontractors and material suppliers and other data requested by the owner to substantiate the Contractor's right to payment

4. Made examination to ascertain how or for what purpose the contractor has used money previously paid on account of the contract sum

The qualifying clause within the certificate that the opinion is based on "the architect's knowledge, information and belief" is not likely to relieve the architect from making the necessary site inspection, being acquainted with the contract documents, and exercising reasonable care in on-site observation and administration of the contract.

Architect's Certificate of Substantial Completion

It is of fundamental significance that the date of substantial completion be conclusively established, as this is the starting date for the warranty period and the ending date of the construction period for computing liquidated damages. The date is to be determined through the independent judgment of the architect and is transmitted to the parties by means of a certificate. (A201, 4.2.9 and 9.8) A convenient form for expressing this determination is the Certificate of Substantial Completion, AIA Document G704.

The certificate should be based on the architect's having made an on-site inspection to ascertain the actual state of the work. G704 contains a definition of the date of substantial completion and also includes a listing of the residual responsibilities of both the owner and contractor and provisions for their signing to indicate acceptance.

Architect's Certificate of Final Completion

The architect is also obligated to determine the date of final completion. Upon reaching this determination, the architect should approve the contractor's final application for payment by executing the architect's certificate for payment on the application form. This constitutes the certificate of final completion. (A201, 4.2.9 and 9.10.1) When a retention has been withheld from the contractor's payments, an additional payment application and certificate will be issued when the retainage is due.

Architect's Certificate of Approval of Owner's Correction of Deficiencies

In the event that the contractor defaults or neglects to carry out the work of the contract, the owner, after observing the requirement of two consecutive 7-day notice periods may correct such deficiencies and charge the costs to the contractor. The owner, however, must obtain the architect's prior approval of both the owner's action and the amount to be charged to the contractor. (A201, 2.4.1) This approval should be in the form of a certificate.

Architect's Certificate of Sufficient Cause

The owner may terminate the contract for cause under conditions listed in Subparagraph 14.2.1 (A201), provided all notice requirements have been complied with. The owner must obtain the architect's prior certificate that sufficient cause exists to justify the owner's action. (A201, 14.2.2)

Considering the easily predictable repercussions of either of the two foregoing certifications, the wording is of extreme importance. The certificates should be based on the architect's independent knowledge and opinion and on evidence of the specified facts. The architect should make a thorough examination of the conditions, should memorialize them in appropriate documentation, notes, and photographs, and should obtain the viewpoints of both the owner and contractor before issuing the certificate. All certificates should be dated.

Others' Reliance on Certificates

The owner and architect will both be relying on the truthfulness and accuracy of the contractor's certificate on applications for payment.

The owner, surety, lender, and others will be relying on the genuineness of the architect's certificates for payment. In fact, the payment would not be made to the contractor at all in the absence of a certificate or if the integrity of the certificate was in any way questionable.

The architect's certificate of substantial completion is relied upon by the owner, contractor, subcontractors and suppliers. It could alter their rights substantially if the certificate is erroneous or improper.

The architect's certificates of approval of the owner's actions in taking over correction of the contractor's faulty work or ousting the contractor from the job will be heavily relied upon by the owner. The contractor's position will be seriously affected by either of these two certificates. In these two latter cases there is no question that legal claims and counterclaims will ensue and the certificates will be minutely examined.

All of the architect's certificates will be relied upon by others to protect their economic and legal positions. Architects must be certain that all their certificates are written in clear, unambiguous language. They must be based on verifiable facts. Each certificate should be backed up in the office files by appropriate documentation, field observation reports, photographs, samples, tests, or whatever is logically needed for authentication.

Certificates should always be dated and issued in a timely fashion. Certificates fraudulently issued after the appropriate time will be completely discredited when the true facts are later discovered.

All opinions expressed by certificates should be the independent judgment of the certifier.

In case of legal action involving an architect's certificate, the certificate will be subjected to a painstaking word for word analysis. All underlying circumstances and facts will be looked into. Anyone who stands to gain by discrediting the certificate will undoubtedly attempt to do so.

Notices

Notices include all communications transmitting information, demands, and claims. Notices required by the AIA General Conditions must be in writing. Occasionally it will be expedient or convenient to give notice orally, and it would legally suffice, although its terms might be difficult to establish later if it became necessary. Therefore, all oral notices should be followed promptly by a confirmation in writing.

All notices should be carefully worded for accurate expression to avoid ambiguity and should always be dated. According to the General Conditions, written notice is deemed to have been duly served when it is delivered in person to the individual, firm, or entity or if delivered or sent by registered or certified mail to the last business address known to the party giving notice. (A201, 13.3.1) This is a checklist of the written notices required or allowed by the AIA General Conditions:

> Owner's order to the contractor to stop the work for any of the specified causes (A201, 2.3.1)
>
> Owner's 7-day notice to the contractor of contractor's default and neglect to carry out the work of the contract (2.4.1)
>
> Owner's second 7-day notice to the contractor of contractor's default and neglect to carry out the work of the contract (2.4.1)
>
> Contractor's report to the architect of errors, inconsistencies, or omissions discovered in the contract drawings and information furnished by the owner (3.2.1)
>
> Contractor's report to the architect of errors, inconsistencies or omissions discovered in field measurements, field conditions, and other information known to the contractor, compared with the contract drawings (3.2.2)
>
> Contractor's giving of notices required by laws, ordinances, rules, regulations, and lawful orders of public authorities bearing on performance of the work (3.7.2)

Written Communication 341

Contractor's notification of architect and owner of violations of applicable laws, statutes, ordinances, building codes, and rules and regulations found in the contract documents (3.7.3)

Contractor's informing the architect of specific deviations in the shop drawings from requirements of the contract documents (3.12.8)

Contractor's direction of architect's specific attention on resubmitted shop drawings to revisions other than those requested by the architect on previous submittals (3.12.9)

Contractor's notification of architect that the required design, process, or product is an infringement of a patent (3.17.1)

Architect's informing the owner of the progress of the work (4.2.2)

Request of owner or contractor for architect's interpretation and decision concerning performance under and requirements of the contract documents (4.2.11)

Architect's response to owner's or contractor's request for interpretation and decision concerning performance under and requirements of the contract documents (4.2.11)

Owner's or contractor's claims for adjustment or interpretation of contract terms, payment of money, extension of time, or other relief (4.3.1)

Notice by the observing party of concealed or unknown conditions differing materially from those indicated in the contract documents or those ordinarily found to exist (4.3.6)

Contractor's claims for an increase in the contract sum (4.3.7)

Contractor's claim for an increase in the contract time (4.3.8.1)

Contractor's or owner's notice of injury or damage to person or property because of act or omission of the other party (4.3.9)

Architect's initial response to contractor's or owner's claims (4.4.1)

Architect's notification of surety (not obligatory) of the nature and amount of claim (4.4.1)

Architect's notification of owner and contractor that decision on claim will be made within 7 days (4.4.4)

Architect's written decision on claim of owner or contractor (4.4.4)

Architect's notification of surety (not obligatory) of decision on claim of owner or contractor and request for the surety's assistance in resolving the controversy (4.4.4)

Owner's or contractor's demand for arbitration (4.5.1)

Owner's or contractor's amendment of demand for arbitration (4.5.6)

Contractor's report to the architect of apparent discrepancies or defects in other construction by owner or separate contractors which would render it unsuitable for contractor's further work (6.2.2)

Owner's notice to contractor to proceed with construction (8.2.2)

Contractor's notice to owner to proceed with construction (in absence of owner's notice) (8.2.2)

Architect's notification to the contractor and owner of the architect's reasons for withholding of certificate for contractor's payment (9.4.1. and 9.5.1)

Owner's notification to the architect that the certified payment to the contractor has been made (9.6.1)

Architect's notification of contractor to complete or correct items remaining on the punch list prior to substantial completion (9.8.2)

Contractor's notice that the work is ready for architect's final inspection (9.10.1)

Architect's confirmation that material delay of final completion of the work is not the fault of the contractor (9.10.3)

Contractor's report to the owner and architect of the encountering of asbestos or PCB in the work area. (10.1.2)

Architect's final determination in respect to asbestos or PCB (10.1.2)

Contractor's giving of notices complying with applicable laws, ordinances, rules, regulations, and lawful orders of public authorities bearing on safety of persons or property or their protection from damage, injury, or loss (10.2.2)

Contractor's posting of danger signs and other warnings against hazards, promulgating safety regulations, and notifying owners and users of adjacent sites and utilities (10.2.3)

Insurer's notice to owner 30 days prior to canceling insurance coverage (11.1.3)

Owner's notice to contractor that it does not intend to purchase property insurance with all of the specified coverages required by the contract (11.3.1.2)

Contractor's request for additional property insurance coverage (11.3.4)

Architect's request for contractor to uncover work which was specified to be available for observation (12.1.1)

Architect's request for contractor to uncover work which was not specified to be available for observation (12.1.2)

Owner's notice to contractor during warranty period to rectify work not in accordance with the contract documents (12.2.2)

Architect's notice to contractor fixing reasonable time to correct nonconforming work during warranty period (12.2.4)

Contractor's timely notice to architect of when and where tests and inspections are to be made (13.5.1 and 13.5.2)

Contractor's notice to the owner and architect of contractor's termination of the contract for any reason listed in subparagraph 14.1.1 (14.1.2 and 9.7.1)

Contractor's notice to the owner and architect of contractor's termination of the contract because of owner's persistent failure to fulfill its obligations with respect to matters important to the progress of the work (14.1.3)

Owner's notice to the contractor and surety of owner's termination of the contract for any reason listed in subparagraph 14.2.1 (14.2.2)

Owner's order to the contractor to suspend, delay, or interrupt the work in whole or in part for such period of time as the owner may determine, with cause, for the convenience of the owner (14.3.1)

Notices required by the Performance Bond and Payment Bond, AIA document A312

Submittals

Submittals are written information presented by the contractor or owner to the other such as site information, shop drawings, product data, samples, schedules, and lists. This is a checklist of submittals required by the AIA General Conditions:

Contractor's returning to the architect or suitably accounting for all of the contract documents, except one record set, upon completion of the work (A201, 1.3.1)

Owner's furnishing to the contractor information which is necessary and relevant for the contractor to evaluate, give notice of, or enforce mechanic's lien rights (2.1.2)

Owner's furnishing to the contractor of reasonable evidence that financial arrangements have been made to fulfill the owner's obligations under the contract (2.2.1)

Owner's furnishing to the contractor of surveys, soil tests, legal limitations, utility locations, and legal description of the site (2.2.2)

Owner's furnishing to the contractor of such copies of drawings and project manuals as are reasonably necessary for execution of the work (2.2.5)

Contractor's furnishing of satisfactory evidence as to the kind and quality of materials and equipment (3.5.1)

Contractor's submission of a construction schedule for the work (3.10.1)

Contractor's submission of a schedule of submittals (3.10.2)

Contractor's delivery to the architect at completion of the work of one record copy of the drawings, specifications, addenda, change orders, and other modifications marked to record changes and selections made during constructions, and in addition approved shop drawings, product data, samples, and similar required submittals (3.11.1)

Architect's submittal to owner of preceding items (3.11.1)

Contractor's submission of shop drawings, product data, samples, and similar submittals required by the contract documents (3.12.5)

Contractor's submission of professional certification of performance criteria of materials, systems, or equipment required by the contract documents (3.12.11)

Architect's forwarding to the owner of written warranties and related documents required by the contract when received from the contractor (4.2.9)

Contractor's or owner's submission to the architect of additional data supporting their claim for architect's decision (4.4.3)

Contractor's submission to the architect of list of names of persons or entities proposed for each principal portion of the work (subcontractors and suppliers) (5.2.1)

Contractor's submission to the architect of names of subcontractors or suppliers to replace those rejected by the owner or architect (5.2.3)

Contractor's submission to the architect of an itemized accounting and supporting data to substantiate pricing of construction change directives (7.3.6)

Contractor's submission to the architect of a schedule of values (9.2.1)

Contractor's submission of application for payment and its supporting data (9.3.1)

Owner's making of payments to the contractor (9.6.1)

Contractor's submission to the architect of a request for inspection accompanied by a comprehensive list of items to be completed or corrected (punch list) to achieve substantial completion (9.8.2 and 9.9.1)

Contractor's request for reinspection to determine substantial completion (9.8.2)

Written Communication

Contractor's submittal to architect of prerequisites to final payment itemized in 9.10.2

Contractor's submittal to architect of prerequisites to partial final payment itemized in 9.10.3

Contractor's designation of a responsible member of the contractor's organization at the site, other than the superintendent, whose duty shall be the prevention of accidents (10.2.6)

Contractor's submission of certificates of required insurance (11.1.3)

Contractor's submission of certificates of insurance required to remain in effect after the final payment (11.1.3)

Owner's filing with contractor of certificates of required insurance (11.3.6)

Owner's furnishing of a surety bond for proper performance of owner's duties as fiduciary in case of an insured loss (11.3.9)

Contractor's furnishing of bonds to the owner covering faithful performance of the contract and payment for labor and material (11.4.1)

Contractor's submitting to the architect of required certificates of testing, inspection, or approval (13.5.4)

Additional Agreements

From time to time during construction, unexpected conditions or circumstances will require the parties to make further agreements in order to facilitate construction progress. As it would be nearly impossible to write a contract that would anticipate all possible occurrences, the AIA General Conditions require the parties to further agree in writing when necessary. In the event the parties cannot agree, the architect will provide the decision subject to arbitration if the determination is unacceptable to either or both parties.

This checklist summarizes further written agreements anticipated by the AIA General Conditions:

Contractor's, owner's and architect's agreement on time limits for architect's decision on claims submitted by contractor and owner (A201, 4.2.11)

Contractor's and owner's agreement to suspend contractor's performance and/or owner's contract payments pending final resolution of a claim (4.3.4)

Owner's, contractor's, and architect's mutual agreement to allow consolidation or joinder of the architect and the architect's employees or consultants in an arbitration between owner and contractor (4.5.5)

Owner's, contractor's, and architect's mutual agreement for change orders (7.2)

Owner's and architect's agreement to proceed with a construction change directive in the absence of contractor's concurrence (7.3)

Contractor's and owner's agreement to start construction prior to effective date of insurance (8.2.2)

Contractor's and owner's agreement of an off site location where materials and equipment may be stored to qualify for progress payments (9.3.2)

Contractor's and owner's written acceptance of responsibilities assigned to them in the architect's certificate of substantial completion (9.8.2 and 9.9.1)

Contractor's and owner's agreement to owner's occupancy of any portion of the work (9.9.1)

Contractor's and owner's agreement for terms of contractor's resumption of work after asbestos or PCB has been encountered in the work area (10.1.2)

Contractor's and owner's agreement in respect to changes in insurance requirements prior to final payment (11.3.1)

Contractor's and owner's agreement as to distribution of insurance proceeds received in insured loss (11.3.9)

Contractor's and owner's agreement to permit assignment of the contract as a whole (13.2.1)

Contractor's and owner's agreement to a breach of the contract (13.4.2)

Contractor's and owner's agreement to the amount of a fixed or percentage fee for the increased cost of work caused by owner's suspension for convenience (14.3.3)

Orders, Authorizations, Approvals, and Objections

In the process of carrying out the contract or administering it, the owner, contractor, and architect all must exercise their rights of approval or disapproval and powers to consent to various proposals or submissions. These decisions and determinations should be presented to the affected parties in writing and should be dated.

The various written authorizations, approvals and objections provided for in the AIA General Conditions are summarized in this checklist:

Architect's and owner's specific written consent to the contractor for use of the contract documents on other work (1.3.1)

Architect's approval of contractor's schedule of submittals (3.10.2)

Written Communication

Contractor's approval of shop drawings, product data, samples, and similar submittals prior to submission to architect (3.12.5 and 3.12.7)

Architect's approval of contractor's deviations in shop drawings, product data or samples (3.12.8)

Owner's and separate contractor's consent for contractor to cut or alter work of the owner or separate contractor (3.14.2)

Contractor's consent for owner or separate contractor to cut or alter work of the contractor (3.14.2)

Owner's, contractor's, and architect's mutual consent to modify, restrict, or extend the duties, responsibilities, and limitations of authority of the architect (4.1.2)

Contractor's reasonable objection to appointment of replacement architect (4.1.3)

Owner's, contractor's, or architect's special authorization of direct communications rather than using the formal communications channels described in Subparagraph 4.2.4

Architect's review and approval or other appropriate action upon the contractor's submittals (4.2.7)

Architect's interpretations of and decisions on matters concerning performance under and requirements of the contract documents (4.2.11 and 4.2.12)

Architect's and owner's approval of or reasonable objection to contractor's list of subcontractors and suppliers (5.2.1)

Architect's ordering of minor changes in the work (7.4.1)

Owner's approval in advance for payment to the contractor for materials stored off the site (9.3.2)

Insurer's consent to owner's partial occupancy of the work (9.9.1 and 11.3.11)

Public authorities' authorization of owner's partial occupancy of the work (9.9.1)

Surety's consent to final payment (9.10.2)

Surety's consent to partial final payment (9.10.3)

Contractor's objection to owner's acting as fiduciary in adjusting insurance proceeds in an insured loss (11.3.10)

Additional Written Communications

Other contract documents such as the supplementary or other conditions or the trade sections of the specifications may contain additional requirements for written submissions, notices, authorizations, approvals, or objections. At any time that any of the interested parties wish to communicate with each other for any purpose, the contractual communications channels should be respected. Whenever casual communication such as face-to-face discussions or telephone conversations are used, the importance of the matter should be evaluated. In the very least, a note in the architect's file will memorialize the particulars and the date. If the subject matter is of greater importance, oral communications should be confirmed to all interested parties in writing.

30

Graphic Communication

The Graphic Part of a Construction Contract

The most voluminous portion of the contract documents in any construction contract is made up of the drawings and specifications.

The specifications, being written in English should be comprehensible to anyone who understands the language and is conversant with the technical subject matter. However, the drawings are only understandable to those who fully comprehend the graphic language as well as the technical subject matter.

Construction contract drawings should be drafted with as much legal precision and craftsmanship as any other part of the contract. They must communicate a precise message as well as be technically complete and correct. The drawings must promote specific and clear interpretation. They must be completely understandable to all who are bound by their requirements. There is little possibility of a meeting of the minds when any of the contracting parties do not understand the drawings.

Understanding Technical Drawings

People who comprehend construction drawings are said to be able to "read" them. Architects and engineers and their technical staffs not only understand how to read drawings but also how to create them and use them as a medium for communicating their ideas and expectations to others. The highest level of conversance with graphic communication includes the facility of three-dimensional visualization even though the drawings are flat, two-dimensional representations. The graphic communication faculty has two complementary factors: the ability to communicate with others and the ability to understand communication originated by others.

Although most members of the construction industry are able to use and interpret construction drawings, there are a significant number who cannot. Some outstandingly capable artisans and craftspersons, although highly skilled in the nuances of their own trades, simply do not understand drawings. They are, therefore, utterly dependent on their supervisors, coworkers, or other trusted advisors for accurate interpretation of contract requirements.

People Who Cannot Read Graphics

Those who have never learned how to read drawings are what could be termed *graphically illiterate*. Presumably they could learn if they spent the time and effort. However, some simply cannot be taught.

Producing technical drawings to be understood by people who cannot read, or who cannot be taught to read plans, is no more possible than creating written contracts to be understood by people who cannot read.

Rational and refined judgment of the meaning and communication value of construction drawings can be made only by persons who are themselves graphically literate.

Superficially, it is not possible to identify people who cannot read plans any more easily than it would be to discern if they are color blind, lack taste, or harbor bias. This is particularly true when an individual is outstandingly literate and articulate in oral or written English; one would not presume to question their graphic communication faculty.

Client Approval of Drawings

Architects often ask their clients to sign the drawings at various stages to signify their approval and acceptance of the work up to that point. These clients often do not really understand the graphic materials but they will sign anyway thus signifying their belief and confidence in the architect's word descriptions and as an act of faith in their architect's professional ability. In this case, the architect may not realize when there is a lapse in communication. It will not become apparent until some time later when the project is under construction. Only then is it possible for this client to fully appreciate the three dimensional implications, but then it may be too late to make changes economically.

Graphic Communication

In situations where owners rely on others who possibly cannot read plans, such as some accountants, lawyers, and bankers, they may not recognize that these consultants are unable to interpret graphic communications, visualize the physical implications, and the economic consequences.

Drafting Conventions

Orthographic Projection. Construction drawings are based on the use of certain commonly accepted and understood drafting conventions, principles, and procedures. One of these is the use of orthographic projection. This is a drafting method utilizing right angle or straight-on views of the various surfaces and sections of an object. Orthographic drawings, being flat and two dimensional, are not as realistic-looking as perspective or isometric projections, but they have the advantages of drafting simplicity, ease of scaling and dimensioning, and preservation of relationships and proportions.

Scale. Another significant drafting convention is the use of scale. Buildings and their parts are much too large to fit on a convenient size piece of paper if drawn at their full size, so they are drawn instead reduced to a fraction of their actual size. This proportionate reduction in size is called the scale of the drawing.

Although construction drawings are drafted to scale it is considered poor practice for contractors to obtain dimensions for field use by measuring the drawings. Only figured dimensions should be followed and used for construction. If the drawings contain insufficient dimensional information for construction, the contractor should request the missing data from the architect responsible for the drawings. The contractor, when scaling the drawings, needlessly incurs additional liability by assuming the architect's duties.

The choice of scale for any particular drawing is decided by the drafter. Various relevant factors must be taken into account. The scale must be small enough so the drawing will fit on the size of paper to be used, but large enough to accommodate the necessary detail. If the choice of scale to satisfy the need for detail produces a drawing too large for the selected paper size, the drawing will have to be drawn in 2 or more parts on separate sheets. Some buildings lend themselves conveniently to a single small scale drawing all on one sheet, with selected areas enlarged sufficiently to show the necessary detail.

Drafting Standards. Unfortunately, there is no accepted written national standard for the composition and production of construction drawings. Oddly, the AIA has never issued any comprehensive national guidelines or standards for this important area of architectural communication. The tacit understandings are based on decades of practical usage, word-of-mouth, and drafting room tradition passed down from one generation to another of architects, engineers, and their drafters.

Many architectural and engineering firms write their own drafting room manuals, developed over years of practical evolution. These unpublished, privately produced manuals enable each firm to produce work of a consistent standard even though numerous designers and drafters may work on various portions of a single set of drawings.

When evaluating drawings produced by different firms and different drafters, there is not a wide difference in communication value even though the drafting standards may be somewhat inconsistent. Contractors and others in the industry who utilize construction drawings for estimating and construction routinely see the work of various architects' offices and are generally able to interpret them satisfactorily without having to question the authors.

Drafter's Prerogative. Some drafting techniques are a matter of the drafter's personal preference, technique, or judgment. For example, a floor plan is understood to be a horizontal section through a building, taken just above the floor line. Drafters, for clarity, will raise and lower the horizontal cutting plane (section) to pass through all wall openings (mainly doors and windows) regardless of their height above the floor. This is a recognized and accepted drafter's prerogative that does not interfere with communication. If anything, it enhances communication.

Legends. A legend is a schedule that defines the arbitrary signs and symbols used on a particular drawing or throughout the set of drawings. Although legends in various sets of drawings appear to be somewhat similar, they are not all the same. There is no nationally accepted list of symbols and their meanings which can be relied upon as one would on a dictionary for uniform definitions of words. It is the drafter's prerogative to assume symbols to be used on any drawing. However all symbols must be defined in the legend, the important objective being unambiguous and accurate communication.

There is no national standard for abbreviations so they also must be defined. A drafter's failure to define all arbitrary signs, symbols, and abbreviations seriously diminishes the certainty and communication value of the drawings.

Measuring Systems in Construction

The U. S. construction industry is lagging behind the rest of the world in its failure to embrace metric measurements, now referred to as SI, for the French Systeme International d'Unites. Although we have been extremely slow to adopt the International System of Units, starting in 2000 all U.S. Federal project construction drawings and specifications must be in SI units of measurement. Private work will undoubtedly eventually follow.

The metric measuring system is extremely versatile and simple to manipulate as it is based entirely on decimals, multiples of ten. Only 2 units of length need be used in construction: meters and millimeters. One meter (m) equals 1000 millimeters (mm). (To convert any linear measurement from the English system to metric, the key relationship is 1 inch equals 25.4 millimeters.)

The archaic English system (now called U.S. Customary Units), which we still use, although familiar, is cumbersome and difficult to manipulate. (The mainstream United Kingdom construction industry no longer uses the English system, having abandoned it several years ago in favor of the simpler more universally accepted metric system.)

The smaller subdivisions of feet and inches involve complicated fractions. Various parts of the drawings use different fractions of length units to accommodate the differing practices of the involved trades. For example, all of these various subdivisions of feet and inches will be found on a typical set of construction drawings:

> Feet are divided into *tenths* and *hundredths* for surveying and civil engineering work and inches are not used.

> Cubic yards (27 cubic feet) are used as the basic measuring unit for volumes of earthwork. Fractions of cubic yards are expressed decimally, not as cubic feet.

> Feet are divided into *twelfths* (inches) for the carpentry and general construction trades, but decimal fractions of feet and inches are not used.

> Inches are further divided into *halves*, *quarters*, *eighths*, and *sixteenths* for the carpentry and general construction trades.

> Inches are divided into *thousandths* for the machinery and metal trades.

Engineering and Architectural Scales

Measurement procedures are further complicated in the various scales used on the drawings. The civil engineering portions of the drawings commonly employ the scales of 1 inch equals 10, 20, 30, 40, 50, or 60 feet and 1 inch equals 100, 200, 300, 400, 500, or 600 feet, proceeding further by multiples of 10. The architectural drawings commonly use the scales of 1/16, 3/32, 1/8, 3/16, 1/4, 3/8, 1/2, 3/4, 1, 1-1/2, 3, or 6 inches equals 1 foot and full size. All drafters must have scale rules in all of these denominations.

These engineering and architectural scales are not a direct expression of proportionality to reality. For example, the architectural scale of 1/4 inch equals 1 foot is 1/48th of actual size. The engineering scale of 1 inch equals 20 feet is 1/240th of actual size.

The scales needed for metric drawings are considerably fewer and directly express a proportionality to reality, such as: 1:20, 1:50, or 1:100. The same scale rule can be used for decimal multiples. For example, the 1:20 scale can be used for scales of 1:2, 1:20, 1:200, 1:2000 and so on. 1:20 means simply that 1 unit on the drawing equals 20 units in reality. For example, 1 millimeter on the drawing equals 20 millimeters in reality. The metric scale closest to the familiar 1/4 inch equals one foot is 1:50.

Conversion from One Scale to Another

Frequently, it is found necessary to convert from one of the English systems to another. A commonly encountered problem is to convert from an engineering dimension to an architectural dimension. For example, converting from feet and hundredths (1.75 feet) to feet and inches. (The answer is 1 foot, 9 inches.) That was easy and most people can handle it mentally. But try converting any odd decimal to feet and inches. This will require computation with a calculator or reference to a chart, if one can be found, and to interpret it properly. Architects, engineers, and contractors know how to perform this computation, but the opportunity for embarrassing or costly error is always present.

There are several moderately priced calculators which make the conversions easier. These can be purchased from any architectural supply store or construction bookstore.

Referencing

There must be a system for relating one part of the drawings to another and for knowing where to find needed details. First, to keep one's mind properly oriented, the cardinal directions should be shown on all site and floor plans to be used in identifying faces or parts of the building or site. If the actual directions on the site fall between the cardinal points, an arbitrary reference north is assumed for convenience. It is usually parallel or perpendicular to a principal axis of the building.

All sections and details are keyed into the general drawings by a referencing system that will identify the applicable location in the building as well as to the proper sheet in the drawings.

Drawings versus Specifications

A convention has developed through the years as to the information that should appear on the drawings rather than in the specifications. (Uniform Location of Subject Matter, AIA Document A521, suggests where in the contract documents a particular matter should be properly covered.) The understandings are based on some broad general principles:

1. The drawings should be used to convey information that is most easily and effectively expressed graphically by means of drawings and diagrams. This will include data such as sizes, gauges, proportions, interrelationships, arrangements, locations, and dimensions.
2. The specifications should be used to convey information that is most easily and economically transmitted by words, such as descriptions, standards, guarantees, procedures, and manufacturers' names.
3. Drawings should be used to express *quantity*, while the specifications should describe *quality*.
4. Drawings should be used to denote *genus* (for example, wood) while the specifications will clarify the *species* (for example, Douglas Fir).

There is a notable exception to these fundamental understandings. The building departments of most municipalities will accept only drawings with applications for building permits. They will not accept a book of specifications or a construction contract. All specifications demonstrating compliance with the building code must be on the drawings. For example, a digest of the roofing materials specification must be on the drawings to illustrate compliance with the roofing section of the building code. This is also required of specifications for various structural materials and fire resistive elements.

This required repetition of the same information on drawings and in the specifications creates the possibility of error and inconsistency.

Graphically Illiterate Lawyers, Judges, and Arbitrators

Lawyers who represent segments of the building industry in matters involving interpretation of construction drawings are severely handicapped if they happen to be graphically illiterate. Lack of knowledge of the technical content is not unusual or unexpected and can be compensated for by resort to expert advisors. However, the graphically illiterate lawyer will find it difficult to proceed with confidence and should, at the very least, disclose this condition to prospective clients. Failure to disclose this material fact is not only intellectually dishonest but unfair to the client.

Graphically illiterate judges who preside over litigation involving interpretation of construction drawings are derelict if they fail to disclose this inherent lack to the parties and their counsel, who could then decide whether or not they wish to proceed. Better yet, graphically illiterate judges should voluntarily disqualify themselves from sitting in such matters.

Arbitration panels that include architects, engineers, or contractors will partially compensate for other panel members who may be graphically illiterate, although the better course would be elimination of such arbitrators. However, some arbitrators or judges may be in such high demand for their other favorable qualities that their graphic illiteracy would be waived upon disclosure.

Appendix A:

AIA Document B141--Standard Form of Agreement Between Owner and Architect, 1987 Edition

Appendix

THE AMERICAN INSTITUTE OF ARCHITECTS

Reproduced with permission of The American Institute of Architects under license number #96051. This license expires September 30, 1997. FURTHER REPRODUCTION IS PROHIBITED. Because AIA Documents are revised from time to time, users should ascertain from the AIA the current edition of this document.

Copies of the current edition of this AIA document may be purchased from The American Institute of Architects or its local distributors. The text of this document is not "model language" and is not intended for use in other documents without permission of the AIA.

AIA Document B141

Standard Form of Agreement Between Owner and Architect

1987 EDITION

THIS DOCUMENT HAS IMPORTANT LEGAL CONSEQUENCES; CONSULTATION WITH AN ATTORNEY IS ENCOURAGED WITH RESPECT TO ITS COMPLETION OR MODIFICATION.

SAMPLE

AGREEMENT

made as of the _____ day of _____ in the year of Nineteen Hundred and _____

BETWEEN the Owner:
(Name and address)

and the Architect:
(Name and address)

For the following Project:
(Include detailed description of Project, location, address and scope.)

The Owner and Architect agree as set forth below.

Copyright 1917, 1926, 1948, 1951, 1953, 1958, 1961, 1963, 1966, 1967, 1970, 1974, 1977, ©1987 by The American Institute of Architects, 1735 New York Avenue, N.W., Washington, D.C. 20006. Reproduction of the material herein or substantial quotation of its provisions without written permission of the AIA violates the copyright laws of the United States and will be subject to legal prosecution.

AIA DOCUMENT B141 • OWNER-ARCHITECT AGREEMENT • FOURTEENTH EDITION • AIA® • ©1987
THE AMERICAN INSTITUTE OF ARCHITECTS, 1735 NEW YORK AVENUE, N.W., WASHINGTON, D.C. 20006

B141-1987

TERMS AND CONDITIONS OF AGREEMENT BETWEEN OWNER AND ARCHITECT

ARTICLE 1
ARCHITECT'S RESPONSIBILITIES

1.1 ARCHITECT'S SERVICES

1.1.1 The Architect's services consist of those services performed by the Architect, Architect's employees and Architect's consultants as enumerated in Articles 2 and 3 of this Agreement and any other services included in Article 12.

1.1.2 The Architect's services shall be performed as expeditiously as is consistent with professional skill and care and the orderly progress of the Work. Upon request of the Owner, the Architect shall submit for the Owner's approval a schedule for the performance of the Architect's services which may be adjusted as the Project proceeds, and shall include allowances for periods of time required for the Owner's review and for approval of submissions by authorities having jurisdiction over the Project. Time limits established by this schedule approved by the Owner shall not, except for reasonable cause, be exceeded by the Architect or Owner.

1.1.3 The services covered by this Agreement are subject to the time limitations contained in Subparagraph 11.5.1.

ARTICLE 2
SCOPE OF ARCHITECT'S BASIC SERVICES

2.1 DEFINITION

2.1.1 The Architect's Basic Services consist of those described in Paragraphs 2.2 through 2.6 and any other services identified in Article 12 as part of Basic Services, and include normal structural, mechanical and electrical engineering services.

2.2 SCHEMATIC DESIGN PHASE

2.2.1 The Architect shall review the program furnished by the Owner to ascertain the requirements of the Project and shall arrive at a mutual understanding of such requirements with the Owner.

2.2.2 The Architect shall provide a preliminary evaluation of the Owner's program, schedule and construction budget requirements, each in terms of the other, subject to the limitations set forth in Subparagraph 5.2.1.

2.2.3 The Architect shall review with the Owner alternative approaches to design and construction of the Project.

2.2.4 Based on the mutually agreed-upon program, schedule and construction budget requirements, the Architect shall prepare, for approval by the Owner, Schematic Design Documents consisting of drawings and other documents illustrating the scale and relationship of Project components.

2.2.5 The Architect shall submit to the Owner a preliminary estimate of Construction Cost based on current area, volume or other unit costs.

2.3 DESIGN DEVELOPMENT PHASE

2.3.1 Based on the approved Schematic Design Documents and any adjustments authorized by the Owner in the program, schedule or construction budget, the Architect shall prepare, for approval by the Owner, Design Development Documents consisting of drawings and other documents to fix and describe the size and character of the Project as to architectural, structural, mechanical and electrical systems, materials and such other elements as may be appropriate.

2.3.2 The Architect shall advise the Owner of any adjustments to the preliminary estimate of Construction Cost.

2.4 CONSTRUCTION DOCUMENTS PHASE

2.4.1 Based on the approved Design Development Documents and any further adjustments in the scope or quality of the Project or in the construction budget authorized by the Owner, the Architect shall prepare, for approval by the Owner, Construction Documents consisting of Drawings and Specifications setting forth in detail the requirements for the construction of the Project.

2.4.2 The Architect shall assist the Owner in the preparation of the necessary bidding information, bidding forms, the Conditions of the Contract, and the form of Agreement between the Owner and Contractor.

2.4.3 The Architect shall advise the Owner of any adjustments to previous preliminary estimates of Construction Cost indicated by changes in requirements or general market conditions.

2.4.4 The Architect shall assist the Owner in connection with the Owner's responsibility for filing documents required for the approval of governmental authorities having jurisdiction over the Project.

2.5 BIDDING OR NEGOTIATION PHASE

2.5.1 The Architect, following the Owner's approval of the Construction Documents and of the latest preliminary estimate of Construction Cost, shall assist the Owner in obtaining bids or negotiated proposals and assist in awarding and preparing contracts for construction.

2.6 CONSTRUCTION PHASE—ADMINISTRATION OF THE CONSTRUCTION CONTRACT

2.6.1 The Architect's responsibility to provide Basic Services for the Construction Phase under this Agreement commences with the award of the Contract for Construction and terminates at the earlier of the issuance to the Owner of the final Certificate for Payment or 60 days after the date of Substantial Completion of the Work, unless extended under the terms of Subparagraph 10.3.3.

2.6.2 The Architect shall provide administration of the Contract for Construction as set forth below and in the edition of AIA Document A201, General Conditions of the Contract for Construction, current as of the date of this Agreement, unless otherwise provided in this Agreement.

2.6.3 Duties, responsibilities and limitations of authority of the Architect shall not be restricted, modified or extended without written agreement of the Owner and Architect with consent of the Contractor, which consent shall not be unreasonably withheld.

2.6.4 The Architect shall be a representative of and shall advise and consult with the Owner (1) during construction until final payment to the Contractor is due, and (2) as an Additional Service at the Owner's direction from time to time during the correction period described in the Contract for Construction. The Architect shall have authority to act on behalf of the Owner only to the extent provided in this Agreement unless otherwise modified by written instrument.

2.6.5 The Architect shall visit the site at intervals appropriate to the stage of construction or as otherwise agreed by the Owner and Architect in writing to become generally familiar with the progress and quality of the Work completed and to determine in general if the Work is being performed in a manner indicating that the Work when completed will be in accordance with the Contract Documents. However, the Architect shall not be required to make exhaustive or continuous on-site inspections to check the quality or quantity of the Work. On the basis of on-site observations as an architect, the Architect shall keep the Owner informed of the progress and quality of the Work, and shall endeavor to guard the Owner against defects and deficiencies in the Work. *(More extensive site representation may be agreed to as an Additional Service, as described in Paragraph 3.2.)*

2.6.6 The Architect shall not have control over or charge of and shall not be responsible for construction means, methods, techniques, sequences or procedures, or for safety precautions and programs in connection with the Work, since these are solely the Contractor's responsibility under the Contract for Construction. The Architect shall not be responsible for the Contractor's schedules or failure to carry out the Work in accordance with the Contract Documents. The Architect shall not have control over or charge of acts or omissions of the Contractor, Subcontractors, or their agents or employees, or of any other persons performing portions of the Work.

2.6.7 The Architect shall at all times have access to the Work wherever it is in preparation or progress.

2.6.8 Except as may otherwise be provided in the Contract Documents or when direct communications have been specially authorized, the Owner and Contractor shall communicate through the Architect. Communications by and with the Architect's consultants shall be through the Architect.

2.6.9 Based on the Architect's observations and evaluations of the Contractor's Applications for Payment, the Architect shall review and certify the amounts due the Contractor.

2.6.10 The Architect's certification for payment shall constitute a representation to the Owner, based on the Architect's observations at the site as provided in Subparagraph 2.6.5 and on the data comprising the Contractor's Application for Payment, that the Work has progressed to the point indicated and that, to the best of the Architect's knowledge, information and belief, quality of the Work is in accordance with the Contract Documents. The foregoing representations are subject to an evaluation of the Work for conformance with the Contract Documents upon Substantial Completion, to results of subsequent tests and inspections, to minor deviations from the Contract Documents correctable prior to completion and to specific qualifications expressed by the Architect. The issuance of a Certificate for Payment shall further constitute a representation that the Contractor is entitled to payment in the amount certified. However, the issuance of a Certificate for Payment shall not be a representation that the Architect has (1) made exhaustive or continuous on-site inspections to check the quality or quantity of the Work, (2) reviewed construction means, methods, techniques, sequences or procedures, (3) reviewed copies of requisitions received from Subcontractors and material suppliers and other data requested by the Owner to substantiate the Contractor's right to payment or (4) ascertained how or for what purpose the Contractor has used money previously paid on account of the Contract Sum.

2.6.11 The Architect shall have authority to reject Work which does not conform to the Contract Documents. Whenever the Architect considers it necessary or advisable for implementation of the intent of the Contract Documents, the Architect will have authority to require additional inspection or testing of the Work in accordance with the provisions of the Contract Documents, whether or not such Work is fabricated, installed or completed. However, neither this authority of the Architect nor a decision made in good faith either to exercise or not to exercise such authority shall give rise to a duty or responsibility of the Architect to the Contractor, Subcontractors, material and equipment suppliers, their agents or employees or other persons performing portions of the Work.

2.6.12 The Architect shall review and approve or take other appropriate action upon Contractor's submittals such as Shop Drawings, Product Data and Samples, but only for the limited purpose of checking for conformance with information given and the design concept expressed in the Contract Documents. The Architect's action shall be taken with such reasonable promptness as to cause no delay in the Work or in the construction of the Owner or of separate contractors, while allowing sufficient time in the Architect's professional judgment to permit adequate review. Review of such submittals is not conducted for the purpose of determining the accuracy and completeness of other details such as dimensions and quantities or for substantiating instructions for installation or performance of equipment or systems designed by the Contractor, all of which remain the responsibility of the Contractor to the extent required by the Contract Documents. The Architect's review shall not constitute approval of safety precautions or, unless otherwise specifically stated by the Architect, of construction means, methods, techniques, sequences or procedures. The Architect's approval of a specific item shall not indicate approval of an assembly of which the item is a component. When professional certification of performance characteristics of materials, systems or equipment is required by the Contract Documents, the Architect shall be entitled to rely upon such certification to establish that the materials, systems or equipment will meet the performance criteria required by the Contract Documents.

2.6.13 The Architect shall prepare Change Orders and Construction Change Directives, with supporting documentation and data if deemed necessary by the Architect as provided in Subparagraphs 3.1.1 and 3.3.3, for the Owner's approval and execution in accordance with the Contract Documents, and may authorize minor changes in the Work not involving an adjustment in the Contract Sum or an extension of the Contract Time which are not inconsistent with the intent of the Contract Documents.

2.6.14 The Architect shall conduct inspections to determine the date or dates of Substantial Completion and the date of final completion, shall receive and forward to the Owner for the Owner's review and records written warranties and related documents required by the Contract Documents and assembled by the Contractor, and shall issue a final Certificate for Payment upon compliance with the requirements of the Contract Documents.

2.6.15 The Architect shall interpret and decide matters concerning performance of the Owner and Contractor under the requirements of the Contract Documents on written request of either the Owner or Contractor. The Architect's response to such requests shall be made with reasonable promptness and within any time limits agreed upon.

2.6.16 Interpretations and decisions of the Architect shall be consistent with the intent of and reasonably inferable from the Contract Documents and shall be in writing or in the form of drawings. When making such interpretations and initial decisions, the Architect shall endeavor to secure faithful performance by both Owner and Contractor, shall not show partiality to either, and shall not be liable for results of interpretations or decisions so rendered in good faith.

2.6.17 The Architect's decisions on matters relating to aesthetic effect shall be final if consistent with the intent expressed in the Contract Documents.

2.6.18 The Architect shall render written decisions within a reasonable time on all claims, disputes or other matters in question between the Owner and Contractor relating to the execution or progress of the Work as provided in the Contract Documents.

2.6.19 The Architect's decisions on claims, disputes or other matters, including those in question between the Owner and Contractor, except for those relating to aesthetic effect as provided in Subparagraph 2.6.17, shall be subject to arbitration as provided in this Agreement and in the Contract Documents.

ARTICLE 3
ADDITIONAL SERVICES

3.1 GENERAL

3.1.1 The services described in this Article 3 are not included in Basic Services unless so identified in Article 12, and they shall be paid for by the Owner as provided in this Agreement, in addition to the compensation for Basic Services. The services described under Paragraphs 3.2 and 3.4 shall only be provided if authorized or confirmed in writing by the Owner. If services described under Contingent Additional Services in Paragraph 3.3 are required due to circumstances beyond the Architect's control, the Architect shall notify the Owner prior to commencing such services. If the Owner deems that such services described under Paragraph 3.3 are not required, the Owner shall give prompt written notice to the Architect. If the Owner indicates in writing that all or part of such Contingent Additional Services are not required, the Architect shall have no obligation to provide those services.

3.2 PROJECT REPRESENTATION BEYOND BASIC SERVICES

3.2.1 If more extensive representation at the site than is described in Subparagraph 2.6.5 is required, the Architect shall provide one or more Project Representatives to assist in carrying out such additional on-site responsibilities.

3.2.2 Project Representatives shall be selected, employed and directed by the Architect, and the Architect shall be compensated therefor as agreed by the Owner and Architect. The duties, responsibilities and limitations of authority of Project Representatives shall be as described in the edition of AIA Document B352 current as of the date of this Agreement, unless otherwise agreed.

3.2.3 Through the observations by such Project Representatives, the Architect shall endeavor to provide further protection for the Owner against defects and deficiencies in the Work, but the furnishing of such project representation shall not modify the rights, responsibilities or obligations of the Architect as described elsewhere in this Agreement.

3.3 CONTINGENT ADDITIONAL SERVICES

3.3.1 Making revisions in Drawings, Specifications or other documents when such revisions are:

.1 inconsistent with approvals or instructions previously given by the Owner, including revisions made necessary by adjustments in the Owner's program or Project budget;

.2 required by the enactment or revision of codes, laws or regulations subsequent to the preparation of such documents; or

.3 due to changes required as a result of the Owner's failure to render decisions in a timely manner.

3.3.2 Providing services required because of significant changes in the Project including, but not limited to, size, quality, complexity, the Owner's schedule, or the method of bidding or negotiating and contracting for construction, except for services required under Subparagraph 5.2.5.

3.3.3 Preparing Drawings, Specifications and other documentation and supporting data, evaluating Contractor's proposals, and providing other services in connection with Change Orders and Construction Change Directives.

3.3.4 Providing services in connection with evaluating substitutions proposed by the Contractor and making subsequent revisions to Drawings, Specifications and other documentation resulting therefrom.

3.3.5 Providing consultation concerning replacement of Work damaged by fire or other cause during construction, and furnishing services required in connection with the replacement of such Work.

3.3.6 Providing services made necessary by the default of the Contractor, by major defects or deficiencies in the Work of the Contractor, or by failure of performance of either the Owner or Contractor under the Contract for Construction.

3.3.7 Providing services in evaluating an extensive number of claims submitted by the Contractor or others in connection with the Work.

3.3.8 Providing services in connection with a public hearing, arbitration proceeding or legal proceeding except where the Architect is party thereto.

3.3.9 Preparing documents for alternate, separate or sequential bids or providing services in connection with bidding, negotiation or construction prior to the completion of the Construction Documents Phase.

3.4 OPTIONAL ADDITIONAL SERVICES

3.4.1 Providing analyses of the Owner's needs and programming the requirements of the Project.

3.4.2 Providing financial feasibility or other special studies.

3.4.3 Providing planning surveys, site evaluations or comparative studies of prospective sites.

3.4.4 Providing special surveys, environmental studies and submissions required for approvals of governmental authorities or others having jurisdiction over the Project.

3.4.5 Providing services relative to future facilities, systems and equipment.

3.4.6 Providing services to investigate existing conditions or facilities or to make measured drawings thereof.

3.4.7 Providing services to verify the accuracy of drawings or other information furnished by the Owner.

3.4.8 Providing coordination of construction performed by separate contractors or by the Owner's own forces and coordination of services required in connection with construction performed and equipment supplied by the Owner.

3.4.9 Providing services in connection with the work of a construction manager or separate consultants retained by the Owner.

3.4.10 Providing detailed estimates of Construction Cost.

3.4.11 Providing detailed quantity surveys or inventories of material, equipment and labor.

3.4.12 Providing analyses of owning and operating costs.

3.4.13 Providing interior design and other similar services required for or in connection with the selection, procurement or installation of furniture, furnishings and related equipment.

3.4.14 Providing services for planning tenant or rental spaces.

3.4.15 Making investigations, inventories of materials or equipment, or valuations and detailed appraisals of existing facilities.

3.4.16 Preparing a set of reproducible record drawings showing significant changes in the Work made during construction based on marked-up prints, drawings and other data furnished by the Contractor to the Architect.

3.4.17 Providing assistance in the utilization of equipment or systems such as testing, adjusting and balancing, preparation of operation and maintenance manuals, training personnel for operation and maintenance, and consultation during operation.

3.4.18 Providing services after issuance to the Owner of the final Certificate for Payment, or in the absence of a final Certificate for Payment, more than 60 days after the date of Substantial Completion of the Work.

3.4.19 Providing services of consultants for other than architectural, structural, mechanical and electrical engineering portions of the Project provided as a part of Basic Services.

3.4.20 Providing any other services not otherwise included in this Agreement or not customarily furnished in accordance with generally accepted architectural practice.

ARTICLE 4
OWNER'S RESPONSIBILITIES

4.1 The Owner shall provide full information regarding requirements for the Project, including a program which shall set forth the Owner's objectives, schedule, constraints and criteria, including space requirements and relationships, flexibility, expandability, special equipment, systems and site requirements.

4.2 The Owner shall establish and update an overall budget for the Project, including the Construction Cost, the Owner's other costs and reasonable contingencies related to all of these costs.

4.3 If requested by the Architect, the Owner shall furnish evidence that financial arrangements have been made to fulfill the Owner's obligations under this Agreement.

4.4 The Owner shall designate a representative authorized to act on the Owner's behalf with respect to the Project. The Owner or such authorized representative shall render decisions in a timely manner pertaining to documents submitted by the Architect in order to avoid unreasonable delay in the orderly and sequential progress of the Architect's services.

4.5 The Owner shall furnish surveys describing physical characteristics, legal limitations and utility locations for the site of the Project, and a written legal description of the site. The surveys and legal information shall include, as applicable, grades and lines of streets, alleys, pavements and adjoining property and structures; adjacent drainage; rights-of-way, restrictions, easements, encroachments, zoning, deed restrictions, boundaries and contours of the site; locations, dimensions and necessary data pertaining to existing buildings, other improvements and trees; and information concerning available utility services and lines, both public and private, above and below grade, including inverts and depths. All the information on the survey shall be referenced to a project benchmark.

4.6 The Owner shall furnish the services of geotechnical engineers when such services are requested by the Architect. Such services may include but are not limited to test borings, test pits, determinations of soil bearing values, percolation tests, evaluations of hazardous materials, ground corrosion and resistivity tests, including necessary operations for anticipating subsoil conditions, with reports and appropriate professional recommendations.

4.6.1 The Owner shall furnish the services of other consultants when such services are reasonably required by the scope of the Project and are requested by the Architect.

4.7 The Owner shall furnish structural, mechanical, chemical, air and water pollution tests, tests for hazardous materials, and other laboratory and environmental tests, inspections and reports required by law or the Contract Documents.

4.8 The Owner shall furnish all legal, accounting and insurance counseling services as may be necessary at any time for the Project, including auditing services the Owner may require to verify the Contractor's Applications for Payment or to ascertain how or for what purposes the Contractor has used the money paid by or on behalf of the Owner.

4.9 The services, information, surveys and reports required by Paragraphs 4.5 through 4.8 shall be furnished at the Owner's expense, and the Architect shall be entitled to rely upon the accuracy and completeness thereof.

4.10 Prompt written notice shall be given by the Owner to the Architect if the Owner becomes aware of any fault or defect in the Project or nonconformance with the Contract Documents.

4.11 The proposed language of certificates or certifications requested of the Architect or Architect's consultants shall be submitted to the Architect for review and approval at least 14 days prior to execution. The Owner shall not request certifications that would require knowledge or services beyond the scope of this Agreement.

ARTICLE 5
CONSTRUCTION COST

5.1 DEFINITION

5.1.1 The Construction Cost shall be the total cost or estimated cost to the Owner of all elements of the Project designed or specified by the Architect.

5.1.2 The Construction Cost shall include the cost at current market rates of labor and materials furnished by the Owner and equipment designed, specified, selected or specially provided for by the Architect, plus a reasonable allowance for the Contractor's overhead and profit. In addition, a reasonable allowance for contingencies shall be included for market conditions at the time of bidding and for changes in the Work during construction.

5.1.3 Construction Cost does not include the compensation of the Architect and Architect's consultants, the costs of the land, rights-of-way, financing or other costs which are the responsibility of the Owner as provided in Article 4.

5.2 RESPONSIBILITY FOR CONSTRUCTION COST

5.2.1 Evaluations of the Owner's Project budget, preliminary estimates of Construction Cost and detailed estimates of Construction Cost, if any, prepared by the Architect, represent the Architect's best judgment as a design professional familiar with the construction industry. It is recognized, however, that neither the Architect nor the Owner has control over the cost of labor, materials or equipment, over the Contractor's methods of determining bid prices, or over competitive bidding, market or negotiating conditions. Accordingly, the Architect cannot and does not warrant or represent that bids or negotiated prices will not vary from the Owner's Project budget or from any estimate of Construction Cost or evaluation prepared or agreed to by the Architect.

5.2.2 No fixed limit of Construction Cost shall be established as a condition of this Agreement by the furnishing, proposal or establishment of a Project budget, unless such fixed limit has been agreed upon in writing and signed by the parties hereto. If such a fixed limit has been established, the Architect shall be permitted to include contingencies for design, bidding and price escalation, to determine what materials, equipment, component systems and types of construction are to be included in the Contract Documents, to make reasonable adjustments in the scope of the Project and to include in the Contract Documents alternate bids to adjust the Construction Cost to the fixed limit. Fixed limits, if any, shall be increased in the amount of an increase in the Contract Sum occurring after execution of the Contract for Construction.

5.2.3 If the Bidding or Negotiation Phase has not commenced within 90 days after the Architect submits the Construction Documents to the Owner, any Project budget or fixed limit of Construction Cost shall be adjusted to reflect changes in the general level of prices in the construction industry between the date of submission of the Construction Documents to the Owner and the date on which proposals are sought.

5.2.4 If a fixed limit of Construction Cost (adjusted as provided in Subparagraph 5.2.3) is exceeded by the lowest bona fide bid or negotiated proposal, the Owner shall:

.1 give written approval of an increase in such fixed limit;

.2 authorize rebidding or renegotiating of the Project within a reasonable time;

.3 if the Project is abandoned, terminate in accordance with Paragraph 8.3; or

.4 cooperate in revising the Project scope and quality as required to reduce the Construction Cost.

5.2.5 If the Owner chooses to proceed under Clause 5.2.4.4, the Architect, without additional charge, shall modify the Contract Documents as necessary to comply with the fixed limit, if established as a condition of this Agreement. The modification of Contract Documents shall be the limit of the Architect's responsibility arising out of the establishment of a fixed limit. The Architect shall be entitled to compensation in accordance with this Agreement for all services performed whether or not the Construction Phase is commenced.

ARTICLE 6
USE OF ARCHITECT'S DRAWINGS, SPECIFICATIONS AND OTHER DOCUMENTS

6.1 The Drawings, Specifications and other documents prepared by the Architect for this Project are instruments of the Architect's service for use solely with respect to this Project and, unless otherwise provided, the Architect shall be deemed the author of these documents and shall retain all common law, statutory and other reserved rights, including the copyright. The Owner shall be permitted to retain copies, including reproducible copies, of the Architect's Drawings, Specifications and other documents for information and reference in connection with the Owner's use and occupancy of the Project. The Architect's Drawings, Specifications or other documents shall not be used by the Owner or others on other projects, for additions to this Project or for completion of this Project by others, unless the Architect is adjudged to be in default under this Agreement, except by agreement in writing and with appropriate compensation to the Architect.

6.2 Submission or distribution of documents to meet official regulatory requirements or for similar purposes in connection with the Project is not to be construed as publication in derogation of the Architect's reserved rights.

ARTICLE 7
ARBITRATION

7.1 Claims, disputes or other matters in question between the parties to this Agreement arising out of or relating to this Agreement or breach thereof shall be subject to and decided by arbitration in accordance with the Construction Industry Arbitration Rules of the American Arbitration Association currently in effect unless the parties mutually agree otherwise.

7.2 Demand for arbitration shall be filed in writing with the other party to this Agreement and with the American Arbitration Association. A demand for arbitration shall be made within a reasonable time after the claim, dispute or other matter in question has arisen. In no event shall the demand for arbitration be made after the date when institution of legal or equitable proceedings based on such claim, dispute or other matter in question would be barred by the applicable statutes of limitations.

7.3 No arbitration arising out of or relating to this Agreement shall include, by consolidation, joinder or in any other manner, an additional person or entity not a party to this Agreement, except by written consent containing a specific reference to this Agreement signed by the Owner, Architect, and any other person or entity sought to be joined. Consent to arbitration involving an additional person or entity shall not constitute consent to arbitration of any claim, dispute or other matter in question not described in the written consent or with a person or entity not named or described therein. The foregoing agreement to arbitrate and other agreements to arbitrate with an additional person or entity duly consented to by the parties to this Agreement shall be specifically enforceable in accordance with applicable law in any court having jurisdiction thereof.

7.4 The award rendered by the arbitrator or arbitrators shall be final, and judgment may be entered upon it in accordance with applicable law in any court having jurisdiction thereof.

ARTICLE 8
TERMINATION, SUSPENSION OR ABANDONMENT

8.1 This Agreement may be terminated by either party upon not less than seven days' written notice should the other party fail substantially to perform in accordance with the terms of this Agreement through no fault of the party initiating the termination.

8.2 If the Project is suspended by the Owner for more than 30 consecutive days, the Architect shall be compensated for services performed prior to notice of such suspension. When the Project is resumed, the Architect's compensation shall be equitably adjusted to provide for expenses incurred in the interruption and resumption of the Architect's services.

8.3 This Agreement may be terminated by the Owner upon not less than seven days' written notice to the Architect in the event that the Project is permanently abandoned. If the Project is abandoned by the Owner for more than 90 consecutive days, the Architect may terminate this Agreement by giving written notice.

8.4 Failure of the Owner to make payments to the Architect in accordance with this Agreement shall be considered substantial nonperformance and cause for termination.

8.5 If the Owner fails to make payment when due the Architect for services and expenses, the Architect may, upon seven days' written notice to the Owner, suspend performance of services under this Agreement. Unless payment in full is received by the Architect within seven days of the date of the notice, the suspension shall take effect without further notice. In the event of a suspension of services, the Architect shall have no liability to the Owner for delay or damage caused the Owner because of such suspension of services.

8.6 In the event of termination not the fault of the Architect, the Architect shall be compensated for services performed prior to termination, together with Reimbursable Expenses then due and all Termination Expenses as defined in Paragraph 8.7.

8.7 Termination Expenses are in addition to compensation for Basic and Additional Services, and include expenses which are directly attributable to termination. Termination Expenses shall be computed as a percentage of the total compensation for Basic Services and Additional Services earned to the time of termination, as follows:

.1 Twenty percent of the total compensation for Basic and Additional Services earned to date if termination occurs before or during the predesign, site analysis, or Schematic Design Phases; or

.2 Ten percent of the total compensation for Basic and Additional Services earned to date if termination occurs during the Design Development Phase; or

.3 Five percent of the total compensation for Basic and Additional Services earned to date if termination occurs during any subsequent phase.

ARTICLE 9
MISCELLANEOUS PROVISIONS

9.1 Unless otherwise provided, this Agreement shall be governed by the law of the principal place of business of the Architect.

9.2 Terms in this Agreement shall have the same meaning as those in AIA Document A201, General Conditions of the Contract for Construction, current as of the date of this Agreement.

9.3 Causes of action between the parties to this Agreement pertaining to acts or failures to act shall be deemed to have accrued and the applicable statutes of limitations shall commence to run not later than either the date of Substantial Completion for acts or failures to act occurring prior to Substantial Completion, or the date of issuance of the final Certificate for Payment for acts or failures to act occurring after Substantial Completion.

9.4 The Owner and Architect waive all rights against each other and against the contractors, consultants, agents and employees of the other for damages, but only to the extent covered by property insurance during construction, except such rights as they may have to the proceeds of such insurance as set forth in the edition of AIA Document A201, General Conditions of the Contract for Construction, current as of the date of this Agreement. The Owner and Architect each shall require similar waivers from their contractors, consultants and agents.

9.5 The Owner and Architect, respectively, bind themselves, their partners, successors, assigns and legal representatives to the other party to this Agreement and to the partners, successors, assigns and legal representatives of such other party with respect to all covenants of this Agreement. Neither Owner nor Architect shall assign this Agreement without the written consent of the other.

9.6 This Agreement represents the entire and integrated agreement between the Owner and Architect and supersedes all prior negotiations, representations or agreements, either written or oral. This Agreement may be amended only by written instrument signed by both Owner and Architect.

9.7 Nothing contained in this Agreement shall create a contractual relationship with or a cause of action in favor of a third party against either the Owner or Architect.

9.8 Unless otherwise provided in this Agreement, the Architect and Architect's consultants shall have no responsibility for the discovery, presence, handling, removal or disposal of or exposure of persons to hazardous materials in any form at the Project site, including but not limited to asbestos, asbestos products, polychlorinated biphenyl (PCB) or other toxic substances.

9.9 The Architect shall have the right to include representations of the design of the Project, including photographs of the exterior and interior, among the Architect's promotional and professional materials. The Architect's materials shall not include the Owner's confidential or proprietary information if the Owner has previously advised the Architect in writing of the specific information considered by the Owner to be confidential or proprietary. The Owner shall provide professional credit for the Architect on the construction sign and in the promotional materials for the Project.

ARTICLE 10
PAYMENTS TO THE ARCHITECT

10.1 DIRECT PERSONNEL EXPENSE

10.1.1 Direct Personnel Expense is defined as the direct salaries of the Architect's personnel engaged on the Project and the portion of the cost of their mandatory and customary contributions and benefits related thereto, such as employment taxes and other statutory employee benefits, insurance, sick leave, holidays, vacations, pensions and similar contributions and benefits.

10.2 REIMBURSABLE EXPENSES

10.2.1 Reimbursable Expenses are in addition to compensation for Basic and Additional Services and include expenses incurred by the Architect and Architect's employees and consultants in the interest of the Project, as identified in the following Clauses.

10.2.1.1 Expense of transportation in connection with the Project; expenses in connection with authorized out-of-town travel; long-distance communications; and fees paid for securing approval of authorities having jurisdiction over the Project.

10.2.1.2 Expense of reproductions, postage and handling of Drawings, Specifications and other documents.

10.2.1.3 If authorized in advance by the Owner, expense of overtime work requiring higher than regular rates.

10.2.1.4 Expense of renderings, models and mock-ups requested by the Owner.

10.2.1.5 Expense of additional insurance coverage or limits, including professional liability insurance, requested by the Owner in excess of that normally carried by the Architect and Architect's consultants.

10.2.1.6 Expense of computer-aided design and drafting equipment time when used in connection with the Project.

10.3 PAYMENTS ON ACCOUNT OF BASIC SERVICES

10.3.1 An initial payment as set forth in Paragraph 11.1 is the minimum payment under this Agreement.

10.3.2 Subsequent payments for Basic Services shall be made monthly and, where applicable, shall be in proportion to services performed within each phase of service, on the basis set forth in Subparagraph 11.2.2.

10.3.3 If and to the extent that the time initially established in Subparagraph 11.5.1 of this Agreement is exceeded or extended through no fault of the Architect, compensation for any services rendered during the additional period of time shall be computed in the manner set forth in Subparagraph 11.3.2.

10.3.4 When compensation is based on a percentage of Construction Cost and any portions of the Project are deleted or otherwise not constructed, compensation for those portions of the Project shall be payable to the extent services are performed on those portions, in accordance with the schedule set forth in Subparagraph 11.2.2, based on (1) the lowest bona fide bid or negotiated proposal, or (2) if no such bid or proposal is received, the most recent preliminary estimate of Construction Cost or detailed estimate of Construction Cost for such portions of the Project.

10.4 PAYMENTS ON ACCOUNT OF ADDITIONAL SERVICES

10.4.1 Payments on account of the Architect's Additional Services and for Reimbursable Expenses shall be made monthly upon presentation of the Architect's statement of services rendered or expenses incurred.

10.5 PAYMENTS WITHHELD

10.5.1 No deductions shall be made from the Architect's compensation on account of penalty, liquidated damages or other sums withheld from payments to contractors, or on account of the cost of changes in the Work other than those for which the Architect has been found to be liable.

10.6 ARCHITECT'S ACCOUNTING RECORDS

10.6.1 Records of Reimbursable Expenses and expenses pertaining to Additional Services and services performed on the basis of a multiple of Direct Personnel Expense shall be available to the Owner or the Owner's authorized representative at mutually convenient times.

ARTICLE 11
BASIS OF COMPENSATION

The Owner shall compensate the Architect as follows:

11.1 AN INITIAL PAYMENT of   Dollars ($
shall be made upon execution of this Agreement and credited to the Owner's account at final payment.

11.2 BASIC COMPENSATION

11.2.1 FOR BASIC SERVICES, as described in Article 2, and any other services included in Article 12 as part of Basic Services, Basic Compensation shall be computed as follows:

(Insert basis of compensation, including stipulated sums, multiples or percentages, and identify phases to which particular methods of compensation apply, if necessary.)

Appendix

11.2.2 Where compensation is based on a stipulated sum or percentage of Construction Cost, progress payments for Basic Services in each phase shall total the following percentages of the total Basic Compensation payable:

(Insert additional phases as appropriate.)

Schematic Design Phase:	percent (%)
Design Development Phase:	percent (%)
Construction Documents Phase:	percent (%)
Bidding or Negotiation Phase:	percent (%)
Construction Phase:	percent (%)
Total Basic Compensation:	one hundred percent (100%)

11.3 COMPENSATION FOR ADDITIONAL SERVICES

11.3.1 FOR PROJECT REPRESENTATION BEYOND BASIC SERVICES, as described in Paragraph 3.2, compensation shall be computed as follows:

11.3.2 FOR ADDITIONAL SERVICES OF THE ARCHITECT, as described in Articles 3 and 12, other than (1) Additional Project Representation, as described in Paragraph 3.2, and (2) services included in Article 12 as part of Additional Services, but excluding services of consultants, compensation shall be computed as follows:

(Insert basis of compensation, including rates and/or multiples of Direct Personnel Expense for Principals and employees, and identify Principals and classify employees, if required. Identify specific services to which particular methods of compensation apply, if necessary.)

11.3.3 FOR ADDITIONAL SERVICES OF CONSULTANTS, including additional structural, mechanical and electrical engineering services and those provided under Subparagraph 3.4.19 or identified in Article 12 as part of Additional Services, a multiple of ( ) times the amounts billed to the Architect for such services.

(Identify specific types of consultants in Article 12, if required.)

11.4 REIMBURSABLE EXPENSES

11.4.1 FOR REIMBURSABLE EXPENSES, as described in Paragraph 10.2, and any other items included in Article 12 as Reimbursable Expenses, a multiple of ( ) times the expenses incurred by the Architect, the Architect's employees and consultants in the interest of the Project.

11.5 ADDITIONAL PROVISIONS

11.5.1 IF THE BASIC SERVICES covered by this Agreement have not been completed within ( ) months of the date hereof, through no fault of the Architect, extension of the Architect's services beyond that time shall be compensated as provided in Subparagraphs 10.3.3 and 11.3.2.

11.5.2 Payments are due and payable ( ) days from the date of the Architect's invoice. Amounts unpaid ( ) days after the invoice date shall bear interest at the rate entered below, or in the absence thereof at the legal rate prevailing from time to time at the principal place of business of the Architect.

(Insert rate of interest agreed upon.)

(Usury laws and requirements under the Federal Truth in Lending Act, similar state and local consumer credit laws and other regulations at the Owner's and Architect's principal places of business, the location of the Project and elsewhere may affect the validity of this provision. Specific legal advice should be obtained with respect to deletions or modifications, and also regarding requirements such as written disclosures or waivers.)

11.5.3 The rates and multiples set forth for Additional Services shall be annually adjusted in accordance with normal salary review practices of the Architect.

ARTICLE 12
OTHER CONDITIONS OR SERVICES

(Insert descriptions of other services, identify Additional Services included within Basic Compensation and modifications to the payment and compensation terms included in this Agreement.)

This Agreement entered into as of the day and year first written above.

OWNER

ARCHITECT

_____

(Signature)

_____

(Printed name and title)

_____

(Signature)

_____

(Printed name and title)

AIA DOCUMENT B141 • OWNER-ARCHITECT AGREEMENT • FOURTEENTH EDITION • AIA® • ©1987
THE AMERICAN INSTITUTE OF ARCHITECTS, 1735 NEW YORK AVENUE, N.W., WASHINGTON, D.C. 20006

B141-1987 10

Appendix B:

AIA Document A101--Standard Form of Agreement Between Owner and Contractor (where the basis of payment is a Stipulated Sum), 1987 Edition

THE AMERICAN INSTITUTE OF ARCHITECTS

Reproduced with permission of The American Institute of Architects under license number #96051. This license expires September 30, 1997. FURTHER REPRODUCTION IS PROHIBITED. Because AIA Documents are revised from time to time, users should ascertain from the AIA the current edition of this document.

Copies of the current edition of this AIA document may be purchased from The American Institute of Architects or its local distributors. The text of this document is not "model language" and is not intended for use in other documents without permission of the AIA.

AIA Document A101

Standard Form of Agreement Between Owner and Contractor

where the basis of payment is a
STIPULATED SUM

1987 EDITION

THIS DOCUMENT HAS IMPORTANT LEGAL CONSEQUENCES; CONSULTATION WITH AN ATTORNEY IS ENCOURAGED WITH RESPECT TO ITS COMPLETION OR MODIFICATION.
The 1987 Edition of AIA Document A201, General Conditions of the Contract for Construction, is adopted in this document by reference. Do not use with other general conditions unless this document is modified.
This document has been approved and endorsed by The Associated General Contractors of America.

AGREEMENT

made as of the   day of   in the year of
Nineteen Hundred and

BETWEEN the Owner:
(Name and address)

and the Contractor:
(Name and address)

The Project is:
(Name and location)

The Architect is:
(Name and address)

The Owner and Contractor agree as set forth below.

Copyright 1915, 1918, 1925, 1937, 1951, 1958, 1961, 1963, 1967, 1974, 1977, ©1987 by The American Institute of Architects, 1735 New York Avenue, N.W., Washington, D.C. 20006. Reproduction of the material herein or substantial quotation of its provisions without written permission of the AIA violates the copyright laws of the United States and will be subject to legal prosecution.

ARTICLE 1
THE CONTRACT DOCUMENTS

The Contract Documents consist of this Agreement, Conditions of the Contract (General, Supplementary and other Conditions), Drawings, Specifications, Addenda issued prior to execution of this Agreement, other documents listed in this Agreement and Modifications issued after execution of this Agreement; these form the Contract, and are as fully a part of the Contract as if attached to this Agreement or repeated herein. The Contract represents the entire and integrated agreement between the parties hereto and supersedes prior negotiations, representations or agreements, either written or oral. An enumeration of the Contract Documents, other than Modifications, appears in Article 9.

ARTICLE 2
THE WORK OF THIS CONTRACT

The Contractor shall execute the entire Work described in the Contract Documents, except to the extent specifically indicated in the Contract Documents to be the responsibility of others, or as follows:

ARTICLE 3
DATE OF COMMENCEMENT AND SUBSTANTIAL COMPLETION

3.1 The date of commencement is the date from which the Contract Time of Paragraph 3.2 is measured, and shall be the date of this Agreement, as first written above, unless a different date is stated below or provision is made for the date to be fixed in a notice to proceed issued by the Owner.
(Insert the date of commencement, if it differs from the date of this Agreement or, if applicable, state that the date will be fixed in a notice to proceed.)

Unless the date of commencement is established by a notice to proceed issued by the Owner, the Contractor shall notify the Owner in writing not less than five days before commencing the Work to permit the timely filing of mortgages, mechanic's liens and other security interests.

3.2 The Contractor shall achieve Substantial Completion of the entire Work not later than
(Insert the calendar date or number of calendar days after the date of commencement. Also insert any requirements for earlier Substantial Completion of certain portions of the Work, if not stated elsewhere in the Contract Documents.)

, subject to adjustments of this Contract Time as provided in the Contract Documents.
(Insert provisions, if any, for liquidated damages relating to failure to complete on time.)

AIA DOCUMENT A101 • OWNER-CONTRACTOR AGREEMENT • TWELFTH EDITION • AIA® • ©1987
THE AMERICAN INSTITUTE OF ARCHITECTS, 1735 NEW YORK AVENUE, N.W., WASHINGTON, D.C. 20006   A101-1987 **2**

Appendix

ARTICLE 4
CONTRACT SUM

4.1 The Owner shall pay the Contractor in current funds for the Contractor's performance of the Contract the Contract Sum of   Dollars
($), subject to additions and deductions as provided in the Contract Documents.

4.2 The Contract Sum is based upon the following alternates, if any, which are described in the Contract Documents and are hereby accepted by the Owner:

(State the numbers or other identification of accepted alternates. If decisions on other alternates are to be made by the Owner subsequent to the execution of this Agreement, attach a schedule of such other alternates showing the amount for each and the date until which that amount is valid.)

4.3 Unit prices, if any, are as follows:

ARTICLE 5
PROGRESS PAYMENTS

5.1 Based upon Applications for Payment submitted to the Architect by the Contractor and Certificates for Payment issued by the Architect, the Owner shall make progress payments on account of the Contract Sum to the Contractor as provided below and elsewhere in the Contract Documents.

5.2 The period covered by each Application for Payment shall be one calendar month ending on the last day of the month, or as follows:

5.3 Provided an Application for Payment is received by the Architect not later than the   day of a month, the Owner shall make payment to the Contractor not later than the   day of the   month. If an Application for Payment is received by the Architect after the application date fixed above, payment shall be made by the Owner not later than   days after the Architect receives the Application for Payment.

5.4 Each Application for Payment shall be based upon the Schedule of Values submitted by the Contractor in accordance with the Contract Documents. The Schedule of Values shall allocate the entire Contract Sum among the various portions of the Work and be prepared in such form and supported by such data to substantiate its accuracy as the Architect may require. This Schedule, unless objected to by the Architect, shall be used as a basis for reviewing the Contractor's Applications for Payment.

5.5 Applications for Payment shall indicate the percentage of completion of each portion of the Work as of the end of the period covered by the Application for Payment.

5.6 Subject to the provisions of the Contract Documents, the amount of each progress payment shall be computed as follows:

5.6.1 Take that portion of the Contract Sum properly allocable to completed Work as determined by multiplying the percentage completion of each portion of the Work by the share of the total Contract Sum allocated to that portion of the Work in the Schedule of Values, less retainage of   percent (%). Pending final determination of cost to the Owner of changes in the Work, amounts not in dispute may be included as provided in Subparagraph 7.3.7 of the General Conditions even though the Contract Sum has not yet been adjusted by Change Order;

5.6.2 Add that portion of the Contract Sum properly allocable to materials and equipment delivered and suitably stored at the site for subsequent incorporation in the completed construction (or, if approved in advance by the Owner, suitably stored off the site at a location agreed upon in writing), less retainage of   percent (%);

5.6.3 Subtract the aggregate of previous payments made by the Owner; and

5.6.4 Subtract amounts, if any, for which the Architect has withheld or nullified a Certificate for Payment as provided in Paragraph 9.5 of the General Conditions.

5.7 The progress payment amount determined in accordance with Paragraph 5.6 shall be further modified under the following circumstances:

5.7.1 Add, upon Substantial Completion of the Work, a sum sufficient to increase the total payments to   percent (%) of the Contract Sum, less such amounts as the Architect shall determine for incomplete Work and unsettled claims; and

5.7.2 Add, if final completion of the Work is thereafter materially delayed through no fault of the Contractor, any additional amounts payable in accordance with Subparagraph 9.10.3 of the General Conditions.

5.8 Reduction or limitation of retainage, if any, shall be as follows:

(If it is intended, prior to Substantial Completion of the entire Work, to reduce or limit the retainage resulting from the percentages inserted in Subparagraphs 5.6.1 and 5.6.2 above, and this is not explained elsewhere in the Contract Documents, insert here provisions for such reduction or limitation.)

Appendix

ARTICLE 6
FINAL PAYMENT

Final payment, constituting the entire unpaid balance of the Contract Sum, shall be made by the Owner to the Contractor when (1) the Contract has been fully performed by the Contractor except for the Contractor's responsibility to correct nonconforming Work as provided in Subparagraph 12.2.2 of the General Conditions and to satisfy other requirements, if any, which necessarily survive final payment; and (2) a final Certificate for Payment has been issued by the Architect; such final payment shall be made by the Owner not more than 30 days after the issuance of the Architect's final Certificate for Payment, or as follows:

ARTICLE 7
MISCELLANEOUS PROVISIONS

7.1 Where reference is made in this Agreement to a provision of the General Conditions or another Contract Document, the reference refers to that provision as amended or supplemented by other provisions of the Contract Documents.

7.2 Payments due and unpaid under the Contract shall bear interest from the date payment is due at the rate stated below, or in the absence thereof, at the legal rate prevailing from time to time at the place where the Project is located.

(Insert rate of interest agreed upon, if any.)

(Usury laws and requirements under the Federal Truth in Lending Act, similar state and local consumer credit laws and other regulations at the Owner's and Contractor's principal places of business, the location of the Project and elsewhere may affect the validity of this provision. Legal advice should be obtained with respect to deletions or modifications, and also regarding requirements such as written disclosures or waivers.)

7.3 Other provisions:

ARTICLE 8
TERMINATION OR SUSPENSION

8.1 The Contract may be terminated by the Owner or the Contractor as provided in Article 14 of the General Conditions.

8.2 The Work may be suspended by the Owner as provided in Article 14 of the General Conditions.

ARTICLE 9
ENUMERATION OF CONTRACT DOCUMENTS

9.1 The Contract Documents, except for Modifications issued after execution of this Agreement, are enumerated as follows:

9.1.1 The Agreement is this executed Standard Form of Agreement Between Owner and Contractor, AIA Document A101, 1987 Edition.

9.1.2 The General Conditions are the General Conditions of the Contract for Construction, AIA Document A201, 1987 Edition.

9.1.3 The Supplementary and other Conditions of the Contract are those contained in the Project Manual dated   , and are as follows:

Document   **Title**   **Pages**

9.1.4 The Specifications are those contained in the Project Manual dated as in Subparagraph 9.1.3, and are as follows:
(Either list the Specifications here or refer to an exhibit attached to this Agreement.)

Section   **Title**   **Pages**

Appendix 377

9.1.5 The Drawings are as follows, and are dated   unless a different date is shown below:
(Either list the Drawings here or refer to an exhibit attached to this Agreement.)

Number	Title	Date

9.1.6 The Addenda, if any, are as follows:

Number	Date	Pages

Portions of Addenda relating to bidding requirements are not part of the Contract Documents unless the bidding requirements are also enumerated in this Article 9.

AIA DOCUMENT A101 • OWNER-CONTRACTOR AGREEMENT • TWELFTH EDITION • AIA® • ©1987
THE AMERICAN INSTITUTE OF ARCHITECTS, 1735 NEW YORK AVENUE, N.W., WASHINGTON, D.C. 20006

378

9.1.7 Other documents, if any, forming part of the Contract Documents are as follows:

(List here any additional documents which are intended to form part of the Contract Documents. The General Conditions provide that bidding requirements such as advertisement or invitation to bid, Instructions to Bidders, sample forms and the Contractor's bid are not part of the Contract Documents unless enumerated in this Agreement. They should be listed here only if intended to be part of the Contract Documents.)

This Agreement is entered into as of the day and year first written above and is executed in at least three original copies of which one is to be delivered to the Contractor, one to the Architect for use in the administration of the Contract, and the remainder to the Owner.

OWNER   CONTRACTOR

_____   _____
(Signature)   *(Signature)*

_____   _____
(Printed name and title)   *(Printed name and title)*

AIA DOCUMENT A101 • OWNER-CONTRACTOR AGREEMENT • TWELFTH EDITION • AIA® • ©1987
THE AMERICAN INSTITUTE OF ARCHITECTS, 1735 NEW YORK AVENUE, N.W., WASHINGTON, D.C. 20006

Appendix C:

AIA Document A111--Standard Form of Agreement Between Owner and Contractor (where the basis of payment is the Cost of the Work Plus a Fee with or without a Guaranteed Maximum Price), 1987 Edition

Appendix 381

THE AMERICAN INSTITUTE OF ARCHITECTS

Reproduced with permission of The American Institute of Architects under license number #96051. This license expires September 30, 1997. FURTHER REPRODUCTION IS PROHIBITED. Because AIA Documents are revised from time to time, users should ascertain from the AIA the current edition of this document.

Copies of the current edition of this AIA document may be purchased from The American Institute of Architects or its local distributors. The text of this document is not "model language" and is not intended for use in other documents without permission of the AIA.

AIA Document A111

Standard Form of Agreement Between Owner and Contractor

where the basis of payment is the
COST OF THE WORK PLUS A FEE
with or without a Guaranteed Maximum Price

1987 EDITION

THIS DOCUMENT HAS IMPORTANT LEGAL CONSEQUENCES; CONSULTATION WITH AN ATTORNEY IS ENCOURAGED WITH RESPECT TO ITS COMPLETION OR MODIFICATION.

The 1987 Edition of AIA Document A201, General Conditions of the Contract for Construction, is adopted in this document by reference. Do not use with other general conditions unless this document is modified.

This document has been approved and endorsed by The Associated General Contractors of America.

AGREEMENT

made as of the   day of   in the year of
Nineteen Hundred and

BETWEEN the Owner:
(Name and address)

and the Contractor:
(Name and address)

the Project is:
(Name and address)

the Architect is:
(Name and address)

The Owner and Contractor agree as set forth below.

Copyright 1920, 1925, 1951, 1958, 1961, 1963, 1967, 1974, 1978, ©1987 by The American Institute of Architects, 1735 New York Avenue, N.W., Washington, D.C. 20006. Reproduction of the material herein or substantial quotation of its provisions without written permission of the AIA violates the copyright laws of the United States and will be subject to legal prosecution.

AIA DOCUMENT A111 • OWNER-CONTRACTOR AGREEMENT • TENTH EDITION • AIA® • ©1987 • THE AMERICAN INSTITUTE OF ARCHITECTS, 1735 NEW YORK AVENUE, N.W., WASHINGTON, D.C. 20006   **A111-1987   1**

WARNING: Unlicensed photocopying violates U.S. copyright laws and is subject to legal prosecution.

ARTICLE 1
THE CONTRACT DOCUMENTS

1.1 The Contract Documents consist of this Agreement, Conditions of the Contract (General, Supplementary and other Conditions), Drawings, Specifications, addenda issued prior to execution of this Agreement, other documents listed in this Agreement and Modifications issued after execution of this Agreement; these form the Contract, and are as fully a part of the Contract as if attached to this Agreement or repeated herein. The Contract represents the entire and integrated agreement between the parties hereto and supersedes prior negotiations, representations or agreements, either written or oral. An enumeration of the Contract Documents, other than Modifications, appears in Article 16. If anything in the other Contract Documents is inconsistent with this Agreement, this Agreement shall govern.

ARTICLE 2
THE WORK OF THIS CONTRACT

2.1 The Contractor shall execute the entire Work described in the Contract Documents, except to the extent specifically indicated in the Contract Documents to be the responsibility of others, or as follows:

ARTICLE 3
RELATIONSHIP OF THE PARTIES

3.1 The Contractor accepts the relationship of trust and confidence established by this Agreement and covenants with the Owner to cooperate with the Architect and utilize the Contractor's best skill, efforts and judgment in furthering the interests of the Owner; to furnish efficient business administration and supervision; to make best efforts to furnish at all times an adequate supply of workers and materials; and to perform the Work in the best way and most expeditious and economical manner consistent with the interests of the Owner. The Owner agrees to exercise best efforts to enable the Contractor to perform the Work in the best way and most expeditious manner by furnishing and approving in a timely way information required by the Contractor and making payments to the Contractor in accordance with requirements of the Contract Documents.

ARTICLE 4
DATE OF COMMENCEMENT AND SUBSTANTIAL COMPLETION

4.1 The date of commencement is the date from which the Contract Time of Subparagraph 4.2 is measured; it shall be the date of this Agreement, as first written above, unless a different date is stated below or provision is made for the date to be fixed in a notice to proceed issued by the Owner.

(Insert the date of commencement, if it differs from the date of this Agreement or, if applicable, state that the date will be fixed in a notice to proceed.)

Unless the date of commencement is established by a notice to proceed issued by the Owner, the Contractor shall notify the Owner in writing not less than five days before commencing the Work to permit the timely filing of mortgages, mechanic's liens and other security interests.

AIA DOCUMENT A111 • OWNER-CONTRACTOR AGREEMENT • TENTH EDITION • AIA® • ©1987 • THE AMERICAN INSTITUTE OF ARCHITECTS, 1735 NEW YORK AVENUE, N.W., WASHINGTON, D.C. 20006

WARNING: Unlicensed photocopying violates U.S. copyright laws and is subject to legal prosecution.

Appendix

4.2 The Contractor shall achieve Substantial Completion of the entire Work not later than

(Insert the calendar date or number of calendar days after the date of commencement. Also insert any requirements for earlier Substantial Completion of certain portions of the Work, if not stated elsewhere in the Contract Documents.)

, subject to adjustments of this Contract Time as provided in the Contract Documents.

(Insert provisions, if any, for liquidated damages relating to failure to complete on time.)

ARTICLE 5
CONTRACT SUM

5.1 The Owner shall pay the Contractor in current funds for the Contractor's performance of the Contract the Contract Sum consisting of the Cost of the Work as defined in Article 7 and the Contractor's Fee determined as follows:

(State a lump sum, percentage of Cost of the Work or other provision for determining the Contractor's Fee, and explain how the Contractor's Fee is to be adjusted for changes in the Work.)

5.2 GUARANTEED MAXIMUM PRICE (IF APPLICABLE)

5.2.1 The sum of the Cost of the Work and the Contractor's Fee is guaranteed by the Contractor not to exceed
Dollars ($), subject to additions and deductions by Change Order as provided in the Contract Documents. Such maximum sum is referred to in the Contract Documents as the Guaranteed Maximum Price. Costs which would cause the Guaranteed Maximum Price to be exceeded shall be paid by the Contractor without reimbursement by the Owner.

(Insert specific provisions if the Contractor is to participate in any savings.)

5.2.2 The Guaranteed Maximum Price is based upon the following alternates, if any, which are described in the Contract Documents and are hereby accepted by the Owner:

(State the numbers or other identification of accepted alternates, but only if a Guaranteed Maximum Price is inserted in Subparagraph 5.2.1. If decisions on other alternates are to be made by the Owner subsequent to the execution of this Agreement, attach a schedule of such other alternates showing the amount for each and the date until which that amount is valid.)

5.2.3 The amounts agreed to for unit prices, if any, are as follows:
(State unit prices only if a Guaranteed Maximum Price is inserted in Subparagraph 5.2.1.)

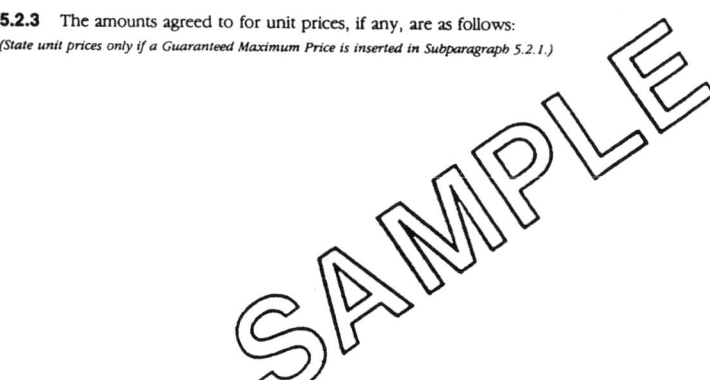

ARTICLE 6
CHANGES IN THE WORK

6.1 CONTRACTS WITH A GUARANTEED MAXIMUM PRICE

6.1.1 Adjustments to the Guaranteed Maximum Price on account of changes in the Work may be determined by any of the methods listed in Subparagraph 7.3.3 of the General Conditions.

6.1.2 In calculating adjustments to subcontracts (except those awarded with the Owner's prior consent on the basis of cost plus a fee), the terms "cost" and "fee" as used in Clause 7.3.3.3 of the General Conditions and the terms "costs" and "a reasonable allowance for overhead and profit" as used in Subparagraph 7.3.6 of the General Conditions shall have the meanings assigned to them in the General Conditions and shall not be modified by Articles 5, 7 and 8 of this Agreement. Adjustments to subcontracts awarded with the Owner's prior consent on the basis of cost plus a fee shall be calculated in accordance with the terms of those subcontracts.

6.1.3 In calculating adjustments to this Contract, the terms "cost" and "costs" as used in the above-referenced provisions of the General Conditions shall mean the Cost of the Work as defined in Article 7 of this Agreement and the terms "fee" and "a reasonable allowance for overhead and profit" shall mean the Contractor's Fee as defined in Paragraph 5.1 of this Agreement.

AIA DOCUMENT A111 • OWNER-CONTRACTOR AGREEMENT • TENTH EDITION • AIA® • ©1987 • THE AMERICAN INSTITUTE OF ARCHITECTS, 1735 NEW YORK AVENUE, N.W., WASHINGTON, D.C. 20006

A111-1987 4

WARNING: Unlicensed photocopying violates U.S. copyright laws and is subject to legal prosecution.

6.2 CONTRACTS WITHOUT A GUARANTEED MAXIMUM PRICE

6.2.1 Increased costs for the items set forth in Article 7 which result from changes in the Work shall become part of the Cost of the Work, and the Contractor's Fee shall be adjusted as provided in Paragraph 5.1.

6.3 ALL CONTRACTS

6.3.1 If no specific provision is made in Paragraph 5.1 for adjustment of the Contractor's Fee in the case of changes in the Work, or if the extent of such changes is such, in the aggregate, that application of the adjustment provisions of Paragraph 5.1 will cause substantial inequity to the Owner or Contractor, the Contractor's Fee shall be equitably adjusted on the basis of the Fee established for the original Work.

ARTICLE 7
COSTS TO BE REIMBURSED

7.1 The term Cost of the Work shall mean costs necessarily incurred by the Contractor in the proper performance of the Work. Such costs shall be at rates not higher than the standard paid at the place of the Project except with prior consent of the Owner. The Cost of the Work shall include only the items set forth in this Article 7.

7.1.1 LABOR COSTS

7.1.1.1 Wages of construction workers directly employed by the Contractor to perform the construction of the Work at the site or, with the Owner's agreement, at off-site workshops.

7.1.1.2 Wages or salaries of the Contractor's supervisory and administrative personnel when stationed at the site with the Owner's agreement.
(If it is intended that the wages or salaries of certain personnel stationed at the Contractor's principal or other offices shall be included in the Cost of the Work, identify in Article 14 the personnel to be included and whether for all or only part of their time.)

7.1.1.3 Wages and salaries of the Contractor's supervisory or administrative personnel engaged, at factories, workshops or on the road, in expediting the production or transportation of materials or equipment required for the Work, but only for that portion of their time required for the Work.

7.1.1.4 Costs paid or incurred by the Contractor for taxes, insurance, contributions, assessments and benefits required by law or collective bargaining agreements and, for personnel not covered by such agreements, customary benefits such as sick leave, medical and health benefits, holidays, vacations and pensions, provided such costs are based on wages and salaries included in the Cost of the Work under Clauses 7.1.1.1 through 7.1.1.3.

7.1.2 SUBCONTRACT COSTS

Payments made by the Contractor to Subcontractors in accordance with the requirements of the subcontracts.

7.1.3 COSTS OF MATERIALS AND EQUIPMENT INCORPORATED IN THE COMPLETED CONSTRUCTION

7.1.3.1 Costs, including transportation, of materials and equipment incorporated or to be incorporated in the completed construction.

7.1.3.2 Costs of materials described in the preceding Clause 7.1.3.1 in excess of those actually installed but required to provide reasonable allowance for waste and for spoilage. Unused excess materials, if any, shall be handed over to the Owner at the completion of the Work or, at the Owner's option, shall be sold by the Contractor; amounts realized, if any, from such sales shall be credited to the Owner as a deduction from the Cost of the Work.

7.1.4 COSTS OF OTHER MATERIALS AND EQUIPMENT, TEMPORARY FACILITIES AND RELATED ITEMS

7.1.4.1 Costs, including transportation, installation, maintenance, dismantling and removal of materials, supplies, temporary facilities, machinery, equipment, and hand tools not customarily owned by the construction workers, which are provided by the Contractor at the site and fully consumed in the performance of the Work; and cost less salvage value on such items if not fully consumed, whether sold to others or retained by the Contractor. Cost for items previously used by the Contractor shall mean fair market value.

7.1.4.2 Rental charges for temporary facilities, machinery, equipment, and hand tools not customarily owned by the construction workers, which are provided by the Contractor at the site, whether rented from the Contractor or others, and costs of transportation, installation, minor repairs and replacements, dismantling and removal thereof. Rates and quantities of equipment rented shall be subject to the Owner's prior approval.

7.1.4.3 Costs of removal of debris from the site.

7.1.4.4 Costs of telegrams and long-distance telephone calls, postage and parcel delivery charges, telephone service at the site and reasonable petty cash expenses of the site office.

7.1.4.5 That portion of the reasonable travel and subsistence expenses of the Contractor's personnel incurred while traveling in discharge of duties connected with the Work.

7.1.5 MISCELLANEOUS COSTS

7.1.5.1 That portion directly attributable to this Contract of premiums for insurance and bonds.

7.1.5.2 Sales, use or similar taxes imposed by a governmental authority which are related to the Work and for which the Contractor is liable.

7.1.5.3 Fees and assessments for the building permit and for other permits, licenses and inspections for which the Contractor is required by the Contract Documents to pay.

7.1.5.4 Fees of testing laboratories for tests required by the Contract Documents, except those related to defective or nonconforming Work for which reimbursement is excluded by Subparagraph 13.5.3 of the General Conditions or other provisions of the Contract Documents and which do not fall within the scope of Subparagraphs 7.2.2 through 7.2.4 below.

7.1.5.5 Royalties and license fees paid for the use of a particular design, process or product required by the Contract Documents; the cost of defending suits or claims for infringement of patent rights arising from such requirement by the Contract Documents; payments made in accordance with legal judgments against the Contractor resulting from such suits or claims and payments of settlements made with the Owner's consent; provided, however, that such costs of legal defenses, judgment and settlements shall not be included in the calculation of the Contractor's Fee or of a Guaranteed Maximum Price, if any, and provided that such royalties, fees and costs are not excluded by the last sentence of Subparagraph 3.17.1 of the General Conditions or other provisions of the Contract Documents.

7.1.5.6 Deposits lost for causes other than the Contractor's fault or negligence.

7.1.6 OTHER COSTS

7.1.6.1 Other costs incurred in the performance of the Work if and to the extent approved in advance in writing by the Owner.

7.2 EMERGENCIES: REPAIRS TO DAMAGED, DEFECTIVE OR NONCONFORMING WORK

The Cost of the Work shall also include costs described in Paragraph 7.1 which are incurred by the Contractor:

7.2.1 In taking action to prevent threatened damage, injury or loss in case of an emergency affecting the safety of persons and property, as provided in Paragraph 10.3 of the General Conditions.

7.2.2 In repairing or correcting Work damaged or improperly executed by construction workers in the employ of the Contractor, provided such damage or improper execution did not result from the fault or negligence of the Contractor or the Contractor's foremen, engineers or superintendents, or other supervisory, administrative or managerial personnel of the Contractor.

7.2.3 In repairing damaged Work other than that described in Subparagraph 7.2.2, provided such damage did not result from the fault or negligence of the Contractor or the Contractor's personnel, and only to the extent that the cost of such repairs is not recoverable by the Contractor from others and the Contractor is not compensated therefor by insurance or otherwise.

7.2.4 In correcting defective or nonconforming Work performed or supplied by a Subcontractor or material supplier and not corrected by them, provided such defective or nonconforming Work did not result from the fault or neglect of the Contractor or the Contractor's personnel adequately to supervise and direct the Work of the Subcontractor or material supplier, and only to the extent that the cost of correcting the defective or nonconforming Work is not recoverable by the Contractor from the Subcontractor or material supplier.

ARTICLE 8
COSTS NOT TO BE REIMBURSED

8.1 The Cost of the Work shall not include:

8.1.1 Salaries and other compensation of the Contractor's personnel stationed at the Contractor's principal office or offices other than the site office, except as specifically provided in Clauses 7.1.1.2 and 7.1.1.3 or as may be provided in Article 14.

8.1.2 Expenses of the Contractor's principal office and offices other than the site office.

8.1.3 Overhead and general expenses, except as may be expressly included in Article 7.

8.1.4 The Contractor's capital expenses, including interest on the Contractor's capital employed for the Work.

8.1.5 Rental costs of machinery and equipment, except as specifically provided in Clause 7.1.4.2.

8.1.6 Except as provided in Subparagraphs 7.2.2 through 7.2.4 and Paragraph 13.5 of this Agreement, costs due to the fault or negligence of the Contractor, Subcontractors, anyone directly or indirectly employed by any of them, or for whose acts any of them may be liable, including but not limited to costs for the correction of damaged, defective or nonconforming Work, disposal and replacement of materials and equipment incorrectly ordered or supplied, and making good damage to property not forming part of the Work.

8.1.7 Any cost not specifically and expressly described in Article 7.

8.1.8 Costs which would cause the Guaranteed Maximum Price, if any, to be exceeded.

Appendix

ARTICLE 9
DISCOUNTS, REBATES AND REFUNDS

9.1 Cash discounts obtained on payments made by the Contractor shall accrue to the Owner if (1) before making the payment, the Contractor included them in an Application for Payment and received payment therefor from the Owner, or (2) the Owner has deposited funds with the Contractor with which to make payments; otherwise, cash discounts shall accrue to the Contractor. Trade discounts, rebates, refunds and amounts received from sales of surplus materials and equipment shall accrue to the Owner, and the Contractor shall make provisions so that they can be secured.

9.2 Amounts which accrue to the Owner in accordance with the provisions of Paragraph 9.1 shall be credited to the Owner as a deduction from the Cost of the Work.

ARTICLE 10
SUBCONTRACTS AND OTHER AGREEMENTS

10.1 Those portions of the Work that the Contractor does not customarily perform with the Contractor's own personnel shall be performed under subcontracts or by other appropriate agreements with the Contractor. The Contractor shall obtain bids from Subcontractors and from suppliers of materials or equipment fabricated especially for the Work and shall deliver such bids to the Architect. The Owner will then determine, with the advice of the Contractor and subject to the reasonable objection of the Architect, which bids will be accepted. The Owner may designate specific persons or entities from whom the Contractor shall obtain bids; however, if a Guaranteed Maximum Price has been established, the Owner may not prohibit the Contractor from obtaining bids from others. The Contractor shall not be required to contract with anyone to whom the Contractor has reasonable objection.

10.2 If a Guaranteed Maximum Price has been established and a specific bidder among those whose bids are delivered by the Contractor to the Architect (1) is recommended to the Owner by the Contractor; (2) is qualified to perform that portion of the Work; and (3) has submitted a bid which conforms to the requirements of the Contract Documents without reservations or exceptions, but the Owner requires that another bid be accepted; then the Contractor may require that a Change Order be issued to adjust the Guaranteed Maximum Price by the difference between the bid of the person or entity recommended to the Owner by the Contractor and the amount of the subcontract or other agreement actually signed with the person or entity designated by the Owner.

10.3 Subcontracts or other agreements shall conform to the payment provisions of Paragraphs 12.7 and 12.8, and shall not be awarded on the basis of cost plus a fee without the prior consent of the Owner.

ARTICLE 11
ACCOUNTING RECORDS

11.1 The Contractor shall keep full and detailed accounts and exercise such controls as may be necessary for proper financial management under this Contract; the accounting and control systems shall be satisfactory to the Owner. The Owner and the Owner's accountants shall be afforded access to the Contractor's records, books, correspondence, instructions, drawings, receipts, subcontracts, purchase orders, vouchers, memoranda and other data relating to this Contract, and the Contractor shall preserve these for a period of three years after final payment, or for such longer period as may be required by law.

ARTICLE 12
PROGRESS PAYMENTS

12.1 Based upon Applications for Payment submitted to the Architect by the Contractor and Certificates for Payment issued by the Architect, the Owner shall make progress payments on account of the Contract Sum to the Contractor as provided below and elsewhere in the Contract Documents.

12.2 The period covered by each Application for Payment shall be one calendar month ending on the last day of the month, or as follows:

12.3 Provided an Application for Payment is received by the Architect not later than the   day of a month, the Owner shall make payment to the Contractor not later than the   day of the month. If an Application for Payment is received by the Architect after the application date fixed above, payment shall be made by the Owner not later than   days after the Architect receives the Application for Payment.

12.4 With each Application for Payment the Contractor shall submit payrolls, petty cash accounts, receipted invoices or invoices with check vouchers attached, and any other evidence required by the Owner or Architect to demonstrate that cash disbursements already made by the Contractor on account of the Cost of the Work equal or exceed (1) progress payments already received by the Contractor; less (2) that portion of those payments attributable to the Contractor's Fee; plus (3) payrolls for the period covered by the present Application for Payment; plus (4) retainage provided in Subparagraph 12.5.4, if any, applicable to prior progress payments.

12.5 CONTRACTS WITH A GUARANTEED MAXIMUM PRICE

12.5.1 Each Application for Payment shall be based upon the most recent schedule of values submitted by the Contractor in accordance with the Contract Documents. The schedule of values shall allocate the entire Guaranteed Maximum Price among the various portions of the Work, except that the Contractor's Fee shall be shown as a single separate item. The schedule of values shall be prepared in such form and supported by such data to substantiate its accuracy as the Architect may require. This schedule, unless objected to by the Architect, shall be used as a basis for reviewing the Contractor's Applications for Payment.

12.5.2 Applications for Payment shall show the percentage completion of each portion of the Work as of the end of the period covered by the Application for Payment. The percentage completion shall be the lesser of (1) the percentage of that portion of the Work which has actually been completed or (2) the percentage obtained by dividing (a) the expense which has actually been incurred by the Contractor on account of that portion of the Work for which the Contractor has made or intends to make actual payment prior to the next Application for Payment by (b) the share of the Guaranteed Maximum Price allocated to that portion of the Work in the schedule of values.

12.5.3 Subject to other provisions of the Contract Documents, the amount of each progress payment shall be computed as follows:

12.5.3.1 Take that portion of the Guaranteed Maximum Price properly allocable to completed Work as determined by multiplying the percentage completion of each portion of the Work by the share of the Guaranteed Maximum Price allocated to that portion of the Work in the schedule of values. Pending final determination of cost to the Owner of changes in the Work, amounts not in dispute may be included as provided in Subparagraph 7.3.7 of the General Conditions, even though the Guaranteed Maximum Price has not yet been adjusted by Change Order.

12.5.3.2 Add that portion of the Guaranteed Maximum Price properly allocable to materials and equipment delivered and suitably stored at the site for subsequent incorporation in the Work or, if approved in advance by the Owner, suitably stored off the site at a location agreed upon in writing.

12.5.3.3 Add the Contractor's Fee, less retainage of   percent (%). The Contractor's Fee shall be computed upon the Cost of the Work described in the two preceding Clauses at the rate stated in Paragraph 5.1 or, if the Contractor's Fee is stated as a fixed sum in that Paragraph, shall be an amount which bears the same ratio to that fixed-sum Fee as the Cost of the Work in the two preceding Clauses bears to a reasonable estimate of the probable Cost of the Work upon its completion.

12.5.3.4 Subtract the aggregate of previous payments made by the Owner.

12.5.3.5 Subtract the shortfall, if any, indicated by the Contractor in the documentation required by Paragraph 12.4 to substantiate prior Applications for Payment, or resulting from errors subsequently discovered by the Owner's accountants in such documentation.

12.5.3.6 Subtract amounts, if any, for which the Architect has withheld or nullified a Certificate for Payment as provided in Paragraph 9.5 of the General Conditions.

12.5.4 Additional retainage, if any, shall be as follows:

(If it is intended to retain additional amounts from progress payments to the Contractor beyond (1) the retainage from the Contractor's Fee provided in Clause 12.5.3.3, (2) the retainage from Subcontractors provided in Paragraph 12.7 below, and (3) the retainage, if any, provided by other provisions of the Contract, insert provision for such additional retainage here. Such provision, if made, should also describe any arrangement for limiting or reducing the amount retained after the Work reaches a certain state of completion.)

12.6 CONTRACTS WITHOUT A GUARANTEED MAXIMUM PRICE

12.6.1 Applications for Payment shall show the Cost of the Work actually incurred by the Contractor through the end of the period covered by the Application for Payment and for which the Contractor has made or intends to make actual payment prior to the next Application for Payment.

12.6.2 Subject to other provisions of the Contract Documents, the amount of each progress payment shall be computed as follows:

12.6.2.1 Take the Cost of the Work as described in Subparagraph 12.6.1.

12.6.2.2 Add the Contractor's Fee, less retainage of   percent (%). The Contractor's Fee shall be computed upon the Cost of the Work described in the preceding Clause 12.6.2.1 at the rate stated in Paragraph 5.1 or, if the Contractor's Fee is stated as a fixed sum in that Paragraph, an amount which bears the same ratio to that fixed-sum Fee as the Cost of the Work in the preceding Clause bears to a reasonable estimate of the probable Cost of the Work upon its completion.

12.6.2.3 Subtract the aggregate of previous payments made by the Owner.

12.6.2.4 Subtract the shortfall, if any, indicated by the Contractor in the documentation required by Paragraph 12.4 or to substantiate prior Applications for Payment or resulting from errors subsequently discovered by the Owner's accountants in such documentation.

Appendix

12.6.2.5 Subtract amounts, if any, for which the Architect has withheld or withdrawn a Certificate for Payment as provided in the Contract Documents.

12.6.3 Additional retainage, if any, shall be as follows:

12.7 Except with the Owner's prior approval, payments to Subcontractors included in the Contractor's Applications for Payment shall not exceed an amount for each Subcontractor calculated as follows:

12.7.1 Take that portion of the Subcontract Sum properly allocable to completed Work as determined by multiplying the percentage completion of each portion of the Subcontractor's Work by the share of the total Subcontract Sum allocated to that portion in the Subcontractor's schedule of values, less retainage of   percent (%). Pending final determination of amounts to be paid to the Subcontractor for changes in the Work, amounts not in dispute may be included as provided in Subparagraph 7.3.7 of the General Conditions even though the Subcontract Sum has not yet been adjusted by Change Order.

12.7.2 Add that portion of the Subcontract Sum properly allocable to materials and equipment delivered and suitably stored at the site for subsequent incorporation in the Work or, if approved in advance by the Owner, suitably stored off the site at a location agreed upon in writing, less retainage of   percent (%).

12.7.3 Subtract the aggregate of previous payments made by the Contractor to the Subcontractor.

12.7.4 Subtract amounts, if any, for which the Architect has withheld or nullified a Certificate for Payment by the Owner to the Contractor for reasons which are the fault of the Subcontractor.

12.7.5 Add, upon Substantial Completion of the entire Work of the Contractor, a sum sufficient to increase the total payments to the Subcontractor to   percent (%) of the Subcontract Sum, less amounts, if any, for incomplete Work and unsettled claims; and, if final completion of the entire Work is thereafter materially delayed through no fault of the Subcontractor, add any additional amounts payable on account of Work of the Subcontractor in accordance with Subparagraph 9.10.3 of the General Conditions.

(If it is intended, prior to Substantial Completion of the entire Work of the Contractor, to reduce or limit the retainage from Subcontractors resulting from the percentages inserted in Subparagraphs 12.7.1 and 12.7.2 above, and this is not explained elsewhere in the Contract Documents, insert here provisions for such reduction or limitation.)

The Subcontract Sum is the total amount stipulated in the subcontract to be paid by the Contractor to the Subcontractor for the Subcontractor's performance of the subcontract.

12.8 Except with the Owner's prior approval, the Contractor shall not make advance payments to suppliers for materials or equipment which have not been delivered and stored at the site.

12.9 In taking action on the Contractor's Applications for Payment, the Architect shall be entitled to rely on the accuracy and completeness of the information furnished by the Contractor and shall not be deemed to represent that the Architect has made a detailed examination, audit or arithmetic verification of the documentation submitted in accordance with Paragraph 12.4 or other supporting data; that the Architect has made exhaustive or continuous on-site inspections or that the Architect has made examinations to ascertain how or for what purposes the Contractor has used amounts previously paid on account of the Contract. Such examinations, audits and verifications, if required by the Owner, will be performed by the Owner's accountants acting in the sole interest of the Owner.

ARTICLE 13
FINAL PAYMENT

13.1 Final payment shall be made by the Owner to the Contractor when (1) the Contract has been fully performed by the Contractor except for the Contractor's responsibility to correct defective or nonconforming Work, as provided in Subparagraph 12.2.2 of the General Conditions, and to satisfy other requirements, if any, which necessarily survive final payment; (2) a final Application for Payment and a final accounting for the Cost of the Work have been submitted by the Contractor and reviewed by the Owner's accountants; and (3) a final Certificate for Payment has then been issued by the Architect; such final payment shall be made by the Owner not more than 30 days after the issuance of the Architect's final Certificate for Payment, or as follows:

13.2 The amount of the final payment shall be calculated as follows:

13.2.1 Take the sum of the Cost of the Work substantiated by the Contractor's final accounting and the Contractor's Fee; but not more than the Guaranteed Maximum Price, if any.

13.2.2 Subtract amounts, if any, for which the Architect withholds, in whole or in part, a final Certificate for Payment as provided in Subparagraph 9.5.1 of the General Conditions or other provisions of the Contract Documents.

13.2.3 Subtract the aggregate of previous payments made by the Owner.

If the aggregate of previous payments made by the Owner exceeds the amount due the Contractor, the Contractor shall reimburse the difference to the Owner.

13.3 The Owner's accountants will review and report in writing on the Contractor's final accounting within 30 days after delivery of the final accounting to the Architect by the Contractor. Based upon such Cost of the Work as the Owner's accountants report to be substantiated by the Contractor's final accounting, and provided the other conditions of Paragraph 13.1 have been met, the Architect will, within seven days after receipt of the written report of the Owner's accountants, either issue to the Owner a final Certificate for Payment with a copy to the Contractor, or notify the Contractor and Owner in writing of the Architect's reasons for withholding a certificate as provided in Subparagraph 9.5.1 of the General Conditions. The time periods stated in this Paragraph 13.3 supersede those stated in Subparagraph 9.4.1 of the General Conditions.

13.4 If the Owner's accountants report the Cost of the Work as substantiated by the Contractor's final accounting to be less than claimed by the Contractor, the Contractor shall be entitled to demand arbitration of the disputed amount without a further decision of the Architect. Such demand for arbitration shall be made by the Contractor within 30 days after the Contractor's receipt of a copy of the Architect's final Certificate for Payment; failure to demand arbitration within this 30-day period shall result in the substantiated amount reported by the Owner's accountants becoming binding on the Contractor. Pending a final resolution by arbitration, the Owner shall pay the Contractor the amount certified in the Architect's final Certificate for Payment.

13.5 If, subsequent to final payment and at the Owner's request, the Contractor incurs costs described in Article 7 and not excluded by Article 8 to correct defective or nonconforming Work, the Owner shall reimburse the Contractor such costs and the Contractor's Fee applicable thereto on the same basis as if such costs had been incurred prior to final payment, but not in excess of the Guaranteed Maximum Price, if any. If the Contractor has participated in savings as provided in Paragraph 5.2, the amount of such savings shall be recalculated and appropriate credit given to the Owner in determining the net amount to be paid by the Owner to the Contractor.

ARTICLE 14
MISCELLANEOUS PROVISIONS

14.1 Where reference is made in this Agreement to a provision of the General Conditions or another Contract Document, the reference refers to that provision as amended or supplemented by other provisions of the Contract Documents.

14.2 Payments due and unpaid under the Contract shall bear interest from the date payment is due at the rate stated below, or in the absence thereof, at the legal rate prevailing from time to time at the place where the Project is located.

(Insert rate of interest agreed upon, if any.)

(Usury laws and requirements under the Federal Truth in Lending Act, similar state and local consumer credit laws and other regulations at the Owner's and Contractor's principal places of business, the location of the Project and elsewhere may affect the validity of this provision. Legal advice should be obtained with respect to deletions or modifications, and also regarding requirements such as written disclosures or waivers.)

14.3 Other provisions:

ARTICLE 15
TERMINATION OR SUSPENSION

15.1 The Contract may be terminated by the Contractor as provided in Article 14 of the General Conditions; however, the amount to be paid to the Contractor under Subparagraph 14.1.2 of the General Conditions shall not exceed the amount the Contractor would be entitled to receive under Paragraph 15.3 below, except that the Contractor's Fee shall be calculated as if the Work had been fully completed by the Contractor, including a reasonable estimate of the Cost of the Work for Work not actually completed.

15.2 If a Guaranteed Maximum Price is established in Article 5, the Contract may be terminated by the Owner for cause as provided in Article 14 of the General Conditions; however, the amount, if any, to be paid to the Contractor under Subparagraph 14.2.4 of the General Conditions shall not cause the Guaranteed Maximum Price to be exceeded, nor shall it exceed the amount the Contractor would be entitled to receive under Paragraph 15.3 below.

15.3 If no Guaranteed Maximum Price is established in Article 5, the Contract may be terminated by the Owner for cause as provided in Article 14 of the General Conditions; however, the Owner shall then pay the Contractor an amount calculated as follows:

15.3.1 Take the Cost of the Work incurred by the Contractor to the date of termination.

15.3.2 Add the Contractor's Fee computed upon the Cost of the Work to the date of termination at the rate stated in Paragraph 5.1 or, if the Contractor's Fee is stated as a fixed sum in that Paragraph, an amount which bears the same ratio to that fixed-sum Fee as the Cost of the Work at the time of termination bears to a reasonable estimate of the probable Cost of the Work upon its completion.

15.3.3 Subtract the aggregate of previous payments made by the Owner.

The Owner shall also pay the Contractor fair compensation, either by purchase or rental at the election of the Owner, for any equipment owned by the Contractor which the Owner elects to retain and which is not otherwise included in the Cost of the Work under Subparagraph 15.3.1. To the extent that the Owner elects to take legal assignment of subcontracts and purchase orders (including rental agreements), the Contractor shall, as a condition of receiving the payments referred to in this Article 15, execute and deliver all such papers and take all such steps, including the legal assignment of such subcontracts and other contractual rights of the Contractor, as the Owner may require for the purpose of fully vesting in the Owner the rights and benefits of the Contractor under such subcontracts or purchase orders.

15.4 The Work may be suspended by the Owner as provided in Article 14 of the General Conditions; in such case, the Guaranteed Maximum Price, if any, shall be increased as provided in Subparagraph 14.3.2 of the General Conditions except that the term "cost of performance of the Contract" in that Subparagraph shall be understood to mean the Cost of the Work and the term "profit" shall be understood to mean the Contractor's Fee as described in Paragraphs 5.1 and 6.3 of this Agreement.

ARTICLE 16
ENUMERATION OF CONTRACT DOCUMENTS

16.1 The Contract Documents, except for Modifications issued after execution of this Agreement, are enumerated as follows:

16.1.1 The Agreement is this executed Standard Form of Agreement Between Owner and Contractor, AIA Document A111, 1987 Edition.

16.1.2 The General Conditions are the General Conditions of the Contract for Construction, AIA Document A201, 1987 Edition.

16.1.3 The Supplementary and other Conditions of the Contract are those contained in the Project Manual dated   , and are as follows:

Document   **Title**   **Pages**

16.1.4 The Specifications are those contained in the Project Manual dated as in Paragraph 16.1.3, and are as follows:
(Either list the Specifications here or refer to an exhibit attached to this Agreement.)

Section   **Title**   **Pages**

Appendix

16.1.5 The Drawings are as follows, and are dated   unless a different date is shown below:
(Either list the Drawings here or refer to an exhibit attached to this Agreement.)

Number   **Title**   **Date**

16.1.6 The addenda, if any, are as follows:

Number   **Date**   **Pages**

Portions of Addenda relating to bidding requirements are not part of the Contract Documents unless the bidding requirements are also enumerated in this Article 16.

16.1.7 Other Documents, if any, forming part of the Contract Documents are as follows:

(List here any additional documents which are intended to form part of the Contract Documents. The General Conditions provide that bidding requirements such as advertisement or invitation to bid, Instructions to Bidders, sample forms and the Contractor's bid are not part of the Contract Documents unless enumerated in this Agreement. They should be listed here only if intended to be part of the Contract Documents.)

This Agreement is entered into as of the day and year first written above and is executed in at least three original copies of which one is to be delivered to the Contractor, one to the Architect for use in the administration of the Contract, and the remainder to the Owner.

OWNER   CONTRACTOR

_____   _____
(Signature)   *(Signature)*

_____   _____
(Printed name and title)   *(Printed name and title)*

 CAUTION: You should sign an original AIA document which has this caution printed in red. An original assures that changes will not be obscured as may occur when documents are reproduced.

AIA DOCUMENT A111 • OWNER-CONTRACTOR AGREEMENT • TENTH EDITION • AIA® • ©1987 • THE AMERICAN INSTITUTE OF ARCHITECTS, 1735 NEW YORK AVENUE, N.W., WASHINGTON, D.C. 20006

WARNING: Unlicensed photocopying violates U.S. copyright laws and is subject to legal prosecution.

Appendix D:

**AIA Document A201--General Conditions
of the Contract for Construction,
1987 Edition**

THE AMERICAN INSTITUTE OF ARCHITECTS

Reproduced with permission of The American Institute of Architects under license number #96051. This license expires September 30, 1997. FURTHER REPRODUCTION IS PROHIBITED. Because AIA Documents are revised from time to time, users should ascertain from the AIA the current edition of this document.

Copies of the current edition of this AIA document may be purchased from The American Institute of Architects or its local distributors. The text of this document is not "model language" and is not intended for use in other documents without permission of the AIA.

AIA Document A201

General Conditions of the Contract for Construction

THIS DOCUMENT HAS IMPORTANT LEGAL CONSEQUENCES; CONSULTATION WITH AN ATTORNEY IS ENCOURAGED WITH RESPECT TO ITS MODIFICATION

1987 EDITION
TABLE OF ARTICLES

1. GENERAL PROVISIONS
2. OWNER
3. CONTRACTOR
4. ADMINISTRATION OF THE CONTRACT
5. SUBCONTRACTORS
6. CONSTRUCTION BY OWNER OR BY SEPARATE CONTRACTORS
7. CHANGES IN THE WORK
8. TIME
9. PAYMENTS AND COMPLETION
10. PROTECTION OF PERSONS AND PROPERTY
11. INSURANCE AND BONDS
12. UNCOVERING AND CORRECTION OF WORK
13. MISCELLANEOUS PROVISIONS
14. TERMINATION OR SUSPENSION OF THE CONTRACT

This document has been approved and endorsed by the Associated General Contractors of America.

Copyright 1911, 1915, 1918, 1925, 1937, 1951, 1958, 1961, 1963, 1966, 1967, 1970, 1976, ©1987 by The American Institute of Architects, 1735 New York Avenue, N.W., Washington, D.C., 20006. Reproduction of the material herein or substantial quotation of its provisions without written permission of the AIA violates the copyright laws of the United States and will be subject to legal prosecutions.

 CAUTION: You should use an original AIA document which has this caution printed in red. An original assures that changes will not be obscured as may occur when documents are reproduced.

AIA DOCUMENT A201 • GENERAL CONDITIONS OF THE CONTRACT FOR CONSTRUCTION • FOURTEENTH EDITION
AIA® • ©1987 THE AMERICAN INSTITUTE OF ARCHITECTS, 1735 NEW YORK AVENUE, N.W., WASHINGTON, D.C. 20006

WARNING: Unlicensed photocopying violates U.S. copyright laws and is subject to legal prosecution.

INDEX

Acceptance of Nonconforming Work 9.6.6, 9.9.3, **12.3**
Acceptance of Work 9.6.6, 9.8.2, 9.9.3, 9.10.1, 9.10.3
Access to Work **3.16**, 6.2.1, 12.1
Accident Prevention 4.2.3, 10
Acts and Omissions ... 3.2.1, 3.2.2, 3.3.2, 3.12.8, 3.18, 4.2.3, 4.3.2,
  4.3.9, 8.3.1, 10.1.4, 10.2.5, 13.4.2, 13.7, 14.1
Addenda 1.1.1, 3.11
Additional Cost, Claims for 4.3.6, 4.3.7, 4.3.9, 6.1.1, 10.3
Additional Inspections and Testing 4.2.6, 9.8.2, 12.2.1, 13.5
Additional Time, Claims for 4.3.6, 4.3.8, 4.3.9, 8.3.2
ADMINISTRATION OF THE CONTRACT 3.3.3, **4**, 9.4, 9.5
Advertisement or Invitation to Bid 1.1.1
Aesthetic Effect 4.2.13, 4.5.1
Allowances **3.8**
All-risk Insurance 11.3.1.1
Applications for Payment 4.2.5, 7.3.7, 9.2, **9.3**, 9.4, 9.5.1, 9.6.3,
  9.8.3, 9.10.1, 9.10.3, 9.10.4, 11.1.3, 14.2.4
Approvals .... 2.4, 3.3.3, 3.5, 3.10.2, 3.12.4 through 3.12.8, 3.18.3,
  4.2.7, 9.3.2, 11.3.1.4, 13.4.2, 13.5
Arbitration 4.1.4, 4.3.2, 4.3.4, 4.4.4, **4.5**,
  8.3.1, 10.1.2, 11.3.9, 11.3.10
Architect **4**
Architect, Definition of 4.1
Architect, Extent of Authority 2.4, 3.12.6, 4.2, 4.3.6, 4.4,
  4.4, 5.2, 6.3, 7.1.2, 7.2.1, 7.3.6, 7.4, 9.2, 9.3.1,
  9.4, 9.5, 9.6.3, 9.8.2, 9.8.3, 9.10.1, 9.10.3, 12.1, 12.2.1,
  13.5.1, 13.5.2, 14.2.2, 14.2.4
Architect, Limitations of Authority and Responsibility .. 3.3.3, 3.12.8,
  3.12.11, 4.1.2, 4.2.1, 4.2.2, 4.2.3, 4.2.6, 4.2.7, 4.2.10, 4.2.12,
  4.2.13, 4.3.2, 5.2.1, 7.4, 9.4.2, 9.6.4, 9.6.6
Architect's Additional Services and Expenses 2.4, 9.8.2,
  11.3.1.1, 12.2.1, 12.2.4, 13.5.2, 13.5.3, 14.2.4
Architect's Administration of the Contract **4.2**, 4.3.6,
  4.3.7, 4.4, 9.4, 9.5
Architect's Approvals   2.4, 3.5.1, 3.10.2, 3.12.6, 3.12.8, 3.18.3, 4.2.7
Architect's Authority to Reject Work 3.5.1, 4.2.6, 12.1.2, 12.2.1
Architect's Copyright 1.3
Architect's Decisions 4.2.6, 4.2.7, 4.2.11, 4.2.12, 4.2.13,
  4.3.2, 4.3.6, 4.4.1, 4.4.4, 4.5, 6.3, 7.3.6, 7.3.8, 8.1.3, 8.3.1,
  9.2, 9.4, 9.5.1, 9.8.2, 9.9.1, 10.1.2, 13.5.2, 14.2.2, 14.2.4
Architect's Inspections 4.2.2, 4.2.9, 4.3.6, 9.4.2, 9.8.2,
  9.9.2, 9.10.1, 13.5
Architect's Instructions .. 4.2.6, 4.2.7, 4.2.8, 4.3.7, 7.4.1, 12.1, 13.5.2
Architect's Interpretations 4.2.11, 4.2.12, 4.3.7
Architect's On-Site Observations 4.2.2, 4.2.5, 4.3.6, 9.4.2,
  9.5.1, 9.10.1, 13.5
Architect's Project Representative 4.2.10
Architect's Relationship with Contractor 1.1.2, 3.2.1, 3.2.2,
  3.3.3, 3.5.1, 3.7.3, 3.11, 3.12.8, 3.12.11, 3.16, 3.18, 4.2.3, 4.2.4,
  4.2.6, 4.2.12, 5.2, 6.2.2, 7.3.4, 9.8.2, 11.3.7, 12.1, 13.5
Architect's Relationship with Subcontractors 1.1.2, 4.2.3, 4.2.4,
  4.2.6, 9.6.3, 9.6.4, 11.3.7
Architect's Representations 9.4.2, 9.5.1, 9.10.1
Architect's Site Visits 4.2.2, 4.2.5, 4.2.9, 4.3.6, 9.4.2, 9.5.1,
  9.8.2, 9.9.2, 9.10.1, 13.5
Asbestos 10.1
Attorneys' Fees 3.18.1, 9.10.2, 10.1.4
Award of Separate Contracts............................. 6.1.1
Award of Subcontracts and Other Contracts for
  Portions of the Work **5.2**
Basic Definitions **1.1**
Bidding Requirements 1.1.1, 1.1.7, 5.2.1, 11.4.1
Boiler and Machinery Insurance **11.3.2**
Bonds, Lien 9.10.2
Bonds, Performance and Payment 7.3.6.4, 9.10.3, 11.3.9, 11.4

Building Permit 3.7.1
Capitalization **1.4**
Certificate of Substantial Completion 9.8.2
Certificates for Payment 4.2.5, 4.2.9, 9.3.3, **9.4**, 9.5, 9.6.1,
  9.6.6, 9.7.1, 9.8.3, 9.10.1, 9.10.3, 13.7, 14.1.1.3, 14.2.4
Certificates of Inspection, Testing or Approval 3.12.11, 13.5.4
Certificates of Insurance 9.3.2, 9.10.2, 11.1.3
Change Orders 1.1.1, 2.4.1, 3.8.2.4, 3.11, 4.2.8, 4.3.3, 5.2.3,
  7.1, **7.2**, 7.3.2, 8.3.1, 9.3.1.1, 9.10.3, 11.3.1.2,
  11.3.4, 11.3.9, 13.1.2, 12.1.2
Change Orders, Definition of 7.2.1
Changes **7.1**
CHANGES IN THE WORK 3.11, 4.2.8, **7**, 8.3.1, 9.3.1.1, 10.1.3
Claim, **Definition of** **4.3.1**
Claims and Disputes **4.3**, 4.4, 4.5, 6.2.5, 8.3.2,
  9.3.1.2, 9.3.3, 9.10.4, 10.1.4
Claims and Timely Assertion of Claims **4.5.6**
Claims for Additional Cost 4.3.6, **4.3.7**, 4.3.9, 6.1.1, 10.3
Claims for Additional Time 4.3.6, **4.3.8**, 4.3.9, 8.3.2
Claims for Concealed or Unknown Conditions **4.3.6**
Claims for Damages ... 3.18, 4.3.9, 6.1.1, 6.2.5, 8.3.2, 9.5.1.2, 10.1.4
Claims Subject to Arbitration 4.3.2, 4.4.4, 4.5.1
Cleaning Up **3.15**, 6.3
Commencement of Statutory Limitation Period **13.7**
Commencement of the Work, Conditions Relating to 2.1.2,
  2.2.1, 3.2.1, 3.2.2, 3.7.1, 3.10.1, 3.12.6, 4.3.7, 5.2.1,
  6.2.2, 8.1.2, 8.2.2, 9.2, 11.1.3, 11.3.6, 11.4.1
Commencement of the Work, Definition of 8.1.2
Communications Facilitating Contract
  Administration 3.9.1, 4.2.4, 5.2.1
Completion, Conditions Relating to 3.11, 3.15, 4.2.2, 4.2.9,
  4.3.2, 9.4.2, 9.8, 9.9.1, 9.10, 11.3.5, 12.2.2, 13.7.1
COMPLETION, PAYMENTS AND **9**
Completion, Substantial 4.2.9, 4.3.5.2, 8.1.1, 8.1.3, 8.2.3,
  9.8, 9.9.1, 12.2, 13.7
Compliance with Laws 1.3, 3.6, 3.7, 3.13, 4.1.1, 10.2.2, 11.1,
  11.3, 13.1, 13.5.1, 13.5.2, 13.6, 14.1.1, 14.2.1.3
Concealed or Unknown Conditions 4.3.6
Conditions of the Contract 1.1.1, 1.1.7, 6.1.1
Consent, Written 1.3.1, 3.12.8, 3.14.2, 4.1.2,
  4.3.4, 4.5.5, 9.3.2, 9.8.2, 9.9.1, 9.10.2, 9.10.3, 10.1.3,
  11.3.1, 11.3.1.4, 11.3.11, 13.2, 13.4.2
CONSTRUCTION BY OWNER OR BY SEPARATE
  CONTRACTORS 1.1.4, **6**
Construction Change Directive, Definition of 7.3.1
Construction Change Directives 1.1.1, 4.2.8, 7.1, **7.3**, 9.3.1.1
Construction Schedules, Contractor's 3.10, 6.1.3
Contingent Assignment of Subcontracts **5.4**
Continuing Contract Performance **4.3.4**
Contract, Definition of 1.1.2
CONTRACT, TERMINATION OR
  SUSPENSION OF THE 4.3.7, 5.4.1.1, **14**
Contract Administration 3.3.3, 4, 9.4, 9.5
Contract Award and Execution, Conditions Relating to 3.7.1,
  3.10, 5.2, 9.2, 11.1.3, 11.3.6, 11.4.1
Contract Documents, The **1.1**, 1.2, 7
Contract Documents, Copies Furnished and Use of ... 1.3, 2.2.5, 5.3
Contract Documents, Definition of 1.1.1
Contract Performance During Arbitration 4.3.4, 4.5.3
Contract Sum 3.8, 4.3.6, 4.3.7, 4.4.4, 5.2.3,
  6.1.3, 7.2, 7.3, **9.1**, 9.7, 11.3.1, 12.2.4, 12.3, 14.2.4
Contract Sum, Definition of............................... **9.1**
Contract Time 4.3.6, 4.3.8, 4.4.4, 7.2.1.3, 7.3,
  8.2.1, 8.3.1, 9.7, 12.1.1
Contract Time, **Definition of** **8.1.1**

Appendix

CONTRACTOR 3
Contractor, **Definition** of **3.1**, 6.1.2
Contractor's Bid 1.1.1
Contractor's Construction Schedules **3.10**, 6.1.3
Contractor's Employees 3.3.2, 3.4.2, 3.8.1, 3.9, 3.18, 4.2.3,
    4.2.6, 8.1.2, 10.2, 10.3, 11.1.1, 14.2.1.1
Contractor's Liability Insurance **11.1**
Contractor's Relationship with Separate Contractors
   and Owner's Forces 2.2.6, 3.12.5, 3.14.2, 4.2.4, 6, 12.2.5
Contractor's Relationship with Subcontractors 1.2.4, 3.3.2,
    3.18.1, 3.18.2, 5.2, 5.3, 5.4, 9.6.2, 11.3.7, 11.3.8, 14.2.1.2
Contractor's Relationship with the Architect 1.1.2, 3.2.1, 3.2.2,
    3.3.3, 3.5.1, 3.7.3, 3.11, 3.12.8 3.16, 3.18, 4.2.3, 4.2.4, 4.2.6,
    4.2.12, 5.2, 6.2.2, 7.3.4, 9.8.2, 11.3.7, 12.1, 13.5
Contractor's Representations . . 1.2.2, 3.5.1, 3.12.7, 6.2.2, 8.2.1, 9.3.3
Contractor's Responsibility for Those
   Performing the Work 3.3.2, 3.18, 4.2.3, 10
Contractor's Review of Contract Documents 1.2.2, 3.2, 3.7.3
Contractor's Right to Stop the Work 9.7
Contractor's Right to Terminate the Contract 14.1
Contractor's Submittals 3.10, 3.11, 3.12, 4.2.7, 5.2.1, 5.2.3,
    7.3.6, 9.2, 9.3.1, 9.8.2, 9.8.3, 9.9.1, 9.10.2,
    9.10.3, 10.1.2, 11.4.2, 11.4.3
Contractor's Superintendent 3.9, 10.2.6
Contractor's Supervision and Construction Procedures ... 1.2.4,
    3.3, 3.4, 4.2.3, 8.2.2, 8.2.3, 10
Contractual Liability Insurance 11.1.1.7, 11.2.1
Coordination and Correlation 1.2.2, 1.2.4, 3.3.1,
    3.10, 3.12.7, 6.1.3, 6.2.1
Copies Furnished of Drawings and Specifications ... 1.3, 2.2.5, 3.11
Correction of Work 2.3, 2.4, 4.2.1, 9.8.2,
    9.9.1, 12.1.2, 12.2, 13.7.1.3
Cost, Definition of 7.3.6, 14.3.5
Costs 2.4, 3.2.1, 3.7.4, 3.8.2, 3.15.2, 4.3.6, 4.3.7, 4.3.8.1, 5.2.3,
    6.1.1, 6.2.3, 6.3, 7.3.3.3, 7.3.6, 7.3.7, 9.7, 9.8.2, 9.10.2, 11.3.1.2,
    11.3.1.3, 11.3.4, 11.3.9, 12.1, 12.2.1, 12.2.4, 12.2.5, 13.5, 14
Cutting and Patching **3.14**, 6.2.6
Damage to Construction of Owner or Separate Contractors 3.14.2,
    6.2.4, 9.5.1.5, 10.2.1.2, 10.2.5, 10.3, 11.1, 11.3, 12.2.5
Damage to the Work 3.14.2, 9.9.1, 10.2.1.2, 10.2.5, 10.3, 11.3
Damages, Claims for . . 3.18, 4.3.9, 6.1.1, 6.2.5, 8.3.2, 9.5.1.2, 10.1.4
Damages for Delay 6.1.1, 8.3.3, 9.5.1.6, 9.7
Date of Commencement of the Work, Definition of 8.1.2
Date of Substantial Completion, Definition of 8.1.3
Day, Definition of 8.1.4
Decisions of the Architect 4.2.6, 4.2.7, 4.2.11, 4.2.12, 4.2.13,
    4.3.2, 4.3.6, 4.4.1, 4.4.4, 4.5, 6.3, 7.3.6, 7.3.8, 8.1.3, 8.3.1, 9.2,
    9.4, 9.5.1, 9.8.2, 9.9.1, 10.1.2, 13.5.2, 14.2.2, 14.2.4
Decisions to Withhold Certification **9.5**, 9.7, 14.1.1.3
Defective or Nonconforming Work, Acceptance,
   Rejection and Correction of 2.3, 2.4, 3.5.1, 4.2.1,
    4.2.6, 4.3.5, 9.5.2, 9.8.2, 9.9.1, 10.2.5, 12, 13.7.1.3
Defective Work, Definition of 3.5.1
Definitions 1.1, 2.1.1, 3.1, 3.5.1, 3.12.1, 3.12.2, 3.12.3, 4.1.1,
    4.3.1, 5.1, 6.1.2, 7.2.1, 7.3.1, 7.3.6, 8.1, 9.1, 9.8.1
Delays and Extensions of Time 4.3.1, 4.3.8.1, 4.3.8.2,
    6.1.1, 6.2.3, 7.2.1, 7.3.1, 7.3.4, 7.3.5, 7.3.8,
    7.3.9, 8.1.1, **8.3**, 10.3.1, 14.1.1.4
Disputes 4.1.4, 4.3, 4.4, 4.5, 6.2.5, 6.3, 7.3.8, 9.3.1.2
Documents and Samples at the Site 3.11
Drawings, Definition of 1.1.5
Drawings and Specifications, Use and Ownership of 1.1.1, 1.3,
    2.2.5, 3.11, 5.3
Duty to Review Contract Documents and Field Conditions 3.2
Effective Date of Insurance 8.2.2, 11.1.2

Emergencies 4.3.7, **10.3**
Employees, Contractor's 3.3.2, 3.4.2, 3.8.1, 3.9, 3.18.1,
    3.18.2, 4.2.3, 4.2.6, 8.1.2, 10.2, 10.3, 11.1.1, 14.2.1.1
Equipment, Labor, Materials and 1.1.3, 1.1.6, 3.4, 3.5.1,
    3.8.2, 3.12.3, 3.12.7, 3.12.11, 3.13, 3.15.1, 4.2.7,
    6.2.1, 7.3.6, 9.3.2, 9.3.3, 11.3, 12.2.4, 14
Execution and Progress of the Work 1.1.3, 1.2.3, 3.2, 3.4.1,
    3.5.1, 4.2.2, 4.2.3, 4.3.4, 4.3.8, 6.2.2, 7.1.3,
    7.3.9, 8.2, 8.3, 9.5, 9.9.1, 10.2, 14.2, 14.3
Execution, Correlation and Intent of the
   Contract Documents **1.2**, 3.7.1
Extensions of Time 4.3.1, 4.3.8, 7.2.1.3, 8.3, 10.3.1
Failure of Payment by Contractor 9.5.1.3, 14.2.1.2
Failure of Payment by Owner 4.3.7, 9.7, 14.1.3
Faulty Work (See Defective or Nonconforming Work)
Final Completion and Final Payment 4.2.1, 4.2.9, 4.3.2,
    4.3.5, **9.10**, 11.1.2, 11.1.3, 11.3.5, 12.3.1, 13.7
Financial Arrangements, Owner's 2.2.1
Fire and Extended Coverage Insurance 11.3
GENERAL PROVISIONS **1**
Governing Law **13.1**
Guarantees (See Warranty and Warranties)
Hazardous Materials 10.1, 10.2.4
Identification of Contract Documents 1.2.1
Identification of Subcontractors and Suppliers 5.2.1
Indemnification 3.17, **3.18**, 9.10.2, 10.1.4, 11.3.1.2, 11.3.7
Information and Services Required of the Owner 2.1.2, **2.2**,
    4.3.4, 6.1.3, 6.1.4, 6.2.6, 9.3.2, 9.6.1, 9.6.4, 9.8.3, 9.9.2,
    9.10.3, 10.1.4, 11.2, 11.3, 13.5.1, 13.5.2
Injury or Damage to Person or Property **4.3.9**
Inspections 3.3.3, 3.3.4, 3.7.1, 4.2.2,
    4.2.6, 4.2.9, 4.3.6, 9.4.2, 9.8.2, 9.9.2, 9.10.1, 13.5
Instructions to Bidders 1.1.1
Instructions to the Contractor 3.8.1, 4.2.8, 5.2.1, 7, 12.1, 13.5.2
Insurance 4.3.9, 6.1.1, 7.3.6.4, 9.3.2, 9.8.2, 9.9.1, 9.10.2, 11
Insurance, Boiler and Machinery **11.3.2**
Insurance, Contractor's Liability **11.1**
Insurance, Effective Date of 8.2.2, 11.1.2
Insurance, Loss of Use **11.3.3**
Insurance, Owner's Liability **11.2**
Insurance, Property 10.2.5, **11.3**
Insurance, Stored Materials 9.3.2, 11.3.1.4
INSURANCE AND BONDS **11**
Insurance Companies, Consent to Partial Occupancy . . 9.9.1, 11.3.11
Insurance Companies, Settlement with 11.3.10
Intent of the Contract Documents 1.2.3, 3.12.4,
    4.2.6, 4.2.7, 4.2.12, 4.2.13, 7.4
Interest **13.6**
Interpretation 1.2.5, 1.4, **1.5**, 4.1.1, 4.3.1, 5.1, 6.1.2, 8.1.4
Interpretations, Written 4.2.11, 4.2.12, 4.3.7
Joinder and Consolidation of Claims Required 4.5.6
Judgment on Final Award 4.5.1, 4.5.4.1, **4.5.7**
Labor and Materials, Equipment 1.1.3, 1.1.6, **3.4**, 3.5.1, 3.8.2,
    3.12.2, 3.12.3, 3.12.7, 3.12.11, 3.13, 3.15.1,
    4.2.7, 6.2.1, 7.3.6, 9.3.2, 9.3.3, 12.2.4, 14
Labor Disputes 8.3.1
Laws and Regulations 1.3, 3.6, 3.7, 3.13, 4.1.1, 4.5.5, 4.5.7,
    9.9.1, 10.2.2, 11.1, 11.3, 13.1, 13.4, 13.5.1, 13.5.2, 13.6
Liens 2.1.2, 4.3.2, 4.3.5.1, 8.2.2, 9.3.3, 9.10.2
Limitation on Consolidation or Joinder **4.5.5**
Limitations, Statutes of 4.5.4.2, 12.2.6, 13.7
Limitations of Authority 3.3.1, 4.1.2, 4.2.1,
    4.2.3, 4.2.7, 4.2.10, 5.2.2, 5.2.4, 7.4, 11.3.10

Limitations of Liability	2.3, 3.2.1, 3.5.1, 3.7.3, 3.12.8, 3.12.11, 3.17, 3.18, 4.2.6, 4.2.7, 4.2.12, 6.2.2, 9.4.2, 9.6.4, 9.10.4, 10.1.4, 10.2.5, 11.1.2, 11.2.1, 11.3.7, 13.4.2, 13.5.2
Limitations of Time, General	2.2.1, 2.2.4, 3.2.1, 3.7.3, 3.8.2, 3.10, 3.12.5, 3.15.1, 4.2.1, 4.2.7, 4.2.11, 4.3.2, 4.3.3, 4.3.4, 4.3.6, 4.3.9, 4.5.4.2, 5.2.1, 5.2.3, 6.2.4, 7.3.4, 7.4, 8.2, 9.5, 9.6.2, 9.8, 9.9, 9.10, 11.1.3, 11.3.1, 11.3.2, 11.3.5, 11.3.6, 12.2.1, 12.2.2, 13.5, 13.7
Limitations of Time, Specific	2.1.2, 2.2.1, 2.4, 3.10, 3.11, 3.15.1, 4.2.1, 4.2.11, 4.3, 4.4, 4.5, 5.3, 5.4, 7.3.5, 7.3.9, 8.2, 9.2, 9.3.1, 9.3.3, 9.4.1, 9.6.1, 9.7, 9.8.2, 9.10.2, 11.1.3, 11.3.6, 11.3.10, 11.3.11, 12.2.2, 12.2.4, 12.2.6, 13.7, 14
Loss of Use Insurance	**11.3.3**
Material Suppliers	1.3.1, 3.12.1, 4.2.4, 4.2.6, 5.2.1, 9.3.1, 9.3.1.2, 9.3.3, 9.4.2, 9.6.5, 9.10.4
Materials, Hazardous	10.1, 10.2.4
Materials, Labor, Equipment and	1.1.3, 1.1.6, 3.4, 3.5.1, 3.8.2, 3.12.2, 3.12.3, 3.12.7, 3.12.11, 3.13, 3.15.1, 4.2.7, 6.2.1, 7.3.6, 9.3.2, 9.3.3, 12.2.4, 14
Means, Methods, Techniques, Sequences and Procedures of Construction	3.3.1, 4.2.3, 4.2.7, 9.4.2
Minor Changes in the Work	1.1.1, 4.2.8, 4.3.7, 7.1, 7.4
MISCELLANEOUS PROVISIONS	**13**
Modifications, Definition of	1.1.1
Modifications to the Contract	1.1.1, 1.1.2, 3.7.3, 3.11, 4.1.2, 4.2.1, 5.2.3, 7, 8.3.1, 9.7
Mutual Responsibility	**6.2**
Nonconforming Work, Acceptance of	**12.3**
Nonconforming Work, Rejection and Correction of	2.3.1, 4.3.5, 9.5.2, 9.8.2, 12, 13.7.1.3
Notice	2.3, 2.4, 3.2.1, 3.2.2, 3.7.3, 3.7.4, 3.9, 3.12.8, 3.12.9, 3.17, 4.3, 4.4.4, 4.5, 5.2.1, 5.5, 5.4.1.1, 8.2.2, 9.4.1, 9.5.1, 9.6.1, 9.7, 9.10, 10.1.2, 10.2.6, 11.1.3, 11.3, 12.2.2, 12.2.4, 13.3, 13.5.1, 13.5.2, 14
Notice, Written	2.3, 2.4, 3.9, 3.12.8, 3.12.9, 4.3, 4.4.4, 4.5, 5.2.1, 5.3, 5.4.1.1, 8.2.2, 9.4.1, 9.5.1, 9.7, 9.10, 10.1.2, 10.2.6, 11.1.3, 11.3, 12.2.2, 12.2.4, **13.3**, 13.5.2, 14
Notice of Testing and Inspections	13.5.1, 13.5.2
Notice to Proceed	8.2.2
Notices, Permits, Fees and	2.2.3, **3.7**, 3.13, 7.3.6.4, 10.2.2
Observations, Architect's On-Site	4.2.2, 4.2.5, 4.3.6, 9.4.2, 9.5.1, 9.10.1, 13.5
Observations, Contractor's	1.2.2, 3.2.2
Occupancy	9.6.6, 9.8.1, 9.9, 11.3.11
On-Site Inspections by the Architect	4.2.2, 4.2.9, 4.3.6, 9.4.2, 9.8.2, 9.9.2, 9.10.1
On-Site Observations by the Architect	4.2.2, 4.2.5, 4.3.6, 9.4.2, 9.5.1, 9.10.1, 13.5
Orders, Written	2.3, 3.9, 4.3.7, 7, 8.2.2, 11.3.9, 12.1, 12.2, 13.5.2, 14.3.1
OWNER	**2**
Owner, **Definition** of	**2.1**
Owner, Information and Services Required of the	2.1.2, **2.2**, 4.3.4, 6, 9, 10.1.4, 11.2, 11.3, 13.5.1, 14.1.1.5, 14.1.3
Owner's Authority	3.8.1, 4.1.3, 4.2.9, 5.2.1, 5.2.4, 5.4.1, 7.3.1, 8.2.2, 9.3.1, 9.3.2, 11.4.1, 12.2.4, 13.5.2, 14.2, 14.3.1
Owner's Financial Capability	2.2.1, 14.1.1.5
Owner's Liability Insurance	**11.2**
Owner's Loss of Use Insurance	11.3.3
Owner's Relationship with Subcontractors	1.1.2, 5.2.1, 5.4.1, 9.6.4
Owner's Right to Carry Out the Work	2.4, 12.2.4, 14.2.2.2
Owner's Right to Clean Up	**6.3**
Owner's Right to Perform Construction and to Award Separate Contracts	**6.1**
Owner's Right to Stop the Work	**2.3**, 4.3.7
Owner's Right to Suspend the Work	14.3
Owner's Right to Terminate the Contract	14.2
Ownership and Use of Architect's Drawings, Specifications and Other Documents	1.1.1, **1.3**, 2.2.5, 5.3
Partial Occupancy or Use	9.6.6, **9.9**, 11.3.11
Patching, Cutting and	**3.14**, 6.2.6
Patents, Royalties and	**3.17**
Payment, Applications for	4.2.5, 9.2, **9.3**, 9.4, 9.5.1, 9.8.3, 9.10.1, 9.10.3, 9.10.4, 14.2.4
Payment, Certificates for	4.2.5, 4.2.9, 9.3.3, **9.4**, 9.5, 9.6.1, 9.6.6, 9.7.1, 9.8.3, 9.10.1, 9.10.3, 13.7, 14.1.1.3, 14.2.4
Payment, Failure of	4.3.7, 9.5.1.3, **9.7**, 9.10.2, 14.1.1.3, 14.2.1.2
Payment, Final	4.2.1, 4.2.9, 4.3.2, 4.3.5, 9.10, 11.1.2, 11.1.3, 11.3.5, 12.3.1
Payment Bond, Performance Bond and	7.3.6.4, 9.10.3, 11.3.9, **11.4**
Payments, Progress	4.3.4, 9.3, 9.6, 9.8.3, 9.10.3, 13.6, 14.2.3
PAYMENTS AND COMPLETION	**9**, 14
Payments to Subcontractors	5.4.2, 9.5.1.3, 9.6.2, 9.6.3, 9.6.4, 11.3.8, 14.2.1.2
PCB	10.1
Performance Bond and Payment Bond	7.3.6.4, 9.10.3, 11.3.9, 11.4
Permits, Fees and Notices	2.2.3, **3.7**, 3.13, 7.3.6.4, 10.2.2
PERSONS AND PROPERTY, PROTECTION OF	**10**
Polychlorinated Biphenyl	10.1
Product Data, Definition of	3.12.2
Product Data and Samples, Shop Drawings	3.11, **3.12**, 4.2.7
Progress and Completion	4.2.2, 4.3.4, **8.2**
Progress Payments	4.3.4, 9.3, **9.6**, 9.8.3, 9.10.3, 13.6, 14.2.3
Project, Definition of the	**1.1.4**
Project Manual, Definition of the	**1.1.7**
Project Manuals	2.2.5
Project Representatives	4.2.10
Property Insurance	10.2.5, **11.3**
PROTECTION OF PERSONS AND PROPERTY	**10**
Regulations and Laws	1.3, 3.6, 3.7, 3.13, 4.1.1, 4.5.5, 4.5.7, 10.2.2, 11.1, 11.3, 13.1, 13.4, 13.5.1, 13.5.2, 13.6, 14
Rejection of Work	3.5.1, 4.2.6, 12.2
Releases of Waivers and Liens	9.10.2
Representations	1.2.2, 3.5.1, 3.12.7, 6.2.2, 8.2.1, 9.3.3, 9.4.2, 9.5.1, 9.8.2, 9.10.1
Representatives	2.1.1, 3.1.1, 3.9, 4.1.1, 4.2.1, 4.2.10, 5.1.1, 5.1.2, 13.2.1
Resolution of Claims and Disputes	**4.4**, 4.5
Responsibility for Those Performing the Work	3.3.2, 4.2.3, 6.1.3, 6.2, 10
Retainage	9.3.1, 9.6.2, 9.8.3, 9.9.1, 9.10.2, 9.10.3
Review of Contract Documents and Field Conditions by Contractor	1.2.2, **3.2**, 3.7.3, 3.12.7
Review of Contractor's Submittals by Owner and Architect	3.10.1, 3.10.2, 3.11, 3.12, 4.2.7, 4.2.9, 5.2.1, 5.2.3, 9.2, 9.8.2
Review of Shop Drawings, Product Data and Samples by Contractor	3.12.5
Rights and Remedies	1.1.2, 2.3, 2.4, 3.5.1, 3.15.2, 4.2.6, 4.3.6, 4.5, 5.3, 6.1, 6.3, 7.3.1, 8.3.1, 9.5.1, 9.7, 10.2.5, 10.3, 12.2.2, 12.2.4, **13.4**, 14
Royalties and Patents	**3.17**

Rules and Notices for Arbitration	**4.5.2**
Safety of Persons and Property	**10.2**
Safety Precautions and Programs	**4.2.3, 4.2.7, 10.1**
Samples, Definition of	3.12.3
Samples, Shop Drawings, Product Data and	3.11, **3.12**, 4.2.7
Samples at the Site, Documents and	**3.11**
Schedule of Values	**9.2**, 9.3.1
Schedules, Construction	3.10
Separate Contracts and Contractors	1.1.4, 3.14.2, 4.2.4, 4.5.5, 6, 11.3.7, 12.1.2, 12.2.5
Shop Drawings, Definition of	3.12.1
Shop Drawings, Product Data and Samples	3.11, **3.12**, 4.2.7
Site, Use of	**3.13**, 6.1.1, 6.2.1
Site Inspections	1.2.2, 3.3.4, 4.2.2, 4.2.9, 4.3.6, 9.8.2, 9.10.1, 13.5
Site Visits, Architect's	4.2.2, 4.2.5, 4.2.9, 4.3.6, 9.4.2, 9.5.1, 9.8.2, 9.9.2, 9.10.1, 13.5
Special Inspections and Testing	4.2.6, 12.2.1, 13.5
Specifications, Definition of the	**1.1.6**
Specifications, The	1.1.1, **1.1.6**, 1.1.7, 1.2.4, 1.3, 3.11
Statutes of Limitations	4.5.4.2, 12.2.6, 13.7
Stopping the Work	2.3, 4.3.7, 9.7, 10.1.2, 10.3, 14.1
Stored Materials	6.2.1, 9.3.2, 10.2.1.2, 11.3.1.4, 12.2.4
Subcontractor, Definition of	5.1.1
SUBCONTRACTORS	**5**
Subcontractors, Work by	1.2.4, 3.3.2, 3.12.1, 4.2.3, 5.3, 5.4
Subcontractual Relations	**5.3**, 5.4, 9.3.1.2, 9.6.2, 9.6.3, 9.6.4, 10.2.1, 11.3.7, 11.3.8, 14.1, 14.2.1.2, 14.3.2
Submittals	1.3, 3.2.3, 3.10, 3.11, 3.12, 4.2.7, 5.2.1, 5.2.3, 7.3.6, 9.2, 9.3.1, 9.8.2, 9.9.1, 9.10.2, 9.10.3, 10.1.2, 11.1.3
Subrogation, Waivers of	6.1.1, 11.3.5, **11.3.7**
Substantial Completion	4.2.9, 4.3.5.2, 8.1.1, 8.1.3, 8.2.3, **9.8**, 9.9.1, 12.2.1, 12.2.2, 13.7
Substantial Completion, Definition of	9.8.1
Substitution of Subcontractors	5.2.3, 5.2.4
Substitution of the Architect	4.1.3
Substitutions of Materials	3.5.1
Sub-subcontractor, Definition of	5.1.2
Subsurface Conditions	4.3.6
Successors and Assigns	**13.2**
Superintendent	**3.9**, 10.2.6
Supervision and Construction Procedures	1.2.4, **3.3**, 3.4, 4.2.3, 4.3.4, 6.1.3, 6.2.4, 7.1.3, 7.3.4, 8.2, 8.3.1, 10, 12, 14
Surety	4.4.1, 4.4.4, 5.4.1.2, 9.10.2, 9.10.3, 14.2.2
Surety, Consent of	9.9.1, 9.10.2, 9.10.3
Surveys	2.2.2, 3.18.3
Suspension by the Owner for Convenience	**14.3**
Suspension of the Work	4.3.7, 5.4.2, 14.1.1.4, 14.3
Suspension or Termination of the Contract	4.3.7, 5.4.1.1, 14
Taxes	**3.6**, 7.3.6.4
Termination by the Contractor	**14.1**
Termination by the Owner for Cause	5.4.1.1, **14.2**
Termination of the Architect	4.1.3
Termination of the Contractor	14.2.2
TERMINATION OR SUSPENSION OF THE CONTRACT	**14**
Tests and Inspections	3.3.3, 4.2.6, 4.2.9, 9.4.2, 12.2.1, **13.5**
TIME	**8**
Time, Delays and Extensions of	4.3.8, 7.2.1, **8.3**
Time Limits, Specific	2.1.2, 2.2.1, 2.4, 3.10, 3.11, 3.15.1, 4.2.1, 4.2.11, 4.3, 4.4, 4.5, 5.3, 5.4, 7.3.5, 7.3.9, 8.2, 9.2, 9.3.1, 9.3.3, 9.4.1, 9.6.1, 9.7, 9.8.2, 9.10.2, 11.1.3, 11.3.6, 11.3.10, 11.3.11, 12.2.2, 12.2.4, 12.2.6, 13.7, 14
Time Limits on Claims	4.3.2, **4.3.3**, 4.3.6, 4.3.9, 4.4, 4.5
Title to Work	9.3.2, 9.3.3
UNCOVERING AND CORRECTION OF WORK	**12**
Uncovering of Work	**12.1**
Unforeseen Conditions	4.3.6, 8.3.1, 10.1
Unit Prices	7.1.4, 7.3.3.2
Use of Documents	1.1.1, 1.3, 2.2.5, 3.12.7, 5.3
Use of Site	**3.13**, 6.1.1, 6.2.1
Values, Schedule of	**9.2**, 9.3.1
Waiver of Claims: Final Payment	**4.3.5**, 4.5.1, 9.10.3
Waiver of Claims by the Architect	13.4.2
Waiver of Claims by the Contractor	9.10.4, 11.3.7, 13.4.2
Waiver of Claims by the Owner	4.3.5, 4.5.1, 9.9.3, 9.10.3, 11.3.3, 11.3.5, 11.3.7, 13.4.2
Waiver of Liens	9.10.2
Waivers of Subrogation	6.1.1, 11.3.5, 11.3.7
Warranty and Warranties	**3.5**, 4.2.9, 4.3.5.3, 9.3.3, 9.8.2, 9.9.1, 12.2.2, 13.7
Weather Delays	4.3.8.2
When Arbitration May Be Demanded	**4.5.4**
Work, Definition of	1:1.3
Written Consent	1.3.1, 3.12.8, 3.14.2, 4.1.2, 4.3.4, 4.5.5, 9.3.2, 9.8.2, 9.9.1, 9.10.2, 9.10.3, 10.1.2, 10.1.3, 11.3.1, 11.3.1.4, 11.3.11, 13.2, 13.4.2
Written Interpretations	4.2.11, 4.2.12, 4.3.7
Written Notice	2.3, 2.4, 3.9, 3.12.8, 3.12.9, 4.3, 4.4.4, 4.5, 5.2.1, 5.3, 5.4.1.1, 8.2.2, 9.4.1, 9.5.1, 9.7, 9.10, 10.1.2, 10.2.6, 11.1.3, 11.3, 12.2.2, 12.2.4, **13.3**, 13.5.2, 14
Written Orders	2.3, 3.9, 4.3.7, 7, 8.2.2, 11.3.9, 12.1, 12.2, 13.5.2, 14.3.1

GENERAL CONDITIONS OF THE CONTRACT FOR CONSTRUCTION

ARTICLE 1
GENERAL PROVISIONS

1.1 BASIC DEFINITIONS

1.1.1 THE CONTRACT DOCUMENTS

The Contract Documents consist of the Agreement between Owner and Contractor (hereinafter the Agreement), Conditions of the Contract (General, Supplementary and other Conditions), Drawings, Specifications, addenda issued prior to execution of the Contract, other documents listed in the Agreement and Modifications issued after execution of the Contract. A Modification is (1) a written amendment to the Contract signed by both parties, (2) a Change Order, (3) a Construction Change Directive or (4) a written order for a minor change in the Work issued by the Architect. Unless specifically enumerated in the Agreement, the Contract Documents do not include other documents such as bidding requirements (advertisement or invitation to bid, Instructions to Bidders, sample forms, the Contractor's bid or portions of addenda relating to bidding requirements).

1.1.2 THE CONTRACT

The Contract Documents form the Contract for Construction. The Contract represents the entire and integrated agreement between the parties hereto and supersedes prior negotiations, representations or agreements, either written or oral. The Contract may be amended or modified only by a Modification. The Contract Documents shall not be construed to create a contractual relationship of any kind (1) between the Architect and Contractor, (2) between the Owner and a Subcontractor or Sub-subcontractor or (3) between any persons or entities other than the Owner and Contractor. The Architect shall, however, be entitled to performance and enforcement of obligations under the Contract intended to facilitate performance of the Architect's duties.

1.1.3 THE WORK

The term "Work" means the construction and services required by the Contract Documents, whether completed or partially completed, and includes all other labor, materials, equipment and services provided or to be provided by the Contractor to fulfill the Contractor's obligations. The Work may constitute the whole or a part of the Project.

1.1.4 THE PROJECT

The Project is the total construction of which the Work performed under the Contract Documents may be the whole or a part and which may include construction by the Owner or by separate contractors.

1.1.5 THE DRAWINGS

The Drawings are the graphic and pictorial portions of the Contract Documents, wherever located and whenever issued, showing the design, location and dimensions of the Work, generally including plans, elevations, sections, details, schedules and diagrams.

1.1.6 THE SPECIFICATIONS

The Specifications are that portion of the Contract Documents consisting of the written requirements for materials, equipment, construction systems, standards and workmanship for the Work, and performance of related services.

1.1.7 THE PROJECT MANUAL

The Project Manual is the volume usually assembled for the Work which may include the bidding requirements, sample forms, Conditions of the Contract and Specifications.

1.2 EXECUTION, CORRELATION AND INTENT

1.2.1 The Contract Documents shall be signed by the Owner and Contractor as provided in the Agreement. If either the Owner or Contractor or both do not sign all the Contract Documents, the Architect shall identify such unsigned Documents upon request.

1.2.2 Execution of the Contract by the Contractor is a representation that the Contractor has visited the site, become familiar with local conditions under which the Work is to be performed and correlated personal observations with requirements of the Contract Documents.

1.2.3 The intent of the Contract Documents is to include all items necessary for the proper execution and completion of the Work by the Contractor. The Contract Documents are complementary, and what is required by one shall be as binding as if required by all; performance by the Contractor shall be required only to the extent consistent with the Contract Documents and reasonably inferable from them as being necessary to produce the intended results.

1.2.4 Organization of the Specifications into divisions, sections and articles, and arrangement of Drawings shall not control the Contractor in dividing the Work among Subcontractors or in establishing the extent of Work to be performed by any trade.

1.2.5 Unless otherwise stated in the Contract Documents, words which have well-known technical or construction industry meanings are used in the Contract Documents in accordance with such recognized meanings.

1.3 OWNERSHIP AND USE OF ARCHITECT'S DRAWINGS, SPECIFICATIONS AND OTHER DOCUMENTS

1.3.1 The Drawings, Specifications and other documents prepared by the Architect are instruments of the Architect's service through which the Work to be executed by the Contractor is described. The Contractor may retain one contract record set. Neither the Contractor nor any Subcontractor, Sub-subcontractor or material or equipment supplier shall own or claim a copyright in the Drawings, Specifications and other documents prepared by the Architect, and unless otherwise indicated the Architect shall be deemed the author of them and will retain all common law, statutory and other reserved rights, in addition to the copyright. All copies of them, except the Contractor's record set, shall be returned or suitably accounted for to the Architect, on request, upon completion of the Work. The Drawings, Specifications and other documents prepared by the Architect, and copies thereof furnished to the Contractor, are for use solely with respect to this Project. They are not to be used by the Contractor or any Subcontractor, Sub-subcontractor or material or equipment supplier on other projects or for additions to this Project outside the scope of the

Work without the specific written consent of the Owner and Architect. The Contractor, Subcontractors, Sub-subcontractors and material or equipment suppliers are granted a limited license to use and reproduce applicable portions of the Drawings, Specifications and other documents prepared by the Architect appropriate to and for use in the execution of their Work under the Contract Documents. All copies made under this license shall bear the statutory copyright notice, if any, shown on the Drawings, Specifications and other documents prepared by the Architect. Submittal or distribution to meet official regulatory requirements or for other purposes in connection with this Project is not to be construed as publication in derogation of the Architect's copyright or other reserved rights.

1.4 CAPITALIZATION

1.4.1 Terms capitalized in these General Conditions include those which are (1) specifically defined, (2) the titles of numbered articles and identified references to Paragraphs, Subparagraphs and Clauses in the document or (3) the titles of other documents published by the American Institute of Architects.

1.5 INTERPRETATION

1.5.1 In the interest of brevity the Contract Documents frequently omit modifying words such as "all" and "any" and articles such as "the" and "an," but the fact that a modifier or an article is absent from one statement and appears in another is not intended to affect the interpretation of either statement.

ARTICLE 2
OWNER

2.1 DEFINITION

2.1.1 The Owner is the person or entity identified as such in the Agreement and is referred to throughout the Contract Documents as if singular in number. The term "Owner" means the Owner or the Owner's authorized representative.

2.1.2 The Owner upon reasonable written request shall furnish to the Contractor in writing information which is necessary and relevant for the Contractor to evaluate, give notice of or enforce mechanic's lien rights. Such information shall include a correct statement of the record legal title to the property on which the Project is located, usually referred to as the site, and the Owner's interest therein at the time of execution of the Agreement and, within five days after any change, information of such change in title, recorded or unrecorded.

2.2 INFORMATION AND SERVICES REQUIRED OF THE OWNER

2.2.1 The Owner shall, at the request of the Contractor, prior to execution of the Agreement and promptly from time to time thereafter, furnish to the Contractor reasonable evidence that financial arrangements have been made to fulfill the Owner's obligations under the Contract. *[Note: Unless such reasonable evidence were furnished on request prior to the execution of the Agreement, the prospective contractor would not be required to execute the Agreement or to commence the Work.]*

2.2.2 The Owner shall furnish surveys describing physical characteristics, legal limitations and utility locations for the site of the Project, and a legal description of the site.

2.2.3 Except for permits and fees which are the responsibility of the Contractor under the Contract Documents, the Owner shall secure and pay for necessary approvals, easements, assessments and charges required for construction, use or occupancy of permanent structures or for permanent changes in existing facilities.

2.2.4 Information or services under the Owner's control shall be furnished by the Owner with reasonable promptness to avoid delay in orderly progress of the Work.

2.2.5 Unless otherwise provided in the Contract Documents, the Contractor will be furnished, free of charge, such copies of Drawings and Project Manuals as are reasonably necessary for execution of the Work.

2.2.6 The foregoing are in addition to other duties and responsibilities of the Owner enumerated herein and especially those in respect to Article 6 (Construction by Owner or by Separate Contractors), Article 9 (Payments and Completion) and Article 11 (Insurance and Bonds).

2.3 OWNER'S RIGHT TO STOP THE WORK

2.3.1 If the Contractor fails to correct Work which is not in accordance with the requirements of the Contract Documents as required by Paragraph 12.2 or persistently fails to carry out Work in accordance with the Contract Documents, the Owner, by written order signed personally or by an agent specifically so empowered by the Owner in writing, may order the Contractor to stop the Work, or any portion thereof, until the cause for such order has been eliminated; however, the right of the Owner to stop the Work shall not give rise to a duty on the part of the Owner to exercise this right for the benefit of the Contractor or any other person or entity, except to the extent required by Subparagraph 6.1.3.

2.4 OWNER'S RIGHT TO CARRY OUT THE WORK

2.4.1 If the Contractor defaults or neglects to carry out the Work in accordance with the Contract Documents and fails within a seven-day period after receipt of written notice from the Owner to commence and continue correction of such default or neglect with diligence and promptness, the Owner may after such seven-day period give the Contractor a second written notice to correct such deficiencies within a second seven-day period. If the Contractor within such second seven-day period after receipt of such second notice fails to commence and continue to correct any deficiencies, the Owner may, without prejudice to other remedies the Owner may have, correct such deficiencies. In such case an appropriate Change Order shall be issued deducting from payments then or thereafter due the Contractor the cost of correcting such deficiencies, including compensation for the Architect's additional services and expenses made necessary by such default, neglect or failure. Such action by the Owner and amounts charged to the Contractor are both subject to prior approval of the Architect. If payments then or thereafter due the Contractor are not sufficient to cover such amounts, the Contractor shall pay the difference to the Owner.

ARTICLE 3
CONTRACTOR

3.1 DEFINITION

3.1.1 The Contractor is the person or entity identified as such in the Agreement and is referred to throughout the Contract Documents as if singular in number. The term "Contractor" means the Contractor or the Contractor's authorized representative.

3.2 REVIEW OF CONTRACT DOCUMENTS AND FIELD CONDITIONS BY CONTRACTOR

3.2.1 The Contractor shall carefully study and compare the Contract Documents with each other and with information furnished by the Owner pursuant to Subparagraph 2.2.2 and shall at once report to the Architect errors, inconsistencies or omissions discovered. The Contractor shall not be liable to the Owner or Architect for damage resulting from errors, inconsistencies or omissions in the Contract Documents unless the Contractor recognized such error, inconsistency or omission and knowingly failed to report it to the Architect. If the Contractor performs any construction activity knowing it involves a recognized error, inconsistency or omission in the Contract Documents without such notice to the Architect, the Contractor shall assume appropriate responsibility for such performance and shall bear an appropriate amount of the attributable costs for correction.

3.2.2 The Contractor shall take field measurements and verify field conditions and shall carefully compare such field measurements and conditions and other information known to the Contractor with the Contract Documents before commencing activities. Errors, inconsistencies or omissions discovered shall be reported to the Architect at once.

3.2.3 The Contractor shall perform the Work in accordance with the Contract Documents and submittals approved pursuant to Paragraph 3.12.

3.3 SUPERVISION AND CONSTRUCTION PROCEDURES

3.3.1 The Contractor shall supervise and direct the Work, using the Contractor's best skill and attention. The Contractor shall be solely responsible for and have control over construction means, methods, techniques, sequences and procedures and for coordinating all portions of the Work under the Contract, unless Contract Documents give other specific instructions concerning these matters.

3.3.2 The Contractor shall be responsible to the Owner for acts and omissions of the Contractor's employees, Subcontractors and their agents and employees, and other persons performing portions of the Work under a contract with the Contractor.

3.3.3 The Contractor shall not be relieved of obligations to perform the Work in accordance with the Contract Documents either by activities or duties of the Architect in the Architect's administration of the Contract, or by tests, inspections or approvals required or performed by persons other than the Contractor.

3.3.4 The Contractor shall be responsible for inspection of portions of Work already performed under this Contract to determine that such portions are in proper condition to receive subsequent Work.

3.4 LABOR AND MATERIALS

3.4.1 Unless otherwise provided in the Contract Documents, the Contractor shall provide and pay for labor, materials, equipment, tools, construction equipment and machinery, water, heat, utilities, transportation, and other facilities and services necessary for proper execution and completion of the Work, whether temporary or permanent and whether or not incorporated or to be incorporated in the Work.

3.4.2 The Contractor shall enforce strict discipline and good order among the Contractor's employees and other persons carrying out the Contract. The Contractor shall not permit employment of unfit persons or persons not skilled in tasks assigned to them.

3.5 WARRANTY

3.5.1 The Contractor warrants to the Owner and Architect that materials and equipment furnished under the Contract will be of good quality and new unless otherwise required or permitted by the Contract Documents, that the Work will be free from defects not inherent in the quality required or permitted, and that the Work will conform with the requirements of the Contract Documents. Work not conforming to these requirements, including substitutions not properly approved and authorized, may be considered defective. The Contractor's warranty excludes remedy for damage or defect caused by abuse, modifications not executed by the Contractor, improper or insufficient maintenance, improper operation, or normal wear and tear under normal usage. If required by the Architect, the Contractor shall furnish satisfactory evidence as to the kind and quality of materials and equipment.

3.6 TAXES

3.6.1 The Contractor shall pay sales, consumer, use and similar taxes for the Work or portions thereof provided by the Contractor which are legally enacted when bids are received or negotiations concluded, whether or not yet effective or merely scheduled to go into effect.

3.7 PERMITS, FEES AND NOTICES

3.7.1 Unless otherwise provided in the Contract Documents, the Contractor shall secure and pay for the building permit and other permits and governmental fees, licenses and inspections necessary for proper execution and completion of the Work which are customarily secured after execution of the Contract and which are legally required when bids are received or negotiations concluded.

3.7.2 The Contractor shall comply with and give notices required by laws, ordinances, rules, regulations and lawful orders of public authorities bearing on performance of the Work.

3.7.3 It is not the Contractor's responsibility to ascertain that the Contract Documents are in accordance with applicable laws, statutes, ordinances, building codes, and rules and regulations. However, if the Contractor observes that portions of the Contract Documents are at variance therewith, the Contractor shall promptly notify the Architect and Owner in writing, and necessary changes shall be accomplished by appropriate Modification.

3.7.4 If the Contractor performs Work knowing it to be contrary to laws, statutes, ordinances, building codes, and rules and regulations without such notice to the Architect and Owner, the Contractor shall assume full responsibility for such Work and shall bear the attributable costs.

3.8 ALLOWANCES

3.8.1 The Contractor shall include in the Contract Sum all allowances stated in the Contract Documents. Items covered by allowances shall be supplied for such amounts and by such persons or entities as the Owner may direct, but the Contractor shall not be required to employ persons or entities against which the Contractor makes reasonable objection.

3.8.2 Unless otherwise provided in the Contract Documents:

- **.1** materials and equipment under an allowance shall be selected promptly by the Owner to avoid delay in the Work;
- **.2** allowances shall cover the cost to the Contractor of materials and equipment delivered at the site and all required taxes, less applicable trade discounts;

- .3 Contractor's costs for unloading and handling at the site, labor, installation costs, overhead, profit and other expenses contemplated for stated allowance amounts shall be included in the Contract Sum and not in the allowances;
- .4 whenever costs are more than or less than allowances, the Contract Sum shall be adjusted accordingly by Change Order. The amount of the Change Order shall reflect (1) the difference between actual costs and the allowances under Clause 3.8.2.2 and (2) changes in Contractor's costs under Clause 3.8.2.3.

3.9 SUPERINTENDENT

3.9.1 The Contractor shall employ a competent superintendent and necessary assistants who shall be in attendance at the Project site during performance of the Work. The superintendent shall represent the Contractor, and communications given to the superintendent shall be as binding as if given to the Contractor. Important communications shall be confirmed in writing. Other communications shall be similarly confirmed on written request in each case.

3.10 CONTRACTOR'S CONSTRUCTION SCHEDULES

3.10.1 The Contractor, promptly after being awarded the Contract, shall prepare and submit for the Owner's and Architect's information a Contractor's construction schedule for the Work. The schedule shall not exceed time limits current under the Contract Documents, shall be revised at appropriate intervals as required by the conditions of the Work and Project, shall be related to the entire Project to the extent required by the Contract Documents, and shall provide for expeditious and practicable execution of the Work.

3.10.2 The Contractor shall prepare and keep current, for the Architect's approval, a schedule of submittals which is coordinated with the Contractor's construction schedule and allows the Architect reasonable time to review submittals.

3.10.3 The Contractor shall conform to the most recent schedules.

3.11 DOCUMENTS AND SAMPLES AT THE SITE

3.11.1 The Contractor shall maintain at the site for the Owner one record copy of the Drawings, Specifications, addenda, Change Orders and other Modifications, in good order and marked currently to record changes and selections made during construction, and in addition approved Shop Drawings, Product Data, Samples and similar required submittals. These shall be available to the Architect and shall be delivered to the Architect for submittal to the Owner upon completion of the Work.

3.12 SHOP DRAWINGS, PRODUCT DATA AND SAMPLES

3.12.1 Shop Drawings are drawings, diagrams, schedules and other data specially prepared for the Work by the Contractor or a Subcontractor, Sub-subcontractor, manufacturer, supplier or distributor to illustrate some portion of the Work.

3.12.2 Product Data are illustrations, standard schedules, performance charts, instructions, brochures, diagrams and other information furnished by the Contractor to illustrate materials or equipment for some portion of the Work.

3.12.3 Samples are physical examples which illustrate materials, equipment or workmanship and establish standards by which the Work will be judged.

3.12.4 Shop Drawings, Product Data, Samples and similar submittals are not Contract Documents. The purpose of their submittal is to demonstrate for those portions of the Work for which submittals are required the way the Contractor proposes to conform to the information given and the design concept expressed in the Contract Documents. Review by the Architect is subject to the limitations of Subparagraph 4.2.7.

3.12.5 The Contractor shall review, approve and submit to the Architect Shop Drawings, Product Data, Samples and similar submittals required by the Contract Documents with reasonable promptness and in such sequence as to cause no delay in the Work or in the activities of the Owner or of separate contractors. Submittals made by the Contractor which are not required by the Contract Documents may be returned without action.

3.12.6 The Contractor shall perform no portion of the Work requiring submittal and review of Shop Drawings, Product Data, Samples or similar submittals until the respective submittal has been approved by the Architect. Such Work shall be in accordance with approved submittals.

3.12.7 By approving and submitting Shop Drawings, Product Data, Samples and similar submittals, the Contractor represents that the Contractor has determined and verified materials, field measurements and field construction criteria related thereto, or will do so, and has checked and coordinated the information contained within such submittals with the requirements of the Work and of the Contract Documents.

3.12.8 The Contractor shall not be relieved of responsibility for deviations from requirements of the Contract Documents by the Architect's approval of Shop Drawings, Product Data, Samples or similar submittals unless the Contractor has specifically informed the Architect in writing of such deviation at the time of submittal and the Architect has given written approval to the specific deviation. The Contractor shall not be relieved of responsibility for errors or omissions in Shop Drawings, Product Data, Samples or similar submittals by the Architect's approval thereof.

3.12.9 The Contractor shall direct specific attention, in writing or on resubmitted Shop Drawings, Product Data, Samples or similar submittals, to revisions other than those requested by the Architect on previous submittals.

3.12.10 Informational submittals upon which the Architect is not expected to take responsive action may be so identified in the Contract Documents.

3.12.11 When professional certification of performance criteria of materials, systems or equipment is required by the Contract Documents, the Architect shall be entitled to rely upon the accuracy and completeness of such calculations and certifications.

3.13 USE OF SITE

3.13.1 The Contractor shall confine operations at the site to areas permitted by law, ordinances, permits and the Contract Documents and shall not unreasonably encumber the site with materials or equipment.

3.14 CUTTING AND PATCHING

3.14.1 The Contractor shall be responsible for cutting, fitting or patching required to complete the Work or to make its parts fit together properly.

3.14.2 The Contractor shall not damage or endanger a portion of the Work or fully or partially completed construction of the Owner or separate contractors by cutting, patching or otherwise altering such construction, or by excavation. The Contractor shall not cut or otherwise alter such construction by the

Owner or a separate contractor except with written consent of the Owner and of such separate contractor; such consent shall not be unreasonably withheld. The Contractor shall not unreasonably withhold from the Owner or a separate contractor the Contractor's consent to cutting or otherwise altering the Work.

3.15 CLEANING UP

3.15.1 The Contractor shall keep the premises and surrounding area free from accumulation of waste materials or rubbish caused by operations under the Contract. At completion of the Work the Contractor shall remove from and about the Project waste materials, rubbish, the Contractor's tools, construction equipment, machinery and surplus materials.

3.15.2 If the Contractor fails to clean up as provided in the Contract Documents, the Owner may do so and the cost thereof shall be charged to the Contractor.

3.16 ACCESS TO WORK

3.16.1 The Contractor shall provide the Owner and Architect access to the Work in preparation and progress wherever located.

3.17 ROYALTIES AND PATENTS

3.17.1 The Contractor shall pay all royalties and license fees. The Contractor shall defend suits or claims for infringement of patent rights and shall hold the Owner and Architect harmless from loss on account thereof, but shall not be responsible for such defense or loss when a particular design, process or product of a particular manufacturer or manufacturers is required by the Contract Documents. However, if the Contractor has reason to believe that the required design, process or product is an infringement of a patent, the Contractor shall be responsible for such loss unless such information is promptly furnished to the Architect.

3.18 INDEMNIFICATION

3.18.1 To the fullest extent permitted by law, the Contractor shall indemnify and hold harmless the Owner, Architect, Architect's consultants, and agents and employees of any of them from and against claims, damages, losses and expenses, including but not limited to attorneys' fees, arising out of or resulting from performance of the Work, provided that such claim, damage, loss or expense is attributable to bodily injury, sickness, disease or death, or to injury to or destruction of tangible property (other than the Work itself) including loss of use resulting therefrom, but only to the extent caused in whole or in part by negligent acts or omissions of the Contractor, a Subcontractor, anyone directly or indirectly employed by them or anyone for whose acts they may be liable, regardless of whether or not such claim, damage, loss or expense is caused in part by a party indemnified hereunder. Such obligation shall not be construed to negate, abridge, or reduce other rights or obligations of indemnity which would otherwise exist as to a party or person described in this Paragraph 3.18.

3.18.2 In claims against any person or entity indemnified under this Paragraph 3.18 by an employee of the Contractor, a Subcontractor, anyone directly or indirectly employed by them or anyone for whose acts they may be liable, the indemnification obligation under this Paragraph 3.18 shall not be limited by a limitation on amount or type of damages, compensation or benefits payable by or for the Contractor or a Subcontractor under workers' or workmen's compensation acts, disability benefit acts or other employee benefit acts.

3.18.3 The obligations of the Contractor under this Paragraph 3.18 shall not extend to the liability of the Architect, the Architect's consultants, and agents and employees of any of them arising out of (1) the preparation or approval of maps, drawings, opinions, reports, surveys, Change Orders, designs or specifications, or (2) the giving of or the failure to give directions or instructions by the Architect, the Architect's consultants, and agents and employees of any of them provided such giving or failure to give is the primary cause of the injury or damage.

ARTICLE 4
ADMINISTRATION OF THE CONTRACT

4.1 ARCHITECT

4.1.1 The Architect is the person lawfully licensed to practice architecture or an entity lawfully practicing architecture identified as such in the Agreement and is referred to throughout the Contract Documents as if singular in number. The term "Architect" means the Architect or the Architect's authorized representative.

4.1.2 Duties, responsibilities and limitations of authority of the Architect as set forth in the Contract Documents shall not be restricted, modified or extended without written consent of the Owner, Contractor and Architect. Consent shall not be unreasonably withheld.

4.1.3 In case of termination of employment of the Architect, the Owner shall appoint an architect against whom the Contractor makes no reasonable objection and whose status under the Contract Documents shall be that of the former architect.

4.1.4 Disputes arising under Subparagraphs 4.1.2 and 4.1.3 shall be subject to arbitration.

4.2 ARCHITECT'S ADMINISTRATION OF THE CONTRACT

4.2.1 The Architect will provide administration of the Contract as described in the Contract Documents, and will be the Owner's representative (1) during construction, (2) until final payment is due and (3) with the Owner's concurrence, from time to time during the correction period described in Paragraph 12.2. The Architect will advise and consult with the Owner. The Architect will have authority to act on behalf of the Owner only to the extent provided in the Contract Documents, unless otherwise modified by written instrument in accordance with other provisions of the Contract.

4.2.2 The Architect will visit the site at intervals appropriate to the stage of construction to become generally familiar with the progress and quality of the completed Work and to determine in general if the Work is being performed in a manner indicating that the Work, when completed, will be in accordance with the Contract Documents. However, the Architect will not be required to make exhaustive or continuous on-site inspections to check quality or quantity of the Work. On the basis of on-site observations as an architect, the Architect will keep the Owner informed of progress of the Work, and will endeavor to guard the Owner against defects and deficiencies in the Work.

4.2.3 The Architect will not have control over or charge of and will not be responsible for construction means, methods, techniques, sequences or procedures, or for safety precautions and programs in connection with the Work, since these are solely the Contractor's responsibility as provided in Paragraph 3.3. The Architect will not be responsible for the Contractor's failure to carry out the Work in accordance with the Contract Documents. The Architect will not have control over or charge of and will not be responsible for acts or omissions of the Contractor, Subcontractors, or their agents or employees, or of any other persons performing portions of the Work.

4.2.4 Communications Facilitating Contract Administration. Except as otherwise provided in the Contract Documents or when direct communications have been specially authorized, the Owner and Contractor shall endeavor to communicate through the Architect. Communications by and with the Architect's consultants shall be through the Architect. Communications by and with Subcontractors and material suppliers shall be through the Contractor. Communications by and with separate contractors shall be through the Owner.

4.2.5 Based on the Architect's observations and evaluations of the Contractor's Applications for Payment, the Architect will review and certify the amounts due the Contractor and will issue Certificates for Payment in such amounts.

4.2.6 The Architect will have authority to reject Work which does not conform to the Contract Documents. Whenever the Architect considers it necessary or advisable for implementation of the intent of the Contract Documents, the Architect will have authority to require additional inspection or testing of the Work in accordance with Subparagraphs 13.5.2 and 13.5.3, whether or not such Work is fabricated, installed or completed. However, neither this authority of the Architect nor a decision made in good faith either to exercise or not to exercise such authority shall give rise to a duty or responsibility of the Architect to the Contractor, Subcontractors, material and equipment suppliers, their agents or employees, or other persons performing portions of the Work.

4.2.7 The Architect will review and approve or take other appropriate action upon the Contractor's submittals such as Shop Drawings, Product Data and Samples, but only for the limited purpose of checking for conformance with information given and the design concept expressed in the Contract Documents. The Architect's action will be taken with such reasonable promptness as to cause no delay in the Work or in the activities of the Owner, Contractor or separate contractors, while allowing sufficient time in the Architect's professional judgment to permit adequate review. Review of such submittals is not conducted for the purpose of determining the accuracy and completeness of other details such as dimensions and quantities, or for substantiating instructions for installation or performance of equipment or systems, all of which remain the responsibility of the Contractor as required by the Contract Documents. The Architect's review of the Contractor's submittals shall not relieve the Contractor of the obligations under Paragraphs 3.3, 3.5 and 3.12. The Architect's review shall not constitute approval of safety precautions or, unless otherwise specifically stated by the Architect, of any construction means, methods, techniques, sequences or procedures. The Architect's approval of a specific item shall not indicate approval of an assembly of which the item is a component.

4.2.8 The Architect will prepare Change Orders and Construction Change Directives, and may authorize minor changes in the Work as provided in Paragraph 7.4.

4.2.9 The Architect will conduct inspections to determine the date or dates of Substantial Completion and the date of final completion, will receive and forward to the Owner for the Owner's review and records written warranties and related documents required by the Contract and assembled by the Contractor, and will issue a final Certificate for Payment upon compliance with the requirements of the Contract Documents.

4.2.10 If the Owner and Architect agree, the Architect will provide one or more project representatives to assist in carrying out the Architect's responsibilities at the site. The duties, responsibilities and limitations of authority of such project representatives shall be as set forth in an exhibit to be incorporated in the Contract Documents.

4.2.11 The Architect will interpret and decide matters concerning performance under and requirements of the Contract Documents on written request of either the Owner or Contractor. The Architect's response to such requests will be made with reasonable promptness and within any time limits agreed upon. If no agreement is made concerning the time within which interpretations required of the Architect shall be furnished in compliance with this Paragraph 4.2, then delay shall not be recognized on account of failure by the Architect to furnish such interpretations until 15 days after written request is made for them.

4.2.12 Interpretations and decisions of the Architect will be consistent with the intent of and reasonably inferable from the Contract Documents and will be in writing or in the form of drawings. When making such interpretations and decisions, the Architect will endeavor to secure faithful performance by both Owner and Contractor, will not show partiality to either and will not be liable for results of interpretations or decisions so rendered in good faith.

4.2.13 The Architect's decisions on matters relating to aesthetic effect will be final if consistent with the intent expressed in the Contract Documents.

4.3 CLAIMS AND DISPUTES

4.3.1 Definition. A Claim is a demand or assertion by one of the parties seeking, as a matter of right, adjustment or interpretation of Contract terms, payment of money, extension of time or other relief with respect to the terms of the Contract. The term "Claim" also includes other disputes and matters in question between the Owner and Contractor arising out of or relating to the Contract. Claims must be made by written notice. The responsibility to substantiate Claims shall rest with the party making the Claim.

4.3.2 Decision of Architect. Claims, including those alleging an error or omission by the Architect, shall be referred initially to the Architect for action as provided in Paragraph 4.4. A decision by the Architect, as provided in Subparagraph 4.4.4, shall be required as a condition precedent to arbitration or litigation of a Claim between the Contractor and Owner as to all such matters arising prior to the date final payment is due, regardless of (1) whether such matters relate to execution and progress of the Work or (2) the extent to which the Work has been completed. The decision by the Architect in response to a Claim shall not be a condition precedent to arbitration or litigation in the event (1) the position of Architect is vacant, (2) the Architect has not received evidence or has failed to render a decision within agreed time limits, (3) the Architect has failed to take action required under Subparagraph 4.4.4 within 30 days after the Claim is made, (4) 45 days have passed after the Claim has been referred to the Architect or (5) the Claim relates to a mechanic's lien.

4.3.3 Time Limits on Claims. Claims by either party must be made within 21 days after occurrence of the event giving rise to such Claim or within 21 days after the claimant first recognizes the condition giving rise to the Claim, whichever is later. Claims must be made by written notice. An additional Claim made after the initial Claim has been implemented by Change Order will not be considered unless submitted in a timely manner.

4.3.4 Continuing Contract Performance. Pending final resolution of a Claim including arbitration, unless otherwise agreed in writing the Contractor shall proceed diligently with performance of the Contract and the Owner shall continue to make payments in accordance with the Contract Documents.

4.3.5 Waiver of Claims: Final Payment. The making of final payment shall constitute a waiver of Claims by the Owner except those arising from:

.1 liens, Claims, security interests or encumbrances arising out of the Contract and unsettled;

.2 failure of the Work to comply with the requirements of the Contract Documents; or

.3 terms of special warranties required by the Contract Documents.

4.3.6 Claims for Concealed or Unknown Conditions. If conditions are encountered at the site which are (1) subsurface or otherwise concealed physical conditions which differ materially from those indicated in the Contract Documents or (2) unknown physical conditions of an unusual nature, which differ materially from those ordinarily found to exist and generally recognized as inherent in construction activities of the character provided for in the Contract Documents, then notice by the observing party shall be given to the other party promptly before conditions are disturbed and in no event later than 21 days after first observance of the conditions. The Architect will promptly investigate such conditions and, if they differ materially and cause an increase or decrease in the Contractor's cost of, or time required for, performance of any part of the Work, will recommend an equitable adjustment in the Contract Sum or Contract Time, or both. If the Architect determines that the conditions at the site are not materially different from those indicated in the Contract Documents and that no change in the terms of the Contract is justified, the Architect shall so notify the Owner and Contractor in writing, stating the reasons. Claims by either party in opposition to such determination must be made within 21 days after the Architect has given notice of the decision. If the Owner and Contractor cannot agree on an adjustment in the Contract Sum or Contract Time, the adjustment shall be referred to the Architect for initial determination, subject to further proceedings pursuant to Paragraph 4.4.

4.3.7 Claims for Additional Cost. If the Contractor wishes to make Claim for an increase in the Contract Sum, written notice as provided herein shall be given before proceeding to execute the Work. Prior notice is not required for Claims relating to an emergency endangering life or property arising under Paragraph 10.3. If the Contractor believes additional cost is involved for reasons including but not limited to (1) a written interpretation from the Architect, (2) an order by the Owner to stop the Work where the Contractor was not at fault, (3) a written order for a minor change in the Work issued by the Architect, (4) failure of payment by the Owner, (5) termination of the Contract by the Owner, (6) Owner's suspension or (7) other reasonable grounds, Claim shall be filed in accordance with the procedure established herein.

4.3.8 Claims for Additional Time

4.3.8.1 If the Contractor wishes to make Claim for an increase in the Contract Time, written notice as provided herein shall be given. The Contractor's Claim shall include an estimate of cost and of probable effect of delay on progress of the Work. In the case of a continuing delay only one Claim is necessary.

4.3.8.2 If adverse weather conditions are the basis for a Claim for additional time, such Claim shall be documented by data substantiating that weather conditions were abnormal for the period of time and could not have been reasonably anticipated, and that weather conditions had an adverse effect on the scheduled construction.

4.3.9 Injury or Damage to Person or Property. If either party to the Contract suffers injury or damage to person or property because of an act or omission of the other party, of any of the other party's employees or agents, or of others for whose acts such party is legally liable, written notice of such injury or damage, whether or not insured, shall be given to the other party within a reasonable time not exceeding 21 days after first observance. The notice shall provide sufficient detail to enable the other party to investigate the matter. If a Claim for additional cost or time related to this Claim is to be asserted, it shall be filed as provided in Subparagraphs 4.3.7 or 4.3.8.

4.4 RESOLUTION OF CLAIMS AND DISPUTES

4.4.1 The Architect will review Claims and take one or more of the following preliminary actions within ten days of receipt of a Claim: (1) request additional supporting data from the claimant, (2) submit a schedule to the parties indicating when the Architect expects to take action, (3) reject the Claim in whole or in part, stating reasons for rejection, (4) recommend approval of the Claim by the other party or (5) suggest a compromise. The Architect may also, but is not obligated to, notify the surety, if any, of the nature and amount of the Claim.

4.4.2 If a Claim has been resolved, the Architect will prepare or obtain appropriate documentation.

4.4.3 If a Claim has not been resolved, the party making the Claim shall, within ten days after the Architect's preliminary response, take one or more of the following actions: (1) submit additional supporting data requested by the Architect, (2) modify the initial Claim or (3) notify the Architect that the initial Claim stands.

4.4.4 If a Claim has not been resolved after consideration of the foregoing and of further evidence presented by the parties or requested by the Architect, the Architect will notify the parties in writing that the Architect's decision will be made within seven days, which decision shall be final and binding on the parties but subject to arbitration. Upon expiration of such time period, the Architect will render to the parties the Architect's written decision relative to the Claim, including any change in the Contract Sum or Contract Time or both. If there is a surety and there appears to be a possibility of a Contractor's default, the Architect may, but is not obligated to, notify the surety and request the surety's assistance in resolving the controversy.

4.5 ARBITRATION

4.5.1 Controversies and Claims Subject to Arbitration. Any controversy or Claim arising out of or related to the Contract, or the breach thereof, shall be settled by arbitration in accordance with the Construction Industry Arbitration Rules of the American Arbitration Association, and judgment upon the award rendered by the arbitrator or arbitrators may be entered in any court having jurisdiction thereof, except controversies or Claims relating to aesthetic effect and except those waived as provided for in Subparagraph 4.3.5. Such controversies or Claims upon which the Architect has given notice and rendered a decision as provided in Subparagraph 4.4.4 shall be subject to arbitration upon written demand of either party. Arbitration may be commenced when 45 days have passed after a Claim has been referred to the Architect as provided in Paragraph 4.3 and no decision has been rendered.

4.5.2 Rules and Notices for Arbitration. Claims between the Owner and Contractor not resolved under Paragraph 4.4 shall, if subject to arbitration under Subparagraph 4.5.1, be decided by arbitration in accordance with the Construction Industry Arbitration Rules of the American Arbitration Association currently in effect, unless the parties mutually agree otherwise. Notice of demand for arbitration shall be filed in writing with the other party to the Agreement between the Owner and Contractor and with the American Arbitration Association, and a copy shall be filed with the Architect.

4.5.3 Contract Performance During Arbitration. During arbitration proceedings, the Owner and Contractor shall comply with Subparagraph 4.3.4.

4.5.4 When Arbitration May Be Demanded. Demand for arbitration of any Claim may not be made until the earlier of (1) the date on which the Architect has rendered a final written decision on the Claim, (2) the tenth day after the parties have presented evidence to the Architect or have been given reasonable opportunity to do so, if the Architect has not rendered a final written decision by that date, or (3) any of the five events described in Subparagraph 4.3.2.

4.5.4.1 When a written decision of the Architect states that (1) the decision is final but subject to arbitration and (2) a demand for arbitration of a Claim covered by such decision must be made within 30 days after the date on which the party making the demand receives the final written decision, then failure to demand arbitration within said 30 days' period shall result in the Architect's decision becoming final and binding upon the Owner and Contractor. If the Architect renders a decision after arbitration proceedings have been initiated, such decision may be entered as evidence, but shall not supersede arbitration proceedings unless the decision is acceptable to all parties concerned.

4.5.4.2 A demand for arbitration shall be made within the time limits specified in Subparagraphs 4.5.1 and 4.5.4 and Clause 4.5.4.1 as applicable, and in other cases within a reasonable time after the Claim has arisen, and in no event shall it be made after the date when institution of legal or equitable proceedings based on such Claim would be barred by the applicable statute of limitations as determined pursuant to Paragraph 13.7.

4.5.5 Limitation on Consolidation or Joinder. No arbitration arising out of or relating to the Contract Documents shall include, by consolidation or joinder or in any other manner, the Architect, the Architect's employees or consultants, except by written consent containing specific reference to the Agreement and signed by the Architect, Owner, Contractor and any other person or entity sought to be joined. No arbitration shall include, by consolidation or joinder or in any other manner, parties other than the Owner, Contractor, a separate contractor as described in Article 6 and other persons substantially involved in a common question of fact or law whose presence is required if complete relief is to be accorded in arbitration. No person or entity other than the Owner, Contractor or a separate contractor as described in Article 6 shall be included as an original third party or additional third party to an arbitration whose interest or responsibility is insubstantial. Consent to arbitration involving an additional person or entity shall not constitute consent to arbitration of a dispute not described therein or with a person or entity not named or described therein. The foregoing agreement to arbitrate and other agreements to arbitrate with an additional person or entity duly consented to by parties to the Agreement shall be specifically enforceable under applicable law in any court having jurisdiction thereof.

4.5.6 Claims and Timely Assertion of Claims. A party who files a notice of demand for arbitration must assert in the demand all Claims then known to that party on which arbitration is permitted to be demanded. When a party fails to include a Claim through oversight, inadvertence or excusable neglect, or when a Claim has matured or been acquired subsequently, the arbitrator or arbitrators may permit amendment.

4.5.7 Judgment on Final Award. The award rendered by the arbitrator or arbitrators shall be final, and judgment may be entered upon it in accordance with applicable law in any court having jurisdiction thereof.

ARTICLE 5

SUBCONTRACTORS

5.1 DEFINITIONS

5.1.1 A Subcontractor is a person or entity who has a direct contract with the Contractor to perform a portion of the Work at the site. The term "Subcontractor" is referred to throughout the Contract Documents as if singular in number and means a Subcontractor or an authorized representative of the Subcontractor. The term "Subcontractor" does not include a separate contractor or subcontractors of a separate contractor.

5.1.2 A Sub-subcontractor is a person or entity who has a direct or indirect contract with a Subcontractor to perform a portion of the Work at the site. The term "Sub-subcontractor" is referred to throughout the Contract Documents as if singular in number and means a Sub-subcontractor or an authorized representative of the Sub-subcontractor.

5.2 AWARD OF SUBCONTRACTS AND OTHER CONTRACTS FOR PORTIONS OF THE WORK

5.2.1 Unless otherwise stated in the Contract Documents or the bidding requirements, the Contractor, as soon as practicable after award of the Contract, shall furnish in writing to the Owner through the Architect the names of persons or entities (including those who are to furnish materials or equipment fabricated to a special design) proposed for each principal portion of the Work. The Architect will promptly reply to the Contractor in writing stating whether or not the Owner or the Architect, after due investigation, has reasonable objection to any such proposed person or entity. Failure of the Owner or Architect to reply promptly shall constitute notice of no reasonable objection.

5.2.2 The Contractor shall not contract with a proposed person or entity to whom the Owner or Architect has made reasonable and timely objection. The Contractor shall not be required to contract with anyone to whom the Contractor has made reasonable objection.

5.2.3 If the Owner or Architect has reasonable objection to a person or entity proposed by the Contractor, the Contractor shall propose another to whom the Owner or Architect has no reasonable objection. The Contract Sum shall be increased or decreased by the difference in cost occasioned by such change and an appropriate Change Order shall be issued. However, no increase in the Contract Sum shall be allowed for such change unless the Contractor has acted promptly and responsively in submitting names as required.

5.2.4 The Contractor shall not change a Subcontractor, person or entity previously selected if the Owner or Architect makes reasonable objection to such change.

5.3 SUBCONTRACTUAL RELATIONS

5.3.1 By appropriate agreement, written where legally required for validity, the Contractor shall require each Subcontractor, to the extent of the Work to be performed by the Subcontractor, to be bound to the Contractor by terms of the Contract Documents, and to assume toward the Contractor all the obligations and responsibilities which the Contractor, by these Documents, assumes toward the Owner and Architect. Each subcontract agreement shall preserve and protect the rights of the Owner and Architect under the Contract Documents with respect to the Work to be performed by the Subcontractor so that subcontracting thereof will not prejudice such rights, and shall allow to the Subcontractor, unless specifically provided otherwise in the subcontract agreement, the benefit of all rights, remedies and redress against the Contractor that the Contractor, by the Contract Documents, has against the Owner. Where appropriate, the Contractor shall require each Subcontractor to enter into similar agreements with Sub-subcontractors. The Contractor shall make available to each proposed Subcontractor, prior to the execution of the subcontract agreement, copies of the Contract Documents to which the Subcontractor will be bound, and, upon written request of the Subcontractor, identify to the Subcontractor terms and conditions of the proposed subcontract agreement which may be at variance with the Contract Documents. Subcontractors shall similarly make copies of applicable portions of such documents available to their respective proposed Sub-subcontractors.

5.4 CONTINGENT ASSIGNMENT OF SUBCONTRACTS

5.4.1 Each subcontract agreement for a portion of the Work is assigned by the Contractor to the Owner provided that:

.1 assignment is effective only after termination of the Contract by the Owner for cause pursuant to Paragraph 14.2 and only for those subcontract agreements which the Owner accepts by notifying the Subcontractor in writing; and

.2 assignment is subject to the prior rights of the surety, if any, obligated under bond relating to the Contract.

5.4.2 If the Work has been suspended for more than 30 days, the Subcontractor's compensation shall be equitably adjusted.

ARTICLE 6

CONSTRUCTION BY OWNER OR BY SEPARATE CONTRACTORS

6.1 OWNER'S RIGHT TO PERFORM CONSTRUCTION AND TO AWARD SEPARATE CONTRACTS

6.1.1 The Owner reserves the right to perform construction or operations related to the Project with the Owner's own forces, and to award separate contracts in connection with other portions of the Project or other construction or operations on the site under Conditions of the Contract identical or substantially similar to these including those portions related to insurance and waiver of subrogation. If the Contractor claims that delay or additional cost is involved because of such action by the Owner, the Contractor shall make such Claim as provided elsewhere in the Contract Documents.

6.1.2 When separate contracts are awarded for different portions of the Project or other construction or operations on the site, the term "Contractor" in the Contract Documents in each case shall mean the Contractor who executes each separate Owner-Contractor Agreement.

6.1.3 The Owner shall provide for coordination of the activities of the Owner's own forces and of each separate contractor with the Work of the Contractor, who shall cooperate with them. The Contractor shall participate with other separate contractors and the Owner in reviewing their construction schedules when directed to do so. The Contractor shall make any revisions to the construction schedule and Contract Sum deemed necessary after a joint review and mutual agreement. The construction schedules shall then constitute the schedules to be used by the Contractor, separate contractors and the Owner until subsequently revised.

6.1.4 Unless otherwise provided in the Contract Documents, when the Owner performs construction or operations related to the Project with the Owner's own forces, the Owner shall be deemed to be subject to the same obligations and to have the same rights which apply to the Contractor under the Conditions of the Contract, including, without excluding others, those stated in Article 3, this Article 6 and Articles 10, 11 and 12.

6.2 MUTUAL RESPONSIBILITY

6.2.1 The Contractor shall afford the Owner and separate contractors reasonable opportunity for introduction and storage of their materials and equipment and performance of their activities and shall connect and coordinate the Contractor's construction and operations with theirs as required by the Contract Documents.

6.2.2 If part of the Contractor's Work depends for proper execution or results upon construction or operations by the Owner or a separate contractor, the Contractor shall, prior to proceeding with that portion of the Work, promptly report to the Architect apparent discrepancies or defects in such other construction that would render it unsuitable for such proper execution and results. Failure of the Contractor so to report shall constitute an acknowledgment that the Owner's or separate contractors' completed or partially completed construction is fit and proper to receive the Contractor's Work, except as to defects not then reasonably discoverable.

6.2.3 Costs caused by delays or by improperly timed activities or defective construction shall be borne by the party responsible therefor.

6.2.4 The Contractor shall promptly remedy damage wrongfully caused by the Contractor to completed or partially completed construction or to property of the Owner or separate contractors as provided in Subparagraph 10.2.5.

6.2.5 Claims and other disputes and matters in question between the Contractor and a separate contractor shall be subject to the provisions of Paragraph 4.3 provided the separate contractor has reciprocal obligations.

6.2.6 The Owner and each separate contractor shall have the same responsibilities for cutting and patching as are described for the Contractor in Paragraph 3.14.

6.3 OWNER'S RIGHT TO CLEAN UP

6.3.1 If a dispute arises among the Contractor, separate contractors and the Owner as to the responsibility under their respective contracts for maintaining the premises and surrounding area free from waste materials and rubbish as described in Paragraph 3.15, the Owner may clean up and allocate the cost among those responsible as the Architect determines to be just.

ARTICLE 7

CHANGES IN THE WORK

7.1 CHANGES

7.1.1 Changes in the Work may be accomplished after execution of the Contract, and without invalidating the Contract, by Change Order, Construction Change Directive or order for a minor change in the Work, subject to the limitations stated in this Article 7 and elsewhere in the Contract Documents.

7.1.2 A Change Order shall be based upon agreement among the Owner, Contractor and Architect; a Construction Change Directive requires agreement by the Owner and Architect and may or may not be agreed to by the Contractor; an order for a minor change in the Work may be issued by the Architect alone.

7.1.3 Changes in the Work shall be performed under applicable provisions of the Contract Documents, and the Contractor shall proceed promptly, unless otherwise provided in the Change Order, Construction Change Directive or order for a minor change in the Work.

7.1.4 If unit prices are stated in the Contract Documents or subsequently agreed upon, and if quantities originally contemplated are so changed in a proposed Change Order or Construction Change Directive that application of such unit prices to quantities of Work proposed will cause substantial inequity to the Owner or Contractor, the applicable unit prices shall be equitably adjusted.

7.2 CHANGE ORDERS

7.2.1 A Change Order is a written instrument prepared by the Architect and signed by the Owner, Contractor and Architect, stating their agreement upon all of the following:

.1 a change in the Work;

.2 the amount of the adjustment in the Contract Sum, if any; and

.3 the extent of the adjustment in the Contract Time, if any.

7.2.2 Methods used in determining adjustments to the Contract Sum may include those listed in Subparagraph 7.3.3.

7.3 CONSTRUCTION CHANGE DIRECTIVES

7.3.1 A Construction Change Directive is a written order prepared by the Architect and signed by the Owner and Architect, directing a change in the Work and stating a proposed basis for adjustment, if any, in the Contract Sum or Contract Time, or both. The Owner may by Construction Change Directive, without invalidating the Contract, order changes in the Work within the general scope of the Contract consisting of additions, deletions or other revisions, the Contract Sum and Contract Time being adjusted accordingly.

7.3.2 A Construction Change Directive shall be used in the absence of total agreement on the terms of a Change Order.

7.3.3 If the Construction Change Directive provides for an adjustment to the Contract Sum, the adjustment shall be based on one of the following methods:

.1 mutual acceptance of a lump sum properly itemized and supported by sufficient substantiating data to permit evaluation;

.2 unit prices stated in the Contract Documents or subsequently agreed upon;

.3 cost to be determined in a manner agreed upon by the parties and a mutually acceptable fixed or percentage fee; or

.4 as provided in Subparagraph 7.3.6.

7.3.4 Upon receipt of a Construction Change Directive, the Contractor shall promptly proceed with the change in the Work involved and advise the Architect of the Contractor's agreement or disagreement with the method, if any, provided in the Construction Change Directive for determining the proposed adjustment in the Contract Sum or Contract Time.

7.3.5 A Construction Change Directive signed by the Contractor indicates the agreement of the Contractor therewith, including adjustment in Contract Sum and Contract Time or the method for determining them. Such agreement shall be effective immediately and shall be recorded as a Change Order.

7.3.6 If the Contractor does not respond promptly or disagrees with the method for adjustment in the Contract Sum, the method and the adjustment shall be determined by the Architect on the basis of reasonable expenditures and savings of those performing the Work attributable to the change, including, in case of an increase in the Contract Sum, a reasonable allowance for overhead and profit. In such case, and also under Clause 7.3.3.3, the Contractor shall keep and present, in such form as the Architect may prescribe, an itemized accounting together with appropriate supporting data. Unless otherwise provided in the Contract Documents, costs for the purposes of this Subparagraph 7.3.6 shall be limited to the following:

.1 costs of labor, including social security, old age and unemployment insurance, fringe benefits required by agreement or custom, and workers' or workmen's compensation insurance;

.2 costs of materials, supplies and equipment, including cost of transportation, whether incorporated or consumed;

.3 rental costs of machinery and equipment, exclusive of hand tools, whether rented from the Contractor or others;

.4 costs of premiums for all bonds and insurance, permit fees, and sales, use or similar taxes related to the Work; and

.5 additional costs of supervision and field office personnel directly attributable to the change.

7.3.7 Pending final determination of cost to the Owner, amounts not in dispute may be included in Applications for Payment. The amount of credit to be allowed by the Contractor to the Owner for a deletion or change which results in a net decrease in the Contract Sum shall be actual net cost as confirmed by the Architect. When both additions and credits covering related Work or substitutions are involved in a change, the allowance for overhead and profit shall be figured on the basis of net increase, if any, with respect to that change.

7.3.8 If the Owner and Contractor do not agree with the adjustment in Contract Time or the method for determining it, the adjustment or the method shall be referred to the Architect for determination.

7.3.9 When the Owner and Contractor agree with the determination made by the Architect concerning the adjustments in the Contract Sum and Contract Time, or otherwise reach agreement upon the adjustments, such agreement shall be effective immediately and shall be recorded by preparation and execution of an appropriate Change Order.

7.4 MINOR CHANGES IN THE WORK

7.4.1 The Architect will have authority to order minor changes in the Work not involving adjustment in the Contract Sum or extension of the Contract Time and not inconsistent with the intent of the Contract Documents. Such changes shall be effected by written order and shall be binding on the Owner and Contractor. The Contractor shall carry out such written orders promptly.

ARTICLE 8

TIME

8.1 DEFINITIONS

8.1.1 Unless otherwise provided, Contract Time is the period of time, including authorized adjustments, allotted in the Contract Documents for Substantial Completion of the Work.

8.1.2 The date of commencement of the Work is the date established in the Agreement. The date shall not be postponed by the failure to act of the Contractor or of persons or entities for whom the Contractor is responsible.

8.1.3 The date of Substantial Completion is the date certified by the Architect in accordance with Paragraph 9.8.

8.1.4 The term "day" as used in the Contract Documents shall mean calendar day unless otherwise specifically defined.

8.2 PROGRESS AND COMPLETION

8.2.1 Time limits stated in the Contract Documents are of the essence of the Contract. By executing the Agreement the Contractor confirms that the Contract Time is a reasonable period for performing the Work.

8.2.2 The Contractor shall not knowingly, except by agreement or instruction of the Owner in writing, prematurely commence operations on the site or elsewhere prior to the effective date of insurance required by Article 11 to be furnished by the Contractor. The date of commencement of the Work shall not be changed by the effective date of such insurance. Unless the date of commencement is established by a notice to proceed given by the Owner, the Contractor shall notify the Owner in writing not less than five days or other agreed period before commencing the Work to permit the timely filing of mortgages, mechanic's liens and other security interests.

8.2.3 The Contractor shall proceed expeditiously with adequate forces and shall achieve Substantial Completion within the Contract Time.

8.3 DELAYS AND EXTENSIONS OF TIME

8.3.1 If the Contractor is delayed at any time in progress of the Work by an act or neglect of the Owner or Architect, or of an employee of either, or of a separate contractor employed by the Owner, or by changes ordered in the Work, or by labor disputes, fire, unusual delay in deliveries, unavoidable casualties or other causes beyond the Contractor's control, or by delay authorized by the Owner pending arbitration, or by other causes which the Architect determines may justify delay, then the Contract Time shall be extended by Change Order for such reasonable time as the Architect may determine.

8.3.2 Claims relating to time shall be made in accordance with applicable provisions of Paragraph 4.3.

8.3.3 This Paragraph 8.3 does not preclude recovery of damages for delay by either party under other provisions of the Contract Documents.

ARTICLE 9

PAYMENTS AND COMPLETION

9.1 CONTRACT SUM

9.1.1 The Contract Sum is stated in the Agreement and, including authorized adjustments, is the total amount payable by the Owner to the Contractor for performance of the Work under the Contract Documents.

9.2 SCHEDULE OF VALUES

9.2.1 Before the first Application for Payment, the Contractor shall submit to the Architect a schedule of values allocated to various portions of the Work, prepared in such form and supported by such data to substantiate its accuracy as the Architect may require. This schedule, unless objected to by the Architect, shall be used as a basis for reviewing the Contractor's Applications for Payment.

9.3 APPLICATIONS FOR PAYMENT

9.3.1 At least ten days before the date established for each progress payment, the Contractor shall submit to the Architect an itemized Application for Payment for operations completed in accordance with the schedule of values. Such application shall be notarized, if required, and supported by such data substantiating the Contractor's right to payment as the Owner or Architect may require, such as copies of requisitions from Subcontractors and material suppliers, and reflecting retainage if provided for elsewhere in the Contract Documents.

9.3.1.1 Such applications may include requests for payment on account of changes in the Work which have been properly authorized by Construction Change Directives but not yet included in Change Orders.

9.3.1.2 Such applications may not include requests for payment of amounts the Contractor does not intend to pay to a Subcontractor or material supplier because of a dispute or other reason.

9.3.2 Unless otherwise provided in the Contract Documents, payments shall be made on account of materials and equipment delivered and suitably stored at the site for subsequent incorporation in the Work. If approved in advance by the Owner, payment may similarly be made for materials and equipment suitably stored off the site at a location agreed upon in writing. Payment for materials and equipment stored on or off the site shall be conditioned upon compliance by the Contractor with procedures satisfactory to the Owner to establish the Owner's title to such materials and equipment or otherwise protect the Owner's interest, and shall include applicable insurance, storage and transportation to the site for such materials and equipment stored off the site.

9.3.3 The Contractor warrants that title to all Work covered by an Application for Payment will pass to the Owner no later than the time of payment. The Contractor further warrants that upon submittal of an Application for Payment all Work for which Certificates for Payment have been previously issued and payments received from the Owner shall, to the best of the Contractor's knowledge, information and belief, be free and clear of liens, claims, security interests or encumbrances in favor of the Contractor, Subcontractors, material suppliers, or other persons or entities making a claim by reason of having provided labor, materials and equipment relating to the Work.

9.4 CERTIFICATES FOR PAYMENT

9.4.1 The Architect will, within seven days after receipt of the Contractor's Application for Payment, either issue to the

Owner a Certificate for Payment, with a copy to the Contractor, for such amount as the Architect determines is properly due, or notify the Contractor and Owner in writing of the Architect's reasons for withholding certification in whole or in part as provided in Subparagraph 9.5.1.

9.4.2 The issuance of a Certificate for Payment will constitute a representation by the Architect to the Owner, based on the Architect's observations at the site and the data comprising the Application for Payment, that the Work has progressed to the point indicated and that, to the best of the Architect's knowledge, information and belief, quality of the Work is in accordance with the Contract Documents. The foregoing representations are subject to an evaluation of the Work for conformance with the Contract Documents upon Substantial Completion, to results of subsequent tests and inspections, to minor deviations from the Contract Documents correctable prior to completion and to specific qualifications expressed by the Architect. The issuance of a Certificate for Payment will further constitute a representation that the Contractor is entitled to payment in the amount certified. However, the issuance of a Certificate for Payment will not be a representation that the Architect has (1) made exhaustive or continuous on-site inspections to check the quality or quantity of the Work, (2) reviewed construction means, methods, techniques, sequences or procedures, (3) reviewed copies of requisitions received from Subcontractors and material suppliers and other data requested by the Owner to substantiate the Contractor's right to payment or (4) made examination to ascertain how or for what purpose the Contractor has used money previously paid on account of the Contract Sum.

9.5 DECISIONS TO WITHHOLD CERTIFICATION

9.5.1 The Architect may decide not to certify payment and may withhold a Certificate for Payment in whole or in part, to the extent reasonably necessary to protect the Owner, if in the Architect's opinion the representations to the Owner required by Subparagraph 9.4.2 cannot be made. If the Architect is unable to certify payment in the amount of the Application, the Architect will notify the Contractor and Owner as provided in Subparagraph 9.4.1. If the Contractor and Architect cannot agree on a revised amount, the Architect will promptly issue a Certificate for Payment for the amount for which the Architect is able to make such representations to the Owner. The Architect may also decide not to certify payment or, because of subsequently discovered evidence or subsequent observations, may nullify the whole or a part of a Certificate for Payment previously issued, to such extent as may be necessary in the Architect's opinion to protect the Owner from loss because of:

.1 defective Work not remedied;
.2 third party claims filed or reasonable evidence indicating probable filing of such claims;
.3 failure of the Contractor to make payments properly to Subcontractors or for labor, materials or equipment;
.4 reasonable evidence that the Work cannot be completed for the unpaid balance of the Contract Sum;
.5 damage to the Owner or another contractor;
.6 reasonable evidence that the Work will not be completed within the Contract Time, and that the unpaid balance would not be adequate to cover actual or liquidated damages for the anticipated delay; or
.7 persistent failure to carry out the Work in accordance with the Contract Documents.

9.5.2 When the above reasons for withholding certification are removed, certification will be made for amounts previously withheld.

9.6 PROGRESS PAYMENTS

9.6.1 After the Architect has issued a Certificate for Payment, the Owner shall make payment in the manner and within the time provided in the Contract Documents, and shall so notify the Architect.

9.6.2 The Contractor shall promptly pay each Subcontractor, upon receipt of payment from the Owner, out of the amount paid to the Contractor on account of such Subcontractor's portion of the Work, the amount to which said Subcontractor is entitled, reflecting percentages actually retained from payments to the Contractor on account of such Subcontractor's portion of the Work. The Contractor shall, by appropriate agreement with each Subcontractor, require each Subcontractor to make payments to Sub-subcontractors in similar manner.

9.6.3 The Architect will, on request, furnish to a Subcontractor, if practicable, information regarding percentages of completion or amounts applied for by the Contractor and action taken thereon by the Architect and Owner on account of portions of the Work done by such Subcontractor.

9.6.4 Neither the Owner nor Architect shall have an obligation to pay or to see to the payment of money to a Subcontractor except as may otherwise be required by law.

9.6.5 Payment to material suppliers shall be treated in a manner similar to that provided in Subparagraphs 9.6.2, 9.6.3 and 9.6.4.

9.6.6 A Certificate for Payment, a progress payment, or partial or entire use or occupancy of the Project by the Owner shall not constitute acceptance of Work not in accordance with the Contract Documents.

9.7 FAILURE OF PAYMENT

9.7.1 If the Architect does not issue a Certificate for Payment, through no fault of the Contractor, within seven days after receipt of the Contractor's Application for Payment, or if the Owner does not pay the Contractor within seven days after the date established in the Contract Documents the amount certified by the Architect or awarded by arbitration, then the Contractor may, upon seven additional days' written notice to the Owner and Architect, stop the Work until payment of the amount owing has been received. The Contract Time shall be extended appropriately and the Contract Sum shall be increased by the amount of the Contractor's reasonable costs of shut-down, delay and start-up, which shall be accomplished as provided in Article 7.

9.8 SUBSTANTIAL COMPLETION

9.8.1 Substantial Completion is the stage in the progress of the Work when the Work or designated portion thereof is sufficiently complete in accordance with the Contract Documents so the Owner can occupy or utilize the Work for its intended use.

9.8.2 When the Contractor considers that the Work, or a portion thereof which the Owner agrees to accept separately, is substantially complete, the Contractor shall prepare and submit to the Architect a comprehensive list of items to be completed or corrected. The Contractor shall proceed promptly to complete and correct items on the list. Failure to include an item on such list does not alter the responsibility of the Contractor to complete all Work in accordance with the Contract Documents. Upon receipt of the Contractor's list, the Architect will make an inspection to determine whether the Work or designated portion thereof is substantially complete. If the Architect's inspection discloses any item, whether or not included on the Contractor's list, which is not in accordance with the requirements of the Contract Documents, the Contractor shall, before issuance of the Certificate of Substantial Completion, complete or correct such item upon notification by the Architect. The Contractor shall then submit a request for another inspection by the Architect to determine Substantial Completion. When the Work or designated portion thereof is substantially complete, the Architect will prepare a Certificate of Substantial Completion which shall establish the date of Substantial Completion, shall establish responsibilities of the Owner and Contractor for security, maintenance, heat, utilities, damage to the Work and insurance, and shall fix the time within which the Contractor shall finish all items on the list accompanying the Certificate. Warranties required by the Contract Documents shall commence on the date of Substantial Completion of the Work or designated portion thereof unless otherwise provided in the Certificate of Substantial Completion. The Certificate of Substantial Completion shall be submitted to the Owner and Contractor for their written acceptance of responsibilities assigned to them in such Certificate.

9.8.3 Upon Substantial Completion of the Work or designated portion thereof and upon application by the Contractor and certification by the Architect, the Owner shall make payment reflecting adjustment in retainage, if any, for such Work or portion thereof as provided in the Contract Documents.

9.9 PARTIAL OCCUPANCY OR USE

9.9.1 The Owner may occupy or use any completed or partially completed portion of the Work at any stage when such portion is designated by separate agreement with the Contractor, provided such occupancy or use is consented to by the insurer as required under Subparagraph 11.3.11 and authorized by public authorities having jurisdiction over the Work. Such partial occupancy or use may commence whether or not the portion is substantially complete, provided the Owner and Contractor have accepted in writing the responsibilities assigned to each of them for payments, retainage if any, security, maintenance, heat, utilities, damage to the Work and insurance, and have agreed in writing concerning the period for correction of the Work and commencement of warranties required by the Contract Documents. When the Contractor considers a portion substantially complete, the Contractor shall prepare and submit a list to the Architect as provided under Subparagraph 9.8.2. Consent of the Contractor to partial occupancy or use shall not be unreasonably withheld. The stage of the progress of the Work shall be determined by written agreement between the Owner and Contractor or, if no agreement is reached, by decision of the Architect.

9.9.2 Immediately prior to such partial occupancy or use, the Owner, Contractor and Architect shall jointly inspect the area to be occupied or portion of the Work to be used in order to determine and record the condition of the Work.

9.9.3 Unless otherwise agreed upon, partial occupancy or use of a portion or portions of the Work shall not constitute acceptance of Work not complying with the requirements of the Contract Documents.

9.10 FINAL COMPLETION AND FINAL PAYMENT

9.10.1 Upon receipt of written notice that the Work is ready for final inspection and acceptance and upon receipt of a final Application for Payment, the Architect will promptly make such inspection and, when the Architect finds the Work acceptable under the Contract Documents and the Contract fully performed, the Architect will promptly issue a final Certificate for Payment stating that to the best of the Architect's knowledge, information and belief, and on the basis of the Architect's observations and inspections, the Work has been completed in accordance with terms and conditions of the Contract Documents and that the entire balance found to be due the Contractor and noted in said final Certificate is due and payable. The Architect's final Certificate for Payment will constitute a further representation that conditions listed in Subparagraph 9.10.2 as precedent to the Contractor's being entitled to final payment have been fulfilled.

9.10.2 Neither final payment nor any remaining retained percentage shall become due until the Contractor submits to the Architect (1) an affidavit that payrolls, bills for materials and equipment, and other indebtedness connected with the Work for which the Owner or the Owner's property might be responsible or encumbered (less amounts withheld by Owner) have been paid or otherwise satisfied, (2) a certificate evidencing that insurance required by the Contract Documents to remain in force after final payment is currently in effect and will not be cancelled or allowed to expire until at least 30 days' prior written notice has been given to the Owner, (3) a written statement that the Contractor knows of no substantial reason that the insurance will not be renewable to cover the period required by the Contract Documents, (4) consent of surety, if any, to final payment and (5), if required by the Owner, other data establishing payment or satisfaction of obligations, such as receipts, releases and waivers of liens, claims, security interests or encumbrances arising out of the Contract, to the extent and in such form as may be designated by the Owner. If a Subcontractor refuses to furnish a release or waiver required by the Owner, the Contractor may furnish a bond satisfactory to the Owner to indemnify the Owner against such lien. If such lien remains unsatisfied after payments are made, the Contractor shall refund to the Owner all money that the Owner may be compelled to pay in discharging such lien, including all costs and reasonable attorneys' fees.

9.10.3 If, after Substantial Completion of the Work, final completion thereof is materially delayed through no fault of the Contractor or by issuance of Change Orders affecting final completion, and the Architect so confirms, the Owner shall, upon application by the Contractor and certification by the Architect, and without terminating the Contract, make payment of the balance due for that portion of the Work fully completed and accepted. If the remaining balance for Work not fully completed or corrected is less than retainage stipulated in the Contract Documents, and if bonds have been furnished, the written consent of surety to payment of the balance due for that portion of the Work fully completed and accepted shall be submitted by the Contractor to the Architect prior to certification of such payment. Such payment shall be made under terms and conditions governing final payment, except that it shall not constitute a waiver of claims. The making of final payment shall constitute a waiver of claims by the Owner as provided in Subparagraph 4.3.5.

9.10.4 Acceptance of final payment by the Contractor, a Subcontractor or material supplier shall constitute a waiver of claims by that payee except those previously made in writing and identified by that payee as unsettled at the time of final Application for Payment. Such waivers shall be in addition to the waiver described in Subparagraph 4.3.5.

ARTICLE 10

PROTECTION OF PERSONS AND PROPERTY

10.1 SAFETY PRECAUTIONS AND PROGRAMS

10.1.1 The Contractor shall be responsible for initiating, maintaining and supervising all safety precautions and programs in connection with the performance of the Contract.

10.1.2 In the event the Contractor encounters on the site material reasonably believed to be asbestos or polychlorinated biphenyl (PCB) which has not been rendered harmless, the Contractor shall immediately stop Work in the area affected and report the condition to the Owner and Architect in writing. The Work in the affected area shall not thereafter be resumed except by written agreement of the Owner and Contractor if in fact the material is asbestos or polychlorinated biphenyl (PCB) and has not been rendered harmless. The Work in the affected area shall be resumed in the absence of asbestos or polychlorinated biphenyl (PCB), or when it has been rendered harmless, by written agreement of the Owner and Contractor, or in accordance with final determination by the Architect on which arbitration has not been demanded, or by arbitration under Article 4.

10.1.3 The Contractor shall not be required pursuant to Article 7 to perform without consent any Work relating to asbestos or polychlorinated biphenyl (PCB).

10.1.4 To the fullest extent permitted by law, the Owner shall indemnify and hold harmless the Contractor, Architect, Architect's consultants and agents and employees of any of them from and against claims, damages, losses and expenses, including but not limited to attorneys' fees, arising out of or resulting from performance of the Work in the affected area if in fact the material is asbestos or polychlorinated biphenyl (PCB) and has not been rendered harmless, provided that such claim, damage, loss or expense is attributable to bodily injury, sickness, disease or death, or to injury or to destruction of tangible property (other than the Work itself) including loss of use resulting therefrom, but only to the extent caused in whole or in part by negligent acts or omissions of the Owner, anyone directly or indirectly employed by the Owner or anyone for whose acts the Owner may be liable, regardless of whether or not such claim, damage, loss or expense is caused in part by a party indemnified hereunder. Such obligation shall not be construed to negate, abridge, or reduce other rights or obligations of indemnity which would otherwise exist as to a party or person described in this Subparagraph 10.1.4.

10.2 SAFETY OF PERSONS AND PROPERTY

10.2.1 The Contractor shall take reasonable precautions for safety of, and shall provide reasonable protection to prevent damage, injury or loss to:

- .1 employees on the Work and other persons who may be affected thereby;
- .2 the Work and materials and equipment to be incorporated therein, whether in storage on or off the site, under care, custody or control of the Contractor or the Contractor's Subcontractors or Sub-subcontractors; and
- .3 other property at the site or adjacent thereto, such as trees, shrubs, lawns, walks, pavements, roadways, structures and utilities not designated for removal, relocation or replacement in the course of construction.

10.2.2 The Contractor shall give notices and comply with applicable laws, ordinances, rules, regulations and lawful orders of public authorities bearing on safety of persons or property or their protection from damage, injury or loss.

10.2.3 The Contractor shall erect and maintain, as required by existing conditions and performance of the Contract, reasonable safeguards for safety and protection, including posting danger signs and other warnings against hazards, promulgating safety regulations and notifying owners and users of adjacent sites and utilities.

10.2.4 When use or storage of explosives or other hazardous materials or equipment or unusual methods are necessary for execution of the Work, the Contractor shall exercise utmost care and carry on such activities under supervision of properly qualified personnel.

10.2.5 The Contractor shall promptly remedy damage and loss (other than damage or loss insured under property insurance required by the Contract Documents) to property referred to in Clauses 10.2.1.2 and 10.2.1.3 caused in whole or in part by the Contractor, a Subcontractor, a Sub-subcontractor, or anyone directly or indirectly employed by any of them, or by anyone for whose acts they may be liable and for which the Contractor is responsible under Clauses 10.2.1.2 and 10.2.1.3, except damage or loss attributable to acts or omissions of the Owner or Architect or anyone directly or indirectly employed by either of them, or by anyone for whose acts either of them may be liable, and not attributable to the fault or negligence of the Contractor. The foregoing obligations of the Contractor are in addition to the Contractor's obligations under Paragraph 3.18.

10.2.6 The Contractor shall designate a responsible member of the Contractor's organization at the site whose duty shall be the prevention of accidents. This person shall be the Contractor's superintendent unless otherwise designated by the Contractor in writing to the Owner and Architect.

10.2.7 The Contractor shall not load or permit any part of the construction or site to be loaded so as to endanger its safety.

10.3 EMERGENCIES

10.3.1 In an emergency affecting safety of persons or property, the Contractor shall act, at the Contractor's discretion, to prevent threatened damage, injury or loss. Additional compensation or extension of time claimed by the Contractor on account of an emergency shall be determined as provided in Paragraph 4.3 and Article 7.

ARTICLE 11

INSURANCE AND BONDS

11.1 CONTRACTOR'S LIABILITY INSURANCE

11.1.1 The Contractor shall purchase from and maintain in a company or companies lawfully authorized to do business in the jurisdiction in which the Project is located such insurance as will protect the Contractor from claims set forth below which may arise out of or result from the Contractor's operations under the Contract and for which the Contractor may be legally liable, whether such operations be by the Contractor or by a Subcontractor or by anyone directly or indirectly employed by any of them, or by anyone for whose acts any of them may be liable:

- .1 claims under workers' or workmen's compensation, disability benefit and other similar employee benefit acts which are applicable to the Work to be performed;

.2 claims for damages because of bodily injury, occupational sickness or disease, or death of the Contractor's employees;

.3 claims for damages because of bodily injury, sickness or disease, or death of any person other than the Contractor's employees;

.4 claims for damages insured by usual personal injury liability coverage which are sustained (1) by a person as a result of an offense directly or indirectly related to employment of such person by the Contractor, or (2) by another person;

.5 claims for damages, other than to the Work itself, because of injury to or destruction of tangible property, including loss of use resulting therefrom;

.6 claims for damages because of bodily injury, death of a person or property damage arising out of ownership, maintenance or use of a motor vehicle; and

.7 claims involving contractual liability insurance applicable to the Contractor's obligations under Paragraph 3.18.

11.1.2 The insurance required by Subparagraph 11.1.1 shall be written for not less than limits of liability specified in the Contract Documents or required by law, whichever coverage is greater. Coverages, whether written on an occurrence or claims-made basis, shall be maintained without interruption from date of commencement of the Work until date of final payment and termination of any coverage required to be maintained after final payment.

11.1.3 Certificates of Insurance acceptable to the Owner shall be filed with the Owner prior to commencement of the Work. These Certificates and the insurance policies required by this Paragraph 11.1 shall contain a provision that coverages afforded under the policies will not be cancelled or allowed to expire until at least 30 days' prior written notice has been given to the Owner. If any of the foregoing insurance coverages are required to remain in force after final payment and are reasonably available, an additional certificate evidencing continuation of such coverage shall be submitted with the final Application for Payment as required by Subparagraph 9.10.2. Information concerning reduction of coverage shall be furnished by the Contractor with reasonable promptness in accordance with the Contractor's information and belief.

11.2 OWNER'S LIABILITY INSURANCE

11.2.1 The Owner shall be responsible for purchasing and maintaining the Owner's usual liability insurance. Optionally, the Owner may purchase and maintain other insurance for self-protection against claims which may arise from operations under the Contract. The Contractor shall not be responsible for purchasing and maintaining this optional Owner's liability insurance unless specifically required by the Contract Documents.

11.3 PROPERTY INSURANCE

11.3.1 Unless otherwise provided, the Owner shall purchase and maintain, in a company or companies lawfully authorized to do business in the jurisdiction in which the Project is located, property insurance in the amount of the initial Contract Sum as well as subsequent modifications thereto for the entire Work at the site on a replacement cost basis without voluntary deductibles. Such property insurance shall be maintained, unless otherwise provided in the Contract Documents or otherwise agreed in writing by all persons and entities who are beneficiaries of such insurance, until final payment has been made as provided in Paragraph 9.10 or until no person or entity other than the Owner has an insurable interest in the property required by this Paragraph 11.3 to be covered, whichever is earlier. This insurance shall include interests of the Owner, the Contractor, Subcontractors and Sub-subcontractors in the Work.

11.3.1.1 Property insurance shall be on an all-risk policy form and shall insure against the perils of fire and extended coverage and physical loss or damage including, without duplication of coverage, theft, vandalism, malicious mischief, collapse, falsework, temporary buildings and debris removal including demolition occasioned by enforcement of any applicable legal requirements, and shall cover reasonable compensation for Architect's services and expenses required as a result of such insured loss. Coverage for other perils shall not be required unless otherwise provided in the Contract Documents.

11.3.1.2 If the Owner does not intend to purchase such property insurance required by the Contract and with all of the coverages in the amount described above, the Owner shall so inform the Contractor in writing prior to commencement of the Work. The Contractor may then effect insurance which will protect the interests of the Contractor, Subcontractors and Sub-subcontractors in the Work, and by appropriate Change Order the cost thereof shall be charged to the Owner. If the Contractor is damaged by the failure or neglect of the Owner to purchase or maintain insurance as described above, without so notifying the Contractor, then the Owner shall bear all reasonable costs properly attributable thereto.

11.3.1.3 If the property insurance requires minimum deductibles and such deductibles are identified in the Contract Documents, the Contractor shall pay costs not covered because of such deductibles. If the Owner or insurer increases the required minimum deductibles above the amounts so identified or if the Owner elects to purchase this insurance with voluntary deductible amounts, the Owner shall be responsible for payment of the additional costs not covered because of such increased or voluntary deductibles. If deductibles are not identified in the Contract Documents, the Owner shall pay costs not covered because of deductibles.

11.3.1.4 Unless otherwise provided in the Contract Documents, this property insurance shall cover portions of the Work stored off the site after written approval of the Owner at the value established in the approval, and also portions of the Work in transit.

11.3.2 Boiler and Machinery Insurance. The Owner shall purchase and maintain boiler and machinery insurance required by the Contract Documents or by law, which shall specifically cover such insured objects during installation and until final acceptance by the Owner; this insurance shall include interests of the Owner, Contractor, Subcontractors and Sub-subcontractors in the Work, and the Owner and Contractor shall be named insureds.

11.3.3 Loss of Use Insurance. The Owner, at the Owner's option, may purchase and maintain such insurance as will insure the Owner against loss of use of the Owner's property due to fire or other hazards, however caused. The Owner waives all rights of action against the Contractor for loss of use of the Owner's property, including consequential losses due to fire or other hazards however caused.

11.3.4 If the Contractor requests in writing that insurance for risks other than those described herein or for other special hazards be included in the property insurance policy, the Owner shall, if possible, include such insurance, and the cost thereof shall be charged to the Contractor by appropriate Change Order.

11.3.5 If during the Project construction period the Owner insures properties, real or personal or both, adjoining or adjacent to the site by property insurance under policies separate from those insuring the Project, or if after final payment property insurance is to be provided on the completed Project through a policy or policies other than those insuring the Project during the construction period, the Owner shall waive all rights in accordance with the terms of Subparagraph 11.3.7 for damages caused by fire or other perils covered by this separate property insurance. All separate policies shall provide this waiver of subrogation by endorsement or otherwise.

11.3.6 Before an exposure to loss may occur, the Owner shall file with the Contractor a copy of each policy that includes insurance coverages required by this Paragraph 11.3. Each policy shall contain all generally applicable conditions, definitions, exclusions and endorsements related to this Project. Each policy shall contain a provision that the policy will not be cancelled or allowed to expire until at least 30 days' prior written notice has been given to the Contractor.

11.3.7 Waivers of Subrogation. The Owner and Contractor waive all rights against (1) each other and any of their subcontractors, sub-subcontractors, agents and employees, each of the other, and (2) the Architect, Architect's consultants, separate contractors described in Article 6, if any, and any of their subcontractors, sub-subcontractors, agents and employees, for damages caused by fire or other perils to the extent covered by property insurance obtained pursuant to this Paragraph 11.3 or other property insurance applicable to the Work, except such rights as they have to proceeds of such insurance held by the Owner as fiduciary. The Owner or Contractor, as appropriate, shall require of the Architect, Architect's consultants, separate contractors described in Article 6, if any, and the subcontractors, sub-subcontractors, agents and employees of any of them, by appropriate agreements, written where legally required for validity, similar waivers each in favor of other parties enumerated herein. The policies shall provide such waivers of subrogation by endorsement or otherwise. A waiver of subrogation shall be effective as to a person or entity even though that person or entity would otherwise have a duty of indemnification, contractual or otherwise, did not pay the insurance premium directly or indirectly, and whether or not the person or entity had an insurable interest in the property damaged.

11.3.8 A loss insured under Owner's property insurance shall be adjusted by the Owner as fiduciary and made payable to the Owner as fiduciary for the insureds, as their interests may appear, subject to requirements of any applicable mortgagee clause and of Subparagraph 11.3.10. The Contractor shall pay Subcontractors their just shares of insurance proceeds received by the Contractor, and by appropriate agreements, written where legally required for validity, shall require Subcontractors to make payments to their Sub-subcontractors in similar manner.

11.3.9 If required in writing by a party in interest, the Owner as fiduciary shall, upon occurrence of an insured loss, give bond for proper performance of the Owner's duties. The cost of required bonds shall be charged against proceeds received as fiduciary. The Owner shall deposit in a separate account proceeds so received, which the Owner shall distribute in accordance with such agreement as the parties in interest may reach, or in accordance with an arbitration award in which case the procedure shall be as provided in Paragraph 4.5. If after such loss no other special agreement is made, replacement of damaged property shall be covered by appropriate Change Order.

11.3.10 The Owner as fiduciary shall have power to adjust and settle a loss with insurers unless one of the parties in interest shall object in writing within five days after occurrence of loss to the Owner's exercise of this power; if such objection be made, arbitrators shall be chosen as provided in Paragraph 4.5. The Owner as fiduciary shall, in that case, make settlement with insurers in accordance with directions of such arbitrators. If distribution of insurance proceeds by arbitration is required, the arbitrators will direct such distribution.

11.3.11 Partial occupancy or use in accordance with Paragraph 9.9 shall not commence until the insurance company or companies providing property insurance have consented to such partial occupancy or use by endorsement or otherwise. The Owner and the Contractor shall take reasonable steps to obtain consent of the insurance company or companies and shall, without mutual written consent, take no action with respect to partial occupancy or use that would cause cancellation, lapse or reduction of insurance.

11.4 PERFORMANCE BOND AND PAYMENT BOND

11.4.1 The Owner shall have the right to require the Contractor to furnish bonds covering faithful performance of the Contract and payment of obligations arising thereunder as stipulated in bidding requirements or specifically required in the Contract Documents on the date of execution of the Contract.

11.4.2 Upon the request of any person or entity appearing to be a potential beneficiary of bonds covering payment of obligations arising under the Contract, the Contractor shall promptly furnish a copy of the bonds or shall permit a copy to be made.

ARTICLE 12

UNCOVERING AND CORRECTION OF WORK

12.1 UNCOVERING OF WORK

12.1.1 If a portion of the Work is covered contrary to the Architect's request or to requirements specifically expressed in the Contract Documents, it must, if required in writing by the Architect, be uncovered for the Architect's observation and be replaced at the Contractor's expense without change in the Contract Time.

12.1.2 If a portion of the Work has been covered which the Architect has not specifically requested to observe prior to its being covered, the Architect may request to see such Work and it shall be uncovered by the Contractor. If such Work is in accordance with the Contract Documents, costs of uncovering and replacement shall, by appropriate Change Order, be charged to the Owner. If such Work is not in accordance with the Contract Documents, the Contractor shall pay such costs unless the condition was caused by the Owner or a separate contractor in which event the Owner shall be responsible for payment of such costs.

12.2 CORRECTION OF WORK

12.2.1 The Contractor shall promptly correct Work rejected by the Architect or failing to conform to the requirements of the Contract Documents, whether observed before or after Substantial Completion and whether or not fabricated, installed or completed. The Contractor shall bear costs of correcting such rejected Work, including additional testing and inspections and compensation for the Architect's services and expenses made necessary thereby.

12.2.2 If, within one year after the date of Substantial Completion of the Work or designated portion thereof, or after the date for commencement of warranties established under Subparagraph 9.9.1, or by terms of an applicable special warranty required by the Contract Documents, any of the Work is found to be not in accordance with the requirements of the Contract Documents, the Contractor shall correct it promptly after receipt of written notice from the Owner to do so unless the Owner has previously given the Contractor a written acceptance of such condition. This period of one year shall be extended with respect to portions of Work first performed after Substantial Completion by the period of time between Substantial Completion and the actual performance of the Work. This obligation under this Subparagraph 12.2.2 shall survive acceptance of the Work under the Contract and termination of the Contract. The Owner shall give such notice promptly after discovery of the condition.

12.2.3 The Contractor shall remove from the site portions of the Work which are not in accordance with the requirements of the Contract Documents and are neither corrected by the Contractor nor accepted by the Owner.

12.2.4 If the Contractor fails to correct nonconforming Work within a reasonable time, the Owner may correct it in accordance with Paragraph 2.4. If the Contractor does not proceed with correction of such nonconforming Work within a reasonable time fixed by written notice from the Architect, the Owner may remove it and store the salvable materials or equipment at the Contractor's expense. If the Contractor does not pay costs of such removal and storage within ten days after written notice, the Owner may upon ten additional days' written notice sell such materials and equipment at auction or at private sale and shall account for the proceeds thereof, after deducting costs and damages that should have been borne by the Contractor, including compensation for the Architect's services and expenses made necessary thereby. If such proceeds of sale do not cover costs which the Contractor should have borne, the Contract Sum shall be reduced by the deficiency. If payments then or thereafter due the Contractor are not sufficient to cover such amount, the Contractor shall pay the difference to the Owner.

12.2.5 The Contractor shall bear the cost of correcting destroyed or damaged construction, whether completed or partially completed, of the Owner or separate contractors caused by the Contractor's correction or removal of Work which is not in accordance with the requirements of the Contract Documents.

12.2.6 Nothing contained in this Paragraph 12.2 shall be construed to establish a period of limitation with respect to other obligations which the Contractor might have under the Contract Documents. Establishment of the time period of one year as described in Subparagraph 12.2.2 relates only to the specific obligation of the Contractor to correct the Work, and has no relationship to the time within which the obligation to comply with the Contract Documents may be sought to be enforced, nor to the time within which proceedings may be commenced to establish the Contractor's liability with respect to the Contractor's obligations other than specifically to correct the Work.

12.3 ACCEPTANCE OF NONCONFORMING WORK

12.3.1 If the Owner prefers to accept Work which is not in accordance with the requirements of the Contract Documents, the Owner may do so instead of requiring its removal and correction, in which case the Contract Sum will be reduced as appropriate and equitable. Such adjustment shall be effected whether or not final payment has been made.

ARTICLE 13

MISCELLANEOUS PROVISIONS

13.1 GOVERNING LAW

13.1.1 The Contract shall be governed by the law of the place where the Project is located.

13.2 SUCCESSORS AND ASSIGNS

13.2.1 The Owner and Contractor respectively bind themselves, their partners, successors, assigns and legal representatives to the other party hereto and to partners, successors, assigns and legal representatives of such other party in respect to covenants, agreements and obligations contained in the Contract Documents. Neither party to the Contract shall assign the Contract as a whole without written consent of the other. If either party attempts to make such an assignment without such consent, that party shall nevertheless remain legally responsible for all obligations under the Contract.

13.3 WRITTEN NOTICE

13.3.1 Written notice shall be deemed to have been duly served if delivered in person to the individual or a member of the firm or entity or to an officer of the corporation for which it was intended, or if delivered at or sent by registered or certified mail to the last business address known to the party giving notice.

13.4 RIGHTS AND REMEDIES

13.4.1 Duties and obligations imposed by the Contract Documents and rights and remedies available thereunder shall be in addition to and not a limitation of duties, obligations, rights and remedies otherwise imposed or available by law.

13.4.2 No action or failure to act by the Owner, Architect or Contractor shall constitute a waiver of a right or duty afforded them under the Contract, nor shall such action or failure to act constitute approval of or acquiescence in a breach thereunder, except as may be specifically agreed in writing.

13.5 TESTS AND INSPECTIONS

13.5.1 Tests, inspections and approvals of portions of the Work required by the Contract Documents or by laws, ordinances, rules, regulations or orders of public authorities having jurisdiction shall be made at an appropriate time. Unless otherwise provided, the Contractor shall make arrangements for such tests, inspections and approvals with an independent testing laboratory or entity acceptable to the Owner, or with the appropriate public authority, and shall bear all related costs of tests, inspections and approvals. The Contractor shall give the Architect timely notice of when and where tests and inspections are to be made so the Architect may observe such procedures. The Owner shall bear costs of tests, inspections or approvals which do not become requirements until after bids are received or negotiations concluded.

13.5.2 If the Architect, Owner or public authorities having jurisdiction determine that portions of the Work require additional testing, inspection or approval not included under Subparagraph 13.5.1, the Architect will, upon written authorization from the Owner, instruct the Contractor to make arrangements for such additional testing, inspection or approval by an entity acceptable to the Owner, and the Contractor shall give timely notice to the Architect of when and where tests and inspections are to be made so the Architect may observe such procedures.

The Owner shall bear such costs except as provided in Subparagraph 13.5.3.

13.5.3 If such procedures for testing, inspection or approval under Subparagraphs 13.5.1 and 13.5.2 reveal failure of the portions of the Work to comply with requirements established by the Contract Documents, the Contractor shall bear all costs made necessary by such failure including those of repeated procedures and compensation for the Architect's services and expenses.

13.5.4 Required certificates of testing, inspection or approval shall, unless otherwise required by the Contract Documents, be secured by the Contractor and promptly delivered to the Architect.

13.5.5 If the Architect is to observe tests, inspections or approvals required by the Contract Documents, the Architect will do so promptly and, where practicable, at the normal place of testing.

13.5.6 Tests or inspections conducted pursuant to the Contract Documents shall be made promptly to avoid unreasonable delay in the Work.

13.6 INTEREST

13.6.1 Payments due and unpaid under the Contract Documents shall bear interest from the date payment is due at such rate as the parties may agree upon in writing or, in the absence thereof, at the legal rate prevailing from time to time at the place where the Project is located.

13.7 COMMENCEMENT OF STATUTORY LIMITATION PERIOD

13.7.1 As between the Owner and Contractor:

.1 **Before Substantial Completion.** As to acts or failures to act occurring prior to the relevant date of Substantial Completion, any applicable statute of limitations shall commence to run and any alleged cause of action shall be deemed to have accrued in any and all events not later than such date of Substantial Completion;

.2 **Between Substantial Completion and Final Certificate for Payment.** As to acts or failures to act occurring subsequent to the relevant date of Substantial Completion and prior to issuance of the final Certificate for Payment, any applicable statute of limitations shall commence to run and any alleged cause of action shall be deemed to have accrued in any and all events not later than the date of issuance of the final Certificate for Payment; and

.3 **After Final Certificate for Payment.** As to acts or failures to act occurring after the relevant date of issuance of the final Certificate for Payment, any applicable statute of limitations shall commence to run and any alleged cause of action shall be deemed to have accrued in any and all events not later than the date of any act or failure to act by the Contractor pursuant to any warranty provided under Paragraph 3.5, the date of any correction of the Work or failure to correct the Work by the Contractor under Paragraph 12.2, or the date of actual commission of any other act or failure to perform any duty or obligation by the Contractor or Owner, whichever occurs last.

ARTICLE 14

TERMINATION OR SUSPENSION OF THE CONTRACT

14.1 TERMINATION BY THE CONTRACTOR

14.1.1 The Contractor may terminate the Contract if the Work is stopped for a period of 30 days through no act or fault of the Contractor or a Subcontractor, Sub-subcontractor or their agents or employees or any other persons performing portions of the Work under contract with the Contractor, for any of the following reasons:

.1 issuance of an order of a court or other public authority having jurisdiction;

.2 an act of government, such as a declaration of national emergency, making material unavailable;

.3 because the Architect has not issued a Certificate for Payment and has not notified the Contractor of the reason for withholding certification as provided in Subparagraph 9.4.1, or because the Owner has not made payment on a Certificate for Payment within the time stated in the Contract Documents;

.4 if repeated suspensions, delays or interruptions by the Owner as described in Paragraph 14.3 constitute in the aggregate more than 100 percent of the total number of days scheduled for completion, or 120 days in any 365-day period, whichever is less; or

.5 the Owner has failed to furnish to the Contractor promptly, upon the Contractor's request, reasonable evidence as required by Subparagraph 2.2.1.

14.1.2 If one of the above reasons exists, the Contractor may, upon seven additional days' written notice to the Owner and Architect, terminate the Contract and recover from the Owner payment for Work executed and for proven loss with respect to materials, equipment, tools, and construction equipment and machinery, including reasonable overhead, profit and damages.

14.1.3 If the Work is stopped for a period of 60 days through no act or fault of the Contractor or a Subcontractor or their agents or employees or any other persons performing portions of the Work under contract with the Contractor because the Owner has persistently failed to fulfill the Owner's obligations under the Contract Documents with respect to matters important to the progress of the Work, the Contractor may, upon seven additional days' written notice to the Owner and the Architect, terminate the Contract and recover from the Owner as provided in Subparagraph 14.1.2.

14.2 TERMINATION BY THE OWNER FOR CAUSE

14.2.1 The Owner may terminate the Contract if the Contractor:

.1 persistently or repeatedly refuses or fails to supply enough properly skilled workers or proper materials;

.2 fails to make payment to Subcontractors for materials or labor in accordance with the respective agreements between the Contractor and the Subcontractors;

.3 persistently disregards laws, ordinances, or rules, regulations or orders of a public authority having jurisdiction; or

.4 otherwise is guilty of substantial breach of a provision of the Contract Documents.

14.2.2 When any of the above reasons exist, the Owner, upon certification by the Architect that sufficient cause exists to justify such action, may without prejudice to any other rights or remedies of the Owner and after giving the Contractor and the Contractor's surety, if any, seven days' written notice, terminate employment of the Contractor and may, subject to any prior rights of the surety:

- .1 take possession of the site and of all materials, equipment, tools, and construction equipment and machinery thereon owned by the Contractor;
- .2 accept assignment of subcontracts pursuant to Paragraph 5.4; and
- .3 finish the Work by whatever reasonable method the Owner may deem expedient.

14.2.3 When the Owner terminates the Contract for one of the reasons stated in Subparagraph 14.2.1, the Contractor shall not be entitled to receive further payment until the Work is finished.

14.2.4 If the unpaid balance of the Contract Sum exceeds costs of finishing the Work, including compensation for the Architect's services and expenses made necessary thereby, such excess shall be paid to the Contractor. If such costs exceed the unpaid balance, the Contractor shall pay the difference to the Owner. The amount to be paid to the Contractor or Owner, as the case may be, shall be certified by the Architect, upon application, and this obligation for payment shall survive termination of the Contract.

14.3 SUSPENSION BY THE OWNER FOR CONVENIENCE

14.3.1 The Owner may, without cause, order the Contractor in writing to suspend, delay or interrupt the Work in whole or in part for such period of time as the Owner may determine.

14.3.2 An adjustment shall be made for increases in the cost of performance of the Contract, including profit on the increased cost of performance, caused by suspension, delay or interruption. No adjustment shall be made to the extent:

- .1 that performance is, was or would have been so suspended, delayed or interrupted by another cause for which the Contractor is responsible; or
- .2 that an equitable adjustment is made or denied under another provision of this Contract.

14.3.3 Adjustments made in the cost of performance may have a mutually agreed fixed or percentage fee.

Appendix E:

AIA Document B163--Standard Form of Agreement
Between Owner and Architect with Descriptions
of Designated Services and Terms
and Conditions, 1993 Edition

Appendix

Reproduced with permission of The American Institute of Architects under license number #96051. This license expires September 30, 1997. FURTHER REPRODUCTION IS PROHIBITED. Because AIA Documents are revised from time to time, users should ascertain from the AIA the current edition of this document. Copies of the current edition of this AIA document may be purchased from The American Institute of Architects or its local distributors. The text of this document is not "model language" and is not intended for use in other documents without permission of the AIA.

AIA Document B163

Standard Form of Agreement Between Owner and Architect with Descriptions of Designated Services and Terms and Conditions

THIS DOCUMENT HAS IMPORTANT LEGAL CONSEQUENCES; CONSULTATION WITH AN ATTORNEY IS ENCOURAGED WITH RESPECT TO ITS COMPLETION OR MODIFICATION.

1993 EDITION

TABLE OF ARTICLES

PART 1—FORM OF AGREEMENT

ARTICLE 1.1 SCHEDULE OF DESIGNATED SERVICES
ARTICLE 1.2 COMPENSATION
ARTICLE 1.3 PAYMENTS
ARTICLE 1.4 TIME AND COST
ARTICLE 1.5 ENUMERATION OF DOCUMENTS
ARTICLE 1.6 OTHER CONDITIONS OR SERVICES

PART 2—DESCRIPTIONS OF DESIGNATED SERVICES

ARTICLE 2.1 DESIGNATED SERVICES
ARTICLE 2.2 PHASES OF DESIGNATED SERVICES
ARTICLE 2.3 DESCRIPTIONS OF DESIGNATED SERVICES
ARTICLE 2.4 DESCRIPTIONS OF SUPPLEMENTAL SERVICES

PART 3—TERMS AND CONDITIONS

ARTICLE 3.1 ARCHITECT'S RESPONSIBILITIES
ARTICLE 3.2 OWNER'S RESPONSIBILITIES
ARTICLE 3.3 CONTRACT ADMINISTRATION
ARTICLE 3.4 USE OF PROJECT DRAWINGS, SPECIFICATIONS AND OTHER DOCUMENTS
ARTICLE 3.5 COST OF THE WORK
ARTICLE 3.6 PAYMENTS TO THE ARCHITECT
ARTICLE 3.7 DISPUTE RESOLUTION
ARTICLE 3.8 MISCELLANEOUS PROVISIONS
ARTICLE 3.9 TERMINATION, SUSPENSION OR ABANDONMENT

Copyright ©1993 by The American Institute of Architects, 1735 New York Avenue, N.W., Washington, D.C. 20006-5292. Reproduction of the material herein or substantial quotation of its provisions without the written permission of the AIA violates the copyright laws of the United States and will subject the violator to legal prosecution.

AIA DOCUMENT B163 • OWNER-ARCHITECT AGREEMENT FOR DESIGNATED SERVICES
AIA® • ©1993 • THE AMERICAN INSTITUTE OF ARCHITECTS, 1735 NEW YORK AVENUE, N.W., WASHINGTON, D.C. 20006-5292 • **WARNING: Unlicensed photocopying violates U.S. copyright laws and will subject the violator to legal prosecution.**

AIA Document B163—PART 1

FORM OF AGREEMENT
Between Owner and Architect
for Designated Services

AGREEMENT

made as of the   day of   in the year of
(In words, indicate day, month and year.)

BETWEEN the Owner:
(Name and address)

and the Architect:
(Name and address)

For the following Project:
(Include a detailed description of Project, location, address and scope.)

The Owner and the Architect agree as set forth below.

AIA DOCUMENT B163 • OWNER-ARCHITECT AGREEMENT FOR DESIGNATED SERVICES
AIA® • ©1993 • THE AMERICAN INSTITUTE OF ARCHITECTS, 1735 NEW YORK AVENUE,
N.W., WASHINGTON, D.C. 20006-5292 • **WARNING:** Unlicensed photocopying
violates U.S. copyright laws and will subject the violator to legal prosecution.

Appendix

ARTICLE 1.2

COMPENSATION

The Owner shall compensate the Architect as follows.

1.2.1 For Designated Services, as identified in the Schedule of Designated Services, described in the Description of Designated Services, and any other services included in Article 1.6, compensation shall be computed as follows:
(Insert basis of compensation, including stipulated sums, multiples or percentages, and identify phases to which particular methods of compensation apply, if necessary.)

1.2.2 For Contingent Additional Services of the Architect, as described in the Terms and Conditions, but excluding Contingent Additional Services of Consultants, compensation shall be computed as follows:
(Insert basis of compensation, including rates and multiples of Direct Personnel Expense for Principals and employees, and identify Principals and classify employees, if required. Identify specific services to which particular methods of compensation apply, if necessary.)

1.2.3 For Contingent Additional Services of the Architect's Consultants, including additional structural, mechanical and electrical engineering, and those identified in Article 1.6 and in the Schedule of Designated Services or as part of the Architect's Contingent Additional Services under the Terms and Conditions, compensation shall be computed as a multiple of
(   ) times the amounts billed to the Architect for such services.
(Identify specific types of consultants in Article 6, if required.)

1.2.4 For Reimbursable Expenses, as described in Article 3.7 of the Terms and Conditions, and any other items included in Article 1.6 as a Reimbursable Expense, the compensations shall be computed as a multiple of
(   ) times the expense incurred by the Architect, the Architect's employees and consultants in the interest of the Project.

1.2.5 If the Designated Services identified in the Schedule of Designated Services have not been completed within
(   ) months of the date hereof, through no fault of the Architect, extension of the Architect's services beyond that time shall be compensated as provided in Paragraph 1.2.2.

1.2.6 The rates and multiples set forth for Contingent Additional Services shall be annually adjusted in accordance with normal salary review practices of the Architect.

AIA DOCUMENT B163 • OWNER-ARCHITECT AGREEMENT FOR DESIGNATED SERVICES
AIA® • ©1993 • THE AMERICAN INSTITUTE OF ARCHITECTS, 1735 NEW YORK AVENUE,
N.W., WASHINGTON, D.C. 20006-5292 • **WARNING: Unlicensed photocopying violates U.S. copyright laws and will subject the violator to legal prosecution.**

ARTICLE 1.3

PAYMENTS

1.3.1 An initial payment of _____ dollars ($ _____)
shall be made upon execution of this **Agreement** and is the minimum payment under this Agreement. It shall be credited to the Owner's account at final payment. Subsequent payments for Designated Services shall be made monthly, and where applicable, shall be in proportion to services performed within each phase of service, on the basis set forth in the Agreement.

1.3.2 Where compensation is based on a stipulated sum or percentage of Construction Cost, progress payments for Designated Services in each phase shall be made monthly and shall be in proportion to services performed within each Phase of Services, so that Compensation for each Phase shall equal the following amounts or percentages of the total compensation payable for such Designated Services.
(Insert or delete phases as appropriate.)

Phase	Amount or Percentage
Pre-Design Phase:	
Site Analysis Phase:	
Schematic Design Phase:	
Design Development Phase:	
Contract Documents Phase:	
Bidding or Negotiation Phase:	
Contract Administration Phase:	
Post-Contract Phase:	

1.3.3 Payments are due and payable _____ (____) days from the date of the Architect's invoice. Amounts unpaid _____) days after the invoice date shall bear interest at the rate entered below, or in the absence thereof, at the legal rate prevailing from time to time at the principal place of business of the Architect.
(Insert rate of interest agreed upon.)

(Usury laws and requirements under the Federal Truth in Lending Act, similar state and local consumer credit laws and other regulations at the Owner's and Architect's principal places of business, the location of the Project and elsewhere may affect the validity of this provision. Specific legal advice should be obtained with respect to deletion or modifications, and also regarding requirements such as written disclosures or waivers.)

ARTICLE 1.4

TIME AND COST

1.4.1 Unless otherwise indicated, the Owner and the Architect shall perform their respective obligations as expeditiously as is consistent with normal skill and care and the orderly progress of the Project. Upon the request of the Owner, the Architect shall prepare a schedule for the performance of the Designated Services which may be adjusted as the Project proceeds, and shall include allowances for periods of time required for the Owner's review and for approval of submissions by authorities having jurisdiction over the Project. Time limits established by this schedule upon approval by the Owner shall not, except for reasonable cause, be exceeded by the Architect or Owner. If the Architect is delayed in the performance of services under this Agreement by the Owner, the Owner's Consultants, or any other cause not within the control of the Architect, any applicable schedule shall be adjusted accordingly.
(Insert time requirements, if any.)

1.4.2 The Owner shall establish and update an overall budget for the Project, which shall include the Cost of the Work; contingencies for design, bidding and changes in the Work during construction; compensation of the Architect, Architect's consultants and the Owner's other consultants; cost of the land, rights-of-way and financing; and other costs that are the responsibility of the Owner as indicated by the Terms and Conditions or Designated Services. Prior to the establishment of such a budget, the Owner and the Architect may agree on Designated Services that include the utilization of the Architect's or other consultants' services to assist the Owner with market, financing and feasibility studies deemed necessary for development of such a budget for the Project.

1.4.3 No fixed limit of the Cost of the Work shall be established as a condition of this Agreement by the furnishing, proposal or establishment of a Project budget unless such fixed limit has been agreed to below or by separate Amendment made in writing and signed by the parties hereto. Any fixed limit of the Cost of the Work shall be subject to the limitations and definitions contained in the Terms and Conditions under Part 3 of this Agreement.
(If no fixed limit, leave blank.)

ARTICLE 1.5
ENUMERATION OF DOCUMENTS

1.5.1 This Agreement represents the entire and integrated agreement between the Owner and Architect and supersedes all prior negotiations, representations or agreements, either written or oral. This Agreement may be amended only by written instrument signed by both Owner and Architect.

1.5.2 The parts of this Agreement between the Owner and Architect, except for amendments issued after execution of this Agreement, are enumerated as follows:

1.5.2.1 Form of Agreement Between Owner and Architect, AIA Document B163—Part 1, 1993 Edition;

1.5.2.2 Descriptions of Designated Services for AIA Document B163, AIA Document B163—Part 2, 1993 Edition;

1.5.2.3 Terms and Conditions of AIA Document B163, AIA Document B163—Part 3, 1993 Edition.

1.5.2.4 Other Documents, if any, forming a part of the contract are as follows:
(Insert any additional documents, but only if they are intended to be part of the contract between the Owner and the Architect.)

428

ARTICLE 1.6
OTHER CONDITIONS OR SERVICES

(Insert modifications to the Descriptions of Services contained in Part 2 and to the Terms and Conditions contained in Part 3 of this Agreement.)

This Agreement entered into as of the day and year first written above.

OWNER   ARCHITECT

_____   _____
(Signature)   *(Signature)*

_____   _____
(Printed name and title)   *(Printed name and title)*

 CAUTION: You should sign an original AIA document which has this caution printed in red. An original assures that changes will not be obscured as may occur when documents are reproduced. See Instruction Sheet for Limited License for Reproduction of this document.

AIA DOCUMENT B163 • OWNER-ARCHITECT AGREEMENT FOR DESIGNATED SERVICES
AIA® • ©1993 • THE AMERICAN INSTITUTE OF ARCHITECTS, 1735 NEW YORK AVENUE,
N.W., WASHINGTON, D.C. 20006-5292 • **WARNING: Unlicensed photocopying violates U.S. copyright laws and will subject the violator to legal prosecution.**

B163—1993 8

Appendix

AIA Document B163—PART 2

DESCRIPTIONS OF DESIGNATED SERVICES
for the Agreement
Between Owner and Architect

The current edition of AIA Document A201, General Conditions of the Contract for Construction, is adopted by reference under the Construction Phase of this document. Do not use with other general conditions unless this document is modified.

ARTICLE 2.1
DESIGNATED SERVICES

2.1.1 In accordance with the Schedule of Designated Services completed under Part 1 of this Agreement, the Owner and Architect shall provide the phases and services designated therein and described herein. Unless the responsibility for a Project phase or service is specifically allocated in the Schedule of Designated Services to the Owner or Architect, such phase or service shall not be a requirement of this Agreement.

ARTICLE 2.2
PHASES OF DESIGNATED SERVICES

2.2.1 Pre-Design Phase. The Pre-Design Phase is the stage in which the Owner's program, the financial and time requirements, and the scope of the Project are established.

2.2.2 Site Analysis Phase. The Site Analysis Phase is the stage in which site-related limitations and requirements for the Project are established.

2.2.3 Schematic Design Phase. The Schematic Design Phase is the stage in which the general scope, conceptual design, and the scale and relationship of components of the Project are established.

2.2.4 Design Development Phase. The Design Development Phase is the stage in which the size and character of the Project are further refined and described, including architectural, structural, mechanical and electrical systems, materials, and such other elements as may be appropriate.

2.2.5 Contract Documents Phase. The Contract Documents Phase is the stage in which the requirements for the Work are set forth in detail.

2.2.6 Bidding or Negotiations Phase. The Bidding or Negotiation Phase is the stage in which bids or negotiated proposals are solicited and obtained and in which contracts are awarded.

2.2.7 Contract Administration Phase. The Contract Administration Phase is the stage in which the Work is performed by one or more Contractors.

2.2.8 Post-Contract Phase. The Post-Contract Phase is the stage in which assistance in the Owner's use and occupancy of the Project is provided.

2.2.9 Sequence of Phases. The services for the above phases are generally performed in a chronological sequence following the order of phases shown in Paragraphs 2.2.1 through 2.2.8.

2.2.9.1 Normal Sequence. The Owner and Architect shall commence the performance of their respective responsibilities with the services assigned to the foremost sequential phase under the completed Schedule of Designated Services of Part 1 to this Agreement. Except as provided under Subparagraphs 2.2.9.2 and 2.2.9.3, subsequent phases shall not be commenced until the Owner has approved the results of the Architect's services for the preceding phase. Such approvals shall not be unreasonably withheld. When phases or services are to be combined or compressed, their chronology shall continue to follow that shown above, unless otherwise provided in this Agreement.

2.2.9.2 Fast Track. Upon the receipt of the Owner's written authorization for Work to commence prior to completion of the Architect's Contract Documents Phase, the Architect shall provide the services designated in an overlapping manner rather than in the normal chronological sequence in order to expedite the Owner's early occupancy of all or a portion of the Project. The Owner shall furnish to the Architect in a timely manner information obtained from all Contractors and prospective contractors regarding anticipated market conditions and construction cost, availability of labor, materials and equipment, and their proposed methods, sequences and time schedules for construction of the Work. Upon receipt of their proposed Work schedules, the Architect shall prepare a schedule for providing services. In the event of a conflict between the proposed Work schedules and the Architect's proposed schedule, the Architect shall inform the Owner of such conflict.

2.2.9.3 Supplemental Services. Supplemental Services may be provided, however, during a single phase or several phases and may not necessarily follow the normal chronological sequence.

ARTICLE 2.3

DESCRIPTIONS OF DESIGNATED SERVICES

PROJECT ADMINISTRATION AND MANAGEMENT SERVICES

.01 **Project Administration** services consisting of administrative functions including:
- .01 Consultation
- .02 Research
- .03 Conferences
- .04 Communications
- .05 Travel time
- .06 Progress reports
- .07 Direction of the work of in-house architectural personnel
- .08 Coordination of work by the Owner's forces.

.02 **Disciplines Coordination/Document Checking** consisting of:
- .01 Coordination between the architectural work and the work of engineering and other disciplines involved in the Project
- .02 Review and checking of documents prepared for the Project by the Architect and the Architect's Consultants.

.03 **Agency Consulting/Review/Approval** services, including:
- .01 Agency consultations
- .02 Research of critical applicable regulations
- .03 Research of community attitudes
- .04 Preparation of written and graphic explanatory materials
- .05 Appearances on Owner's behalf at agency and community meetings.

The services below apply to applicable laws, statutes, regulations and codes of regulating entities and to reviews required of user or community groups with limited or no statutory authority but significant influence on approving agencies and individuals, including:

- .06 Local political subdivisions
- .07 Planning boards
- .08 County agencies
- .09 Regional agencies
- .10 Federal agencies
- .11 User organizations
- .12 Community organizations
- .13 Consumer interest organizations
- .14 Environmental interest groups.

.04 **Owner-Supplied Data Coordination,** including:
- .01 Review and coordination of data furnished for the Project as a responsibility of the Owner
- .02 Assistance in establishing criteria
- .03 Assistance in obtaining data, including, where applicable, documentation of existing conditions.

.05 **Schedule Development/Monitoring** services, including:
- .01 Establishment of initial schedule for Architect's services, decision-making, design, documentation, contracting and construction, based on determination of scope of Architect's services
- .02 Review and update of previously established schedules during subsequent phases.

.06 **Preliminary Estimate of the Cost of the Work,** including:
- .01 Preparation of a preliminary estimate of the Cost of the Work
- .02 Review and update the preliminary estimate of the Cost of the Work during subsequent phases.

.07 **Presentation** services consisting of presentations and recommendations by the Architect to the following client representatives:
- .01 Owner
- .02 Building committee(s)
- .03 Staff committee (s)
- .04 User group(s)
- .05 Board(s) of Directors
- .06 Financing entity (entities)
- .07 Owner's consultant(s).

PRE-DESIGN SERVICES

.08 **Programming** services consisting of consultation to establish and document the following detailed requirements for the Project:
- .01 Design objectives, limitations and criteria
- .02 Development of initial approximate gross facility areas and space requirements
- .03 Space relations
- .04 Number of functional responsibilities personnel
- .05 Flexibility and expandability
- .06 Special equipment and systems
- .07 Site requirements
- .08 Development of a preliminary budget for the Work based on programming and scheduling studies
- .09 Operating procedures
- .10 Security criteria
- .11 Communications relationships
- .12 Project schedule.

.09 **Space Schematics/Flow Diagrams** consisting of diagrammatic studies and pertinent descriptive text for:
- .01 Conversion of programmed requirements to net area requirements
- .02 Internal functions
- .03 Human, vehicular and material flow patterns
- .04 General space allocations
- .05 Analysis of operating functions
- .06 Adjacency
- .07 Special facilities and equipment
- .08 Flexibility and expandability

.10 **Existing Facilities Surveys** consisting of researching, assembling, reviewing and supplementing information for Projects involving alterations and additions to existing facilities or determining new space usage in conjunction with a new building program and including:
- .01 Photography
- .02 Field measurements
- .03 Review of existing design data

- .04 Analysis of existing structural capabilities
- .05 Analysis of existing mechanical capabilities
- .06 Analysis of existing electrical capabilities
- .07 Review of existing drawings for critical inaccuracies, and the development of required measured drawings.

.11 **Marketing Studies** relating to determination of social, economic and political need for and acceptability of the Project and consisting of:
- .01 Determination with Owner of the scope, parameters, schedule and budget for marketing studies
- .02 Identification, assembly, review and organization of existing pertinent data
- .03 Arrangement of clearances for use of existing data
- .04 Mail survey studies
- .05 Personal survey studies
- .06 Analysis of data
- .07 Assistance in obtaining computerized analysis and modeling
- .08 Computerized analysis and modeling
- .09 Preparation of interim reports
- .10 Preparation of final report
- .11 Assistance in production of final report.

.12 **Economic Feasibility Studies** consisting of the preparation of economic analysis and feasibility evaluation of the Project based on estimates of:
- .01 Total Project cost
- .02 Operation and ownership cost
- .03 Financing requirements
- .04 Cash flow for design, construction and operation
- .05 Return on investment studies
- .06 Equity requirements.

.13 **Project Financing** services as required in connection with:
- .01 Assistance to Owner in preparing and submitting data, supplementary drawings and documentation
- .02 Research of financing availability
- .03 Direct solicitation of financing sources by the Architect.

Project financing services are required for:
- .04 Development costs
- .05 Site control and/or acquisition
- .06 Predesign and site analysis services
- .07 Planning, design, documentation and bidding services
- .08 Interim or construction financing
- .09 Permanent or long-term financing.

SITE DEVELOPMENT SERVICES

.14 **Site Analysis and Selection** consisting of:
- .01 Identification of potential site(s)
- .02 On-site observations
- .03 Movement systems, traffic and parking studies
- .04 Topography analysis
- .05 Analysis of deed, zoning and other legal restrictions
- .06 Studies of availability of labor force to staff Owner's facility
- .07 Studies of availability of construction materials, equipment and labor
- .08 Studies of construction market
- .09 Overall site analysis and evaluation
- .10 Comparative site studies.

Appendix

.15 **Site Development Planning** consisting of preliminary site analysis, and preparation and comparative evaluation of conceptual site development designs, based on:
- .01 Land utilization
- .02 Structures placement
- .03 Facilities development
- .04 Development phasing
- .05 Movement systems, circulation and parking
- .06 Utilities systems
- .07 Surface and subsurface conditions
- .08 Ecological requirements
- .09 Deeds, zoning and other legal restrictions
- .10 Landscape concepts and forms.

.16 **Detailed Site Utilization Studies** consisting of detailed site analyses, based on the approved conceptual site development design, including:
- .01 Land utilization
- .02 Structures placement
- .03 Facilities development
- .04 Development phasing
- .05 Movement systems, circulation and parking
- .06 Utilities systems
- .07 Surface and subsurface conditions
- .08 Review of soils report
- .09 Vegetation
- .10 Slope analysis
- .11 Ecological studies
- .12 Deeds, zoning and other legal restrictions
- .13 Landscape forms and materials.

.17 **On-Site Utility Studies** consisting of establishing requirements and preparing initial designs for on-site:
- .01 Electrical service and distribution
- .02 Gas service and distribution
- .03 Water supply and distribution
- .04 Site drainage
- .05 Sanitary sewer collection and disposal
- .06 Process waste water treatment
- .07 Storm water collection and disposal
- .08 Central-plant mechanical systems
- .09 Fire systems
- .10 Emergency systems
- .11 Security
- .12 Pollution control
- .13 Site illumination
- .14 Communications systems.

.18 **Off-Site Utility Studies** consisting of:
- .01 Confirmation of location, size and adequacy of utilities serving the site
- .02 Determination of requirements for connections to utilities
- .03 Planning for off-site utility extensions and facilities
- .04 Design of off-site utility extensions and facilities.

.19 **Environmental Studies and Reports** consisting of:
  .01 Determination of need or requirements for environmental monitoring, assessment and/or impact statements
  .02 Ecological studies
  .03 Preparation of environmental assessment reports
  .04 Preparation of environmental impact reports
  .05 Attendance at public meetings and hearings
  .06 Presentations to governing authorities.

.20 **Zoning Processing Assistance** consisting of:
  .01 Assistance in preparing applications
  .02 Development of supporting data
  .03 Preparation of presentation materials
  .04 Attendance at public meetings and hearings.

.21 **Geotechnical Engineering** services, including, but not limited to:
  .01 Test borings, test pits, determinations of soil bearing values, percolation tests, evaluations of hazardous materials, ground corrosion and resistivity tests, including necessary operations for anticipating subsoil conditions
  .02 Reports and appropriate professional recommendations.

.22 **Site Surveying** services, to include:
  .01 Furnishing a survey by a licensed surveyor, describing the physical characteristics, legal limitations and utility locations for the site of the Project, including a written legal description of the site
  .02 Include, as applicable, grades and lines of streets, alleys, pavements and adjoining property and structures; adjacent drainage; rights-of-way, restrictions, easements, encroachments, zoning, deed restrictions, boundaries and contours of the site; locations, dimensions and necessary data pertaining to existing buildings, other improvements and trees; and information concerning available utility services and lines, both public and private, above and below grade, including inverts and depths. All information shall be referenced to a project benchmark.

DESIGN SERVICES

.23 **Architectural Design/Documentation:**
  .01 During the Schematic Design Phase, responding to program requirements and preparing:
    .01 Review of Owner's Program and Budget
    .02 Conceptual site and building plans
    .03 Preliminary sections and elevations
    .04 Preliminary selection of building systems and materials
    .05 Development of approximate dimensions, areas and volumes
    .06 Perspective sketch(es)
    .07 Study model(s).

  .02 During the Design Development Phase consisting of continued development and expansion of architectural Schematic Design Documents to establish the final scope, relationships, forms, size and appearance of the Project through:
    .01 Plans, sections and elevations
    .02 Typical construction details
    .03 Three-dimensional sketch(es)
    .04 Study model(s)
    .05 Final materials selection
    .06 Equipment layouts.

  .03 During the Contract Documents Phase consisting of preparation of Drawings based on approved Design Development Documents setting forth in detail the architectural construction requirements for the Project.

.24 Structural Design/Documentation:

.01 During the Schematic Design Phase consisting of recommendations regarding basic structural materials and systems, analyses, and development of conceptual design solutions for:
  .01 A predetermined structural system
  .02 Alternate structural systems.

.02 During the Design Development Phase consisting of continued development of the specific structural system(s) and Schematic Design Documents in sufficient detail to establish:
  .01 Basic structural system and dimensions
  .02 Final structural design criteria
  .03 Foundation design criteria
  .04 Preliminary sizing of major structural components
  .05 Critical coordination clearances
  .06 Outline Specifications or materials lists.

.03 During the Contract Documents Phase consisting of preparation of final structural engineering calculations, Drawings and Specifications based on approved Design Development Documents, setting forth in detail the structural construction requirements for the Project.

.25 Mechanical Design/Documentation:

.01 During the Schematic Design Phase consisting of consideration of alternate materials, systems and equipment, and development of conceptual design solutions for:
  .01 Energy source(s)
  .02 Energy conservation
  .03 Heating and ventilating
  .04 Air conditioning
  .05 Plumbing
  .06 Fire protection
  .07 General space requirements.

.02 During the Design Development Phase consisting of continued development and expansion of mechanical Schematic Design Documents and development of outline Specifications or materials lists to establish:
  .01 Approximate equipment sizes and capacities
  .02 Preliminary equipment layouts
  .03 Required space for equipment
  .04 Required chases and clearances
  .05 Acoustical and vibration control
  .06 Visual impacts
  .07 Energy conservation measures.

.03 During the Contract Documents Phase consisting of preparation of final mechanical engineering calculations, Drawings and Specifications based on approved Design Development Documents, setting forth in detail the mechanical construction requirements for the Project.

.26 Electrical Design/Documentation:

.01 During the Schematic Design Phase consisting of consideration of alternate systems, recommendations regarding basic electrical materials, systems and equipment, analyses, and development of conceptual solutions for:
  .01 Power service and distribution
  .02 Lighting
  .03 Telephones
  .04 Fire detection and alarms

- .05 Security systems
- .06 Electronic communications
- .07 Special electrical systems
- .08 General space requirements.

.02 During the Design Development Phase consisting of continued development and expansion of electrical Schematic Design Documents and development of outline Specifications or materials lists to establish:
- .01 Criteria for lighting, electrical and communications systems
- .02 Approximate sizes and capacities of major components
- .03 Preliminary equipment layouts
- .04 Required space for equipment
- .05 Required chases and clearances.

.03 During the Contract Documents Phase consisting of preparation of final electrical engineering calculations, Drawings and Specifications based on approved Design Development Documents, setting forth in detail the electrical requirements for the Project.

.27 Civil Design/Documentation:

.01 During the Schematic Design Phase consisting of consideration of alternate materials and systems and development of conceptual design solutions for:
- .01 On-site utility systems
- .02 Fire protection systems
- .03 Drainage systems
- .04 Paving.

.02 During the Design Development Phase consisting of continued development and expansion of civil Schematic Design Documents and development of outline Specifications or materials lists to establish the final scope of and preliminary details for on-site and off-site civil engineering work

.03 During the Contract Documents Phase consisting of preparation of final civil engineering calculations, Drawings and Specifications based on approved Design Development Documents, setting forth in detail the civil construction requirements for the Project.

.28 Landscape Design/Documentation:

.01 During the Schematic Design Phase consisting of consideration of alternate materials, systems and equipment and development of conceptual design solutions for land forms, lawns and plantings based on program requirements, physical site characteristics, design objectives and environmental determinants

.02 During the Design Development Phase consisting of continued development and expansion of landscape Schematic Design Documents and development of outline Specifications or materials lists to establish final scope and preliminary details for landscape work

.03 During the Contract Documents Phase consisting of preparation of Drawings and Specifications based on approved Design Development Documents, setting forth in detail the landscape requirements for the Project.

.29 Interior Design/Documentation:

.01 During the Schematic Design Phase consisting of space allocation and utilization plans based on functional relationships, consideration of alternate materials, systems and equipment and development of conceptual design solutions for architectural, mechanical, electrical and equipment requirements in order to establish:
- .01 Partition locations
- .02 Furniture and equipment layouts
- .03 Types and qualities of finishes and materials for furniture, furnishings and equipment.

.02 During the Design Development Phase consisting of continued development and expansion of interior Schematic Design Documents and development of outline Specifications or materials lists to establish final scope and preliminary details relative to:
   .01 Interior construction of the Project
   .02 Special interior design features
   .03 Furniture, furnishings and equipment selections
   .04 Materials, finishes and colors.

.03 During the Contract Documents Phase consisting of preparation of Drawings, Specifications and other documents based on approved Design Development Documents, setting forth in detail the requirements for interior construction and for furniture, furnishings and equipment for the Project.

.30 Special Design/Documentation, including:
   .01 Preparation and coordination of special Drawings and Specifications for obtaining bids or prices on alternate subdivisions of the Work
   .02 Preparation and coordination of special Drawings and Specifications for obtaining alternate bids or prices on changes in the scope of the Work
   .03 Preparation and coordination of Drawings, Specifications, Bidding Documents and schedules for out-of-sequence bidding or pricing of subdivisions of the Work
   .04 Preparation and coordination of Drawings, Specifications and Bidding Documents for multiple prime contracts for subdivisions of the Work.

.31 Materials Research/Specifications:
  .01 During the Schematic Design Phase consisting of:
     .01 Identification of potential architectural materials, systems and equipment and their criteria and quality standards consistent with the conceptual design
     .02 Investigation of availability and suitability of alternative architectural materials, systems and equipment
     .03 Coordination of similar activities of other disciplines.

  .02 During the Design Development Phase consisting of activities by in-house architectural personnel in:
     .01 Presentation of proposed General and Supplementary Conditions of the Contract for Owner's approval
     .02 Development of architectural outline Specifications or itemized lists and brief form identification of significant architectural materials, systems and equipment, including their criteria and quality standards
     .03 Coordination of similar activities of other disciplines
     .04 Production of design manual including design criteria and outline Specifications or materials lists.

  .03 During the Contract Documents Phase consisting of activities of in-house architectural personnel in:
     .01 Assistance to the Owner in development and preparation of bidding and procurement information which describes the time, place and conditions of bidding, bidding forms, and the form(s) of Agreement between the Owner and Contractor(s)
     .02 Assistance to the Owner in development and preparation of the Conditions of the Contract (General, Supplementary and other Conditions)
     .03 Development and preparation of architectural Specifications describing materials, systems and equipment, workmanship, quality and performance criteria required for the construction of the Project
     .04 Coordination of the development of Specifications by other disciplines
     .05 Compilation of Project Manual including Conditions of the Contract, bidding and procurement information and Specifications.

BIDDING OR NEGOTIATION SERVICES

.32 Bidding Materials services consisting of organizing and handling Bidding Documents for:
- .01 Coordination
- .02 Reproduction
- .03 Completeness review
- .04 Distribution
- .05 Distribution records
- .06 Retrieval
- .07 Receipt and return of document deposits
- .08 Review, repair and reassembly of returned materials.

.33 Addenda services consisting of preparation and distribution of Addenda as may be required during bidding or negotiation and including supplementary Drawings, Specifications, instructions and notice(s) of changes in the bidding schedule and procedure.

.34 Bidding/Negotiation services consisting of:
- .01 Assistance to Owner in establishing list of Bidders or proposers
- .02 Prequalification of Bidders or proposers
- .03 Participation in pre-bid conferences
- .04 Responses to questions from Bidders or proposers and clarifications or interpretations of the Bidding Documents
- .05 Attendance at bid opening(s)
- .06 Documentation and distribution of bidding results.

.35 Analysis of Alternates/Substitutions consisting of consideration, analyses, comparisons, and recommendations relative to alternates or substitutions proposed by Bidders or proposers either prior or subsequent to receipt of Bids or proposals.

.36 Special Bidding services consisting of:
- .01 Attendance at bid openings, participation in negotiations, and documentation of decisions for multiple contracts or phased Work
- .02 Technical evaluation of proposals for building systems
- .03 Participation in detailed evaluation procedures for building systems proposals.

.37 Bid Evaluation services consisting of:
- .01 Validation of bids or proposals
- .02 Participation in reviews of bids or proposals
- .03 Evaluation of bids or proposals
- .04 Recommendation on award of Contract(s)
- .05 Participation in negotiations prior to or following decisions on award of the Contract(s).

.38 Contract Award services consisting of:
- .01 Notification of Contract award(s)
- .02 Assistance in preparation of construction contract Agreement forms for approval by Owner
- .03 Preparation and distribution of sets of Contract Documents for execution by parties to the Contract(s)
- .04 Receipt, distribution and processing, for Owner's approval, of required certificates of insurance, bonds and similar documents
- .05 Preparation and distribution to Contractor(s), on behalf of the Owner, of notice(s) to proceed with the Work.

Appendix

CONTRACT ADMINISTRATION SERVICES

.39 **Submittal Services** consisting of:
    .01 Processing of submittals, including receipt, review of, and appropriate action on Shop Drawings, Product Data, Samples and other submittals required by the Contract Documents
    .02 Distribution of submittals to Owner, Contractor and/or Architect's field representative as required
    .03 Maintenance of master file of submittals
    .04 Related communications.

.40 **Observation Services** consisting of visits to the site at intervals appropriate to the stage of the work or as otherwise agreed by the Owner and Architect in writing to become generally familiar with the progress and quality of the Work completed and to determine in general if the Work when completed will be in accordance with Contract Documents; preparing related reports and communications.

.41 **Project Representation** consisting of selection, employment and direction of:
    .01 Project Representative(s) whose specific duties, responsibilities and limitations of authority shall be as described in the edition of AIA Document B352 current as of the date of this Agreement or as set forth in an exhibit to be incorporated in this Agreement under Article 1.6.

.42 **Testing and Inspection Administration** relating to independent inspection and testing agencies, consisting of:
    .01 Administration and coordination of field testing required by the Contract Documents
    .02 Recommending scope, standards, procedures and frequency of testing and inspections
    .03 Arranging for testing and inspection on Owner's behalf
    .04 Notifying inspection and testing agencies of status of Work requiring testing and inspection
    .05 Evaluating compliance by testing and inspection agencies with required scope, standards, procedures and frequency
    .06 Review of reports on inspections and tests and notifications to Owner and Contractor(s) of observed deficiencies in the Work.

.43 **Supplemental Documentation** services consisting of:
    .01 Preparation, reproduction and distribution of supplemental Drawings, Specifications and interpretations in response to requests for clarification by Contractor(s) or the Owner
    .02 Forwarding Owner's instructions and providing guidance to the Contractor(s) on the Owner's behalf relative to changed requirements and schedule revisions.

.44 **Quotation Requests/Change Orders** consisting of:
    .01 Preparation, reproduction and distribution of Drawings and Specifications to describe Work to be added, deleted or modified
    .02 Review of proposals from Contractor(s) for reasonableness of quantities and costs of labor and materials
    .03 Review and recommendations relative to changes in time for Substantial Completion
    .04 Negotiations with Contractor(s) on Owner's behalf relative to costs of Work proposed to be added, deleted or modified
    .05 Assisting in the preparation of appropriate Modifications of the Contract(s) for Construction
    .06 Coordination of communications, approvals, notifications and record-keeping relative to changes in the Work.

.45 **Contract Cost Accounting** services consisting of:
    .01 Maintenance of records of payments on account of the Contract Sum and all changes thereto
    .02 Evaluation of Applications for Payment and certification thereof
    .03 Review and evaluation of expense data submitted by the Contractor(s) for Work performed under cost-plus-fee arrangements.

AIA DOCUMENT B163 • OWNER-ARCHITECT AGREEMENT FOR DESIGNATED SERVICES
AIA® • ©1993 • THE AMERICAN INSTITUTE OF ARCHITECTS, 1735 NEW YORK AVENUE, N.W., WASHINGTON, D.C. 20006-5292 • **WARNING: Unlicensed photocopying violates U.S. copyright laws and will subject the violator to legal prosecution.**

.46 **Furniture, Furnishings and Equipment Installation Administration** consisting of:
  .01 Assistance to the Owner in coordinating schedules for delivery and installation of the Work
  .02 Review of final placement and inspection for damage, quality, assembly and function to determine that furniture, furnishings and equipment are in accordance with the requirements of the Contract Documents.

.47 **Interpretations and Decisions** consisting of:
  .01 Review of claims, disputes or other matters between the Owner and Contractor relating to the execution or progress of the Work as provided in the Contract Documents
  .02 Rendering written decisions within a reasonable time and following the procedures set forth in the General Conditions of the Contract for Construction, AIA Document A201, current as of the date of this Agreeement, or the General Conditions of the Contract for Furniture, Furnishings and Equipment, AIA Document A271, current as of the date of this Agreement, for Resolution of Claims and disputes.

.48 **Project Closeout** services initiated upon notice from the Contractor(s) that the Work, or a designated portion thereof which is acceptable to the Owner, is sufficiently complete, in accordance with the Contract Documents, to permit occupancy or utilization for the use for which it is intended, and consisting of:
  .01 A detailed inspection with the Owner's representative for conformity of the Work to the Contract Documents to verify the list submitted by the Contractor(s) of items to be completed or corrected
  .02 Determination of the amounts to be withheld until final completion
  .03 Securing and receipt of consent of surety or sureties, if any, to reduction in or partial release of retainage or the making of final payment(s)
  .04 Issuance of Certificate(s) of Substantial Completion
  .05 Inspection(s) upon notice by the Contractor(s) that the Work is ready for final inspection and acceptance
  .06 Notification to Owner and Contractor(s) of deficiencies found in follow-up inspection(s), if any
  .07 Final inspection with the Owner's representative to verify final completion of the Work
  .08 Receipt and transmittal of warranties, affidavits, receipts, releases and waivers of liens or bonds indemnifying the Owner against liens
  .09 Securing and receipt of consent of surety or sureties, if any, to the making of final payment(s)
  .10 Issuance of final Certificate(s) for Payment.

POST-CONTRACT SERVICES

.49 **Maintenance and Operational Programming** services consisting of:
  .01 Assistance in the establishment by the Owner of in-house or contract program(s) of operation and maintenance of the physical plant and equipment
  .02 Arranging for and coordinating instructions on operations and maintenance of equipment in conjunction with manufacturer's representatives
  .03 Assistance in the preparation of operations and maintenance manual(s) for the Owner's use.

.50 **Start-Up Assistance** consisting of:
  .01 On-site assistance in the operation of building systems during initial occupancy
  .02 Assistance in the training of the Owner's operation and maintenance personnel in proper operations, schedules and procedures
  .03 Administration and coordination of remedial work by the Contractor(s) after final completion.

.51 **Record Drawing** services consisting of:
  .01 Making arrangements for obtaining from Contractor(s) information in the form of marked-up prints, drawings and other data certified by them on changes made during performance of the Work
  .02 Review of general accuracy of information submitted and certified by the Contractor(s)
  .03 Preparation of record drawings based on certified information furnished by the Contractor(s)
  .04 Transmittal of record drawings and general data, appropriately identified, to the Owner and others as directed.

.52 **Warranty Review** consisting of:
- .01 Consultation with and recommendation to the Owner during the duration of warranties in connection with inadequate performance of materials, systems and equipment under warranty
- .02 Inspection(s) prior to expiration of the warranty period(s) to ascertain adequacy of performance of materials, systems and equipment
- .03 Documenting defects or deficiencies and assisting the Owner in preparing instructions to the Contractor(s) for correction of noted defects.

.53 **Post-Contract Evaluation** consisting of a Project inspection at least one year after completion of the Work; review with appropriate supervisory, operating and maintenance personnel, and analysis of operating costs and related data for evaluation of:
- .01 The initial Project programming versus actual facility use
- .02 The functional effectiveness of planned spaces and relationships
- .03 The operational effectiveness of systems and materials installed.

ARTICLE 2.4
DESCRIPTIONS OF SUPPLEMENTAL SERVICES

SUPPLEMENTAL SERVICES

.54 **Special Studies** consisting of investigation, research and analysis of the Owner's special requirements for the Project and documentation of findings, conclusions and recommendations for:
- .01 Master planning to provide design services relative to future facilities, systems and equipment which are not intended to be constructed as part of the Project during the Construction Phase
- .02 Providing special studies for the project such as analyzing acoustical or lighting requirements, record retention, communications and security systems.

.55 **Tenant-Related Services** consisting of design and documentation services for tenants or potential tenants relating to:
- .01 Space planning, partition and furnishings locations, and furniture and equipment layouts
- .02 Material and color selections and coordination
- .03 Adaptation of mechanical, electrical and other building systems to meet tenant needs
- .04 Preliminary estimate of Construction Cost.

.56 **Special Furnishings Design** services relating to Architect-designed special furnishings and/or equipment incorporated into or provided for the Project and consisting of:
- .01 Design and documentation
- .02 Specifications or standards
- .03 Management of procurement
- .04 Coordination of installation
- .05 Purchase on the Owner's behalf.

.57 **Furniture, Furnishings and Equipment Services** relating to equipment and furnishings not incorporated into the construction of the Project and consisting of:
- .01 Establishment of needs and criteria
- .02 Preparation of requirements, Specifications and bidding or purchasing procedures
- .03 Management of procurement
- .04 Coordination of delivery and installation.

.58 **Special Disciplines Consultation,** which entails retaining, directing and coordinating the work of special disciplines consultants identified from the following list and as more specifically described in Article 1.6, whose specialized training, experience and knowledge relative to specific elements and features of the Project are required for the Project:

.01	Acoustics	.14	Elevators/Escalators	.27	Public Relations
.02	Audio-Visual	.15	Fallout Shelters	.28	Radiation Shielding
.03	CPM Scheduling	.16	Financial	.29	Real Estate
.04	Code Interpretation	.17	Fire Protection	.30	Reprographics
.05	Communications	.18	Food Service	.31	Safety
.06	Computer Technology	.19	Insurance	.32	Sociology
.07	Concrete	.20	Historic Preservation	.33	Soils/Foundations
.08	Cost Estimating	.21	Legal	.34	Space Planning
.09	Demography	.22	Life Safety	.35	Specifications
.10	Display	.23	Lightning	.36	Traffic/Parking
.11	Ecology	.24	Management	.37	Transportation
.12	Economics	.25	Materials Handling	.38	Security
.13	Editorial	.26	Psychology	.39	Record Retention

.59 **Special Building Type Consultation,** which entails retaining, directing and coordinating the work of special building type consultants whose specialized training, experience and knowledge relative to the requirements, planning and design of the Project are required for the Project.

.60 **Fine Arts and Crafts** services relating to acquisition of fine arts or crafts to be a part of the Project and consisting of:

.01 Consultations on selection, commissioning and/or execution
.02 Design integration
.03 Managing procurement
.04 Purchasing fine arts or crafts on the Owner's behalf.

.61 **Graphic Design** services consisting of:

.01 Design and selection of interior and exterior signs and identifying symbols
.02 Material and color selections and coordination
.03 Documentation of requirements for procurement of graphics work
.04 Managing procurement of graphics work
.05 Coordination of delivery and installation.

.62 **Renderings** relating to graphic pictorial representations, as required by the Owner, of the proposed Project and consisting of:

.01 Black and white elevation view(s)
.02 Black and white perspective view(s)
.03 Elevation view(s) in color
.04 Perspective view(s) in color.

.63 **Model Construction** consisting of preparation of:

.01 Small-scale block model(s) showing relationship of structure(s) to site
.02 Moderate-scale block model(s) of structure(s) designed for the Project
.03 Moderate-scale detailed model(s) of structure(s) designed for the Project showing both interior and exterior design
.04 Large-scale models of designated interior or exterior components of the Project.

.64 **Still Photography** consisting of:

.01 Documentation of existing conditions
.02 Aerial site photography
.03 Photographic recording for study purposes of facilities similar to the Project

Appendix 443

  .04 Periscopic photography of models for the Project
  .05 Presentation photography of renderings(s) and model(s) for the Project
  .06 Construction progress photography
  .07 Architectural photography of the completed Project.

.65 **Motion Picture and Videotape** services relating to preparation of promotional or explanatory presentations of the Project during the design and/or construction phases.

.66 **Life Cycle Cost Analysis** consisting of assessment, on the basis of established relevant economic consequences over a given time period, of:
  .01 A given planning and design solution for the Project
  .02 Alternative planning and design solutions for the Project
  .03 Selected systems, subsystems or building components proposed for the Project.

.67 **Value Analysis** consisting of the review during design phases of the cost, quality and time influences of proposed building materials, systems and construction methods relative to design objectives in order to identify options for obtaining value for the Owner.

.68 **Energy Studies** consisting of special analyses of mechanical systems, fuel costs, on-site energy generation and energy conservation options for the Owner's consideration.

.69 **Quantity Surveys** consisting of:
  .01 A detailed determination of the quantities of materials to be used in the Project to establish the basis for price determination by bidding or negotiations.
  .02 Making investigations, inventories of materials or furniture, furnishings and equipment, or valuations and detailed appraisals of existing facilities, furniture, furnishings and equipment, and the relocation thereof.

.70 **Detailed Cost Estimating** services consisting of:
  .01 Development, when the Contract Documents are approximately 90% complete, of a Detailed Estimate of the Cost of the Work based on quantity take-offs and unit-cost pricing of materials, labor, tools, equipment and services required for the Work plus estimates for the Contractor's supervision cost. Work required by General and Supplementary Conditions, and an allowance for reasonable Contractor's overhead and profit; or
  .02 Continuous development during all phases of design and documentation, of an Estimate of the Cost of the Work for the purpose of greater cost control, culminating in a Detailed Estimate of the Cost of the Work or detailed quantity surveys or inventories of material, equipment and labor.

.71 **Environmental Monitoring** services consisting of:
  .01 Monitoring of air, water and other designated components of the environment to establish existing conditions, and the preparation of related analyses and reports.

.72 **Expert Witness** services consisting of preparing to serve and/or serving as an expert witness in connection with any public hearing, arbitration proceeding or legal proceeding.

.73 **Materials and Systems Testing** relating to testing of components of the completed Project for conformance with Contract requirements and consisting of:
  .01 Establishment of requirements
  .02 Procurement of testing services
  .03 Monitoring testing
  .04 Review, analysis and reporting of test results.

.74 **Demolition Services** consisting of:
   .01 Preparation of Contract Documents for demolition of existing structures
   .02 Managing the bidding/negotiation/award process
   .03 Providing field observation and general administration services during demolition.

.75 **Mock-Up Services** relating to the construction of full-size details of components of the Project for study and testing during the design phases and consisting of:
   .01 Design and documentation for the required mock-up(s)
   .02 Management and coordination of pricing and contracting for mock-up services
   .03 Construction administration of mock-up construction activities
   .04 Arrangements for testing and monitoring performance of mock-up(s)
   .05 Administration of testing and monitoring services
   .06 Review, analysis and reporting of results of testing and monitoring services.

.76 **Coordination of Designated Services** with those of non-design professionals, such as economists, sociologists, attorneys and accountants, consiting of:
   .01 Preparation of economic studies
   .02 Condominium documentation
   .03 Sociological impact studies.

.77 **Furniture, Furnishings and Equipment Purchasing/Installation,** consisting of:
   .01 Purchasing furniture, furnishings and equipment on behalf of the Owner with funds provided by the Owner
   .02 Receipt, inspection and acceptance on behalf of the Owner of furniture, furnishings and equipment at the time of their delivery to the premises and installation
   .03 Providing services including travel for the purpose of evaluating materials, furniture, furnishings and equipment proposed for the Project.

.78 **Computer Applications** consisting of computer program development and/or computer program search and acquisition, plus on-line computer time charges, for:
   .01 Programming
   .02 Economic feasibility
   .03 Financial analysis
   .04 Site analysis
   .05 Construction cost estimating
   .06 Detailed Project scheduling
   .07 Market analysis
   .08 Architectural analysis and design
   .09 Structural analysis and design
   .10 Mechanical analysis and design
   .11 Electrical analysis and design
   .12 Production of Drawings
   .13 Construction cost accounting

.79 **Project Promotion/Public Relations** relating to presentation of the Project to the public or identified groups and consisting of:
   .01 Preparation of press releases
   .02 Preparation of special brochures and/or promotional pieces
   .03 Assistance in production and distribution of promotional materials
   .04 Presentations at public relations and/or promotional meetings.

.80 **Leasing Brochures,** including preparation of special materials to assist the Owner in leasing the Project and consisting of:
   .01 Design
   .02 Preparation of illustrations and text
   .03 Arranging for and managing production.

.81 **Pre-Contract Administration/Management,** consisting of:
   .01 Evaluating feasibility of Owner's program, schedule and budget for the Work, each in terms of the other
   .02 Preparing, updating and monitoring Detailed Project Schedule, including services and contract Work, identifying critical and long-lead items

.03 Preparing, updating and monitoring Detailed Estimates of the Cost of the Work prior to completion of each design phase
.04 Assisting the Owner in selecting, retaining and coordinating the professional services of surveyors, testing labs and other special consultants as designated
.05 Assisting the Owner in evaluating relative feasibility of methods of executing the Work, methods of project delivery, availability of materials and labor, time requirements for procurement, installation and delivery, and utilization of the site for mobilization and staging
.06 Assisting the Owner in determining the method of contracting for the Work; evaluating single versus multiple contracts; advising on categories of separate contracts and provisions for coordinating responsibilities.

.82 **Extended Bidding** services, consisting of:
.01 Developing Bidders' interest in the Project and establishing bidding schedules
.02 Receiving and analyzing bids and providing recommendations as to the Owner's acceptance or rejection of bids
.03 Advising the Owner on acceptance of Contractors
.04 Conducting pre-award conferences.

.83 **Extended Contract Administration/Management,** consisting of:
.01 Assisting Owner in obtaining building permits
.02 Updating and monitoring actual costs against estimates of final cost; assisting Owner in monitoring cash flow
.03 Providing a detailed schedule showing time periods for each Contractor, including long-lead items and Owner's occupancy requirements; updating and monitoring periodically; recommending corrective action when required
.04 Endeavoring to achieve satisfactory performance of Contractors through development and implementation of a quality control program; assisting Owner in determining compliance with schedule, cost and Contract Documents
.05 Scheduling and conducting periodic project meetings with the Owner, Contractor and Subcontractors
.06 Assisting Owner in maintaining cost accounting records
.07 Maintaining a daily log including conditions at site and job progress, periodically indicating percentage of completion of each contract
.08 Assisting the Owner in coordinating and scheduling activities of the separate Contractors.
.09 Maintaining and periodically updating a record of all significant changes made during construction; maintaining record copies of Contract Documents; maintaining samples and lay-out drawings at the job site.

AIA Document B163—PART 3

TERMS AND CONDITIONS
of the Agreement Between Owner and Architect for Designated Services

ARTICLE 3.1

ARCHITECT'S RESPONSIBILITIES

3.1.1 Designated Services. Unless otherwise provided, the Architect's designated services consist of those services identified in the Schedule of Designated Services as being performed by the Architect, Architect's employees and Architect's consultants, and as described in the Descriptions of Designated Services.

3.1.2 Contingent Additional Services. Contingent Additional Services described in Subparagraphs 3.1.2.1 through 3.1.2.7 are not included in the Architect's Designated Services, but may be required due to circumstances beyond the Architect's control. The Architect shall notify the Owner prior to commencing such services. If the Owner deems that such services are not required, the Owner shall give prompt written notice to the Architect. If the Owner indicates in writing that all or part of such Contingent Additional Services are not required, the Architect shall have no obligation to provide those services.

3.1.2.1 Document Revisions. Services required to revise Drawings, Specifications or other documents when such revisions are:

.1 inconsistent with approvals or instructions previously given by the Owner, including revisions made necessary by adjustments in the Owner's program or Project budget;

.2 required by the enactment or revision of codes, laws or regulations subsequent to the preparation of such documents; or

.3 due to changes required as a result of the Owner's failure to render decisions in a timely manner.

3.1.2.2 Changes in Project Scope. Services required because of significant changes in the Project including, but not limited to, size, quality, complexity, the Owner's schedule, or the method of bidding or negotiating and contracting for construction, except for services required under Subparagraph 1.6.

3.1.2.3 Replacement of Damaged Work. Consultation concerning replacement of Work damaged by fire or other cause during construction, and furnishing services required in connection with the replacement of such Work.

3.1.2.4 Default by Others. Services made necessary by the default of the Owner's consultants or the Contractor, by major defects or deficiencies in their services or the Work, or by failure of performance of any of them under their respective contracts.

3.1.2.5 Correction Period. Advice and consultation to the Owner during the correction period described in the Contracts for Construction or for Furniture, Furnishings and Equipment.

3.1.2.6 Purchasing of Furniture, Furnishings and Equipment by the Architect. If the Owner and Architect agree that the Architect will purchase furniture, furnishings and equipment on behalf of the Owner with funds provided by the Owner, the duties relating to such services shall be set forth in Article 1.6 of this Agreement. The Owner shall provide and maintain working funds with the Architect, if required, to pay invoices charged to the Project for materials and furnishings, to secure cash discounts and for required deposits.

3.1.2.7 Services Related to Separate Consultants. The Architect shall provide information to and incorporate information received in a timely manner from those separate consultants retained by the Owner and identified in this Agreement whose activities directly relate to the Project.

ARTICLE 3.2

OWNER'S RESPONSIBILITIES

3.2.1 Representative. The Owner shall designate a representative authorized to act on the Owner's behalf with respect to the Project. The Owner or such authorized representative shall render decisions in a timely manner pertaining to documents submitted by the Architect in order to avoid unreasonable delay in the orderly and sequential progress of the Architect's services.

3.2.2 Notice. Prompt written notice shall be given by the Owner to the Architect if the Owner becomes aware of any fault or defect in the Project or nonconformance with the Contract Documents.

3.2.3 Designated Services. The Owner's responsibilities consist of those services identified in the Schedule of Designated Services as being performed by the Owner, Owner's employees and Owner's consultants.

3.2.4 Information. The Owner shall provide full information regarding requirements for the Project.

3.2.5 Owner's Financial Arrangements. If requested by the Architect, the Owner shall furnish evidence that financial arrangements have been made to fulfill the Owner's obligations to the Architect under this Agreement.

Appendix

3.2.6 Tests, Inspections and Reports Furnished by Owner. The Owner shall furnish structural, mechanical, chemical, air and water pollution tests, tests for hazardous materials, and other laboratory and environmental tests, inspections and reports required by law or the Contract Documents, or unless otherwise provided in this Agreement.

3.2.7 Legal, Accounting and Insurance Services Furnished by Owner. The Owner shall furnish all legal, accounting and insurance counseling services required for the Project, including auditing services the Owner may require to verify the Contractor's Applications for Payment or to ascertain how or for what purposes the Contractor has used the money paid by or on behalf of the Owner.

3.2.8 Space Arrangements. The Owner shall provide suitable space for the receipt, inspection and storage of materials, furniture, furnishings and equipment.

3.2.9 Removal of Existing Facilities. The Owner shall be responsible for the relocation or removal of existing facilities, furniture, furnishings and equipment, and the contents thereof, unless otherwise provided by this Agreement.

3.2.10 Responsibility for Services. The drawings, specifications, services, information, surveys and reports required of the Owner under the Agreement shall be furnished at the Owner's expense, and the Architect shall be entitled to rely upon the accuracy and completeness thereof.

3.2.11 Certificates and Certifications. The proposed language of certificates or certifications requested of the Architect or Architect's consultants shall be submitted to the Architect for review and approval at least 14 days prior to execution. The Owner shall not request certifications that would require knowledge or services of the Architect or the Architect's Consultants beyond the scope of this Agreement.

3.2.12 Communications and Security Systems. The Owner shall contract for all temporary and permanent telephone, communications and security systems required for the Project so as not to delay the performance of the Architect's services.

ARTICLE 3.3
CONTRACT ADMINISTRATION

3.3.1 General. The following terms and conditions shall apply to the relevant Contract Administration Phase services, if any, as may be included in the Schedule of Designated Services.

3.3.1.1 Interpretations and Decisions: Timing. To the extent that the following services of the Architect have been designated in the Schedule of Designated Services, the Architect shall interpret and decide matters concerning performance of the Owner and Contractor under the requirements of the Contract Documents on written request of either the Owner or Contractor. The Architect's response to such requests shall be made with reasonable promptness and within any time limits agreed upon.

3.3.1.2 Interpretations and Decisions: Form and Intent. Interpretations and decisions of the Architect shall be consistent with the intent of and reasonably inferable from the Contract Documents and shall be in writing or in the form of drawings. When making such interpretations and initial decisions, the Architect shall endeavor to secure faithful performance by both Owner and Contractor, shall not show partiality to either, and shall not be liable for the results of interpretations or decisions so rendered in good faith.

3.3.1.3 Decisions on Aesthetic Effect. The Architect's decisions on matters relating to aesthetic effect shall be final if consistent with the intent expressed in the Contract Documents.

3.3.1.4 Architect's Decisions Subject to Arbitration. The Architect's decisions on claims, disputes or other matters, including those in question between the Owner and Contractor, except for those relating to aesthetic effect as provided in Clause 3.3.1.3, shall be subject to arbitration as provided in this Agreement and in the Contract Documents.

3.3.2 Duration of Contract Administration Phase. The Architect's responsibility to provide services for the Contract Administration Phase under this Agreement commences with the award of the initial Contract for Construction or for Furniture, Furnishings and Equipment, and terminates at the earlier of the issuance to the Owner of the final Certificate for Payment or 60 days after the date of Substantial Completion of the Work.

3.3.3 Contract(s) for the Work. The Architect shall provide administration of Contract(s) for Construction or Furniture, Furnishings and Equipment as set forth below and in the edition of AIA Document A201, General Conditions of the Contract for Construction, or AIA Document A271, General Conditions of the Contract for Furniture, Furnishings and Equipment, current as of the date of this Agreement.

3.3.4 Modification of Responsibilities. Duties, responsibilities, and limitations of authority of the Architect shall not be restricted, modified or extended without written agreement of the Owner and Architect with the consent of the Contractor; which consent shall not be unreasonably withheld.

3.3.5 Authority of Architect. The Architect shall be a representative of and shall advise and consult with the Owner (1) during the Contract Administration Phase, and (2) by an amendment to this Agreement, from time to time during the correction period described in the Contract for Construction. The Architect shall have authority to act on behalf of the Owner only to the extent provided in this Agreement unless otherwise modified by written instrument.

3.3.6 CONSTRUCTION OBSERVATION SERVICES

3.3.6.1 Architect's Responsibility for Observation. On the basis of on-site observations as an architect, the Architect shall keep the Owner informed of the progress and quality of the Work, and shall endeavor to guard the Owner against defects and deficiencies in the Work. The Architect shall not be required to make exhaustive or continuous on-site inspections to check the quality or quantity of the Work.

3.3.6.2 Project Representation. The furnishing of Project representation services shall not modify the rights, responsibilities or obligations of the Architect as described elsewhere in this Agreement.

3.3.6.3 Means and Methods. The Architect shall not have control over or charge of and shall not be responsible for construction means, methods, techniques, sequences or procedures, or for safety precautions and programs in connection with the Work, since these are solely the Contractor's responsibility under the Contract for Construction and the Contract for Furniture, Furnishings and Equipment. The Architect shall not be responsible for the Contractor's schedules or failure to carry out the Work in accordance with the Contract Documents. The Architect shall not have control over or charge of acts or omissions of the Contractor, Subcontractors, or their agents or employees, or of any other persons performing portions of the Work.

3.3.6.4 Access to Work. The Architect shall at all times have access to the Work wherever it is in preparation or progress.

3.3.6.5 Communications. Except as may otherwise be provided in the Contract Documents or when direct communications have been specially authorized, the Owner and Contractor shall communicate through the Architect. Communications by and with the Architect's consultants shall be through the Architect.

3.3.6.6 Minor Changes. The Architect may authorize minor changes in the Work not involving an adjustment in Contract Sum or an extension of the Contract Time which are not inconsistent with the intent of the Contract Documents.

3.3.6.7 Coordination of Furniture, Furnishings and Equipment Delivery and Installation. When the Architect assists the Owner in coordinating schedules for delivery and installation of furniture, furnishings and equipment, the Architect shall not be responsible for malfeasance, neglect or failure of a Contractor, Subcontractor, Sub-subcontractor or material supplier to meet their schedules for completion or to perform their respective duties and responsibilities.

3.3.7 COST ACCOUNTING SERVICES

3.3.7.1 Certificates for Payment. If certification of the Contractor's Applications for Payment is required by this Agreement, the Architect's certification for payment shall constitute a representation to the Owner, based on the Architect's observations at the site as provided in Subparagraph 3.3.6.1 and on the data comprising the Contractor's Application for Payment, that the Work has progressed to the point indicated and that, to the best of the Architect's knowledge, information and belief, the quality of the Work is in accordance with the Contract Documents. The foregoing representations are subject to an evaluation of the Work for conformance with the Contract Documents upon Substantial Completion, to results of subsequent tests and inspections, to minor deviations from the Contract Documents correctable prior to completion and to specific qualifications expressed by the Architect.

3.3.7.2 Limitations. The issuance of a Certificate for Payment shall not be a representation that the Architect has (1) made exhaustive or continuous on-site inspections to check the quality or quantity of the Work, (2) reviewed means, methods, techniques, sequences or procedures, (3) reviewed copies of requisitions received from Subcontractors and material suppliers and other data requested by the Owner to substantiate the Contractor's right to payment or (4) ascertained how or for what purpose the Contractor has used money previously paid on account of the Contract Sum.

3.3.8 INSPECTION AND TESTING ADMINISTRATION SERVICES

3.3.8.1 Rejection of Work. Except as provided in Subparagraph 3.3.8.3, the Architect shall have authority to reject Work which does not conform to the Contract Documents. Whenever the Architect considers it necessary or advisable for implementation of the intent of the Contract Documents, the Architect will have authority to require additional inspection or testing of the Work in accordance with the provisions of the Contract Documents, whether or not such Work is fabricated, installed or completed. However, neither this authority of the Architect nor a decision made in good faith either to exercise or not to exercise such authority shall give rise to a duty or responsibility of the Architect to the Contractor, Subcontractors, material and equipment suppliers, their agents or employees or other persons performing portions of the Work.

3.3.8.2 Review and Inspection of Work. The Architect shall review final placement and inspect for damage, quality, assembly and function in order to determine that furniture, furnishings and equipment are in accordance with the requirements of the Contract Documents.

3.3.8.3 Rejection of Work Involving Furniture, Furnishings and Equipment. Unless otherwise designated, the Architect's duties shall not extend to the receipt, inspection and acceptance on behalf of the Owner of furniture, furnishings and equipment at the time of their delivery to the premises and installation. The Architect is not authorized to reject nonconforming furniture, furnishings and equipment, sign Change Orders on behalf of the Owner, stop the Work, or terminate a Contract on behalf of the Owner. However, the Architect shall recommend to the Owner rejection of furniture, furnishings and equipment which does not conform to the Contract Documents. Whenever the Architect considers it necessary or advisable for implementation of the intent of the Contract Documents, the Architect will have authority to require additional inspection or testing of furniture, furnishings and equipment in accordance with the provisions of the Contract Documents, whether or not such furniture, furnishings and equipment is fabricated, installed or completed.

3.3.9 SUBMITTAL SERVICES

3.3.9.1 Submittal Review. To the extent required by this Agreement, the Architect shall review and approve or take other appropriate action upon Contractor's submittals such as Shop Drawings, Product Data and Samples, but only for the limited purpose of checking for conformance with information given and the design concept expressed in the Contract Documents. The Architect's action shall be taken with such reasonable promptness as to cause no delay in the Work or in the construction of the Owner or of separate contractors, while allowing sufficient time in the Architect's professional judgment to permit adequate review.

3.3.9.2 Limitations. Review of such submittals is not conducted for the purpose of determining the accuracy and completeness of other details such as dimensions and quantities or for substantiating instructions for installation or performance of equipment or systems designed by the Contractor, all of which remain the responsibility of the Contractor to the extent required by the Contract Documents. The Architect's review shall not constitute approval of safety precautions or, unless otherwise specifically stated by the Architect, of construction means, methods, techniques, sequences or procedures. The Architect's approval of a specific item shall not indicate approval of an assembly of which the item is a component.

3.3.10 Reliance on Professional Certification. When professional certification of performance characteristics of materials, systems or equipment is required by the Contract Documents, the Architect shall be entitled to rely upon such certification to establish that the materials, systems or equipment will meet the performance criteria required by the Contract Documents.

ARTICLE 3.4

USE OF PROJECT DRAWINGS, SPECIFICATIONS AND OTHER DOCUMENTS

3.4.1 Architect's Reserved Rights. The Drawings, Specifications and other documents prepared by the Architect for this Project are instruments of the Architect's service for use solely with respect to this Project and, unless otherwise provided, the Architect shall be deemed the author of these documents and shall retain all common law, statutory and other reserved rights, including the copyright.

3.4.2 Limitations on Use. The Owner shall be permitted to retain copies, including reproducible copies, of the Project Drawings, Specifications and other documents for information and reference in connection with the Owner's use and occupancy of the Project. The Project Drawings, Specifications or other documents shall not be used by the Owner or others on other projects, for additions to this Project or for completion of this Project by others, unless the Architect is adjudged to be in default under this Agreement, except by agreement in writing and with appropriate compensation to the Architect.

3.4.3 Unpublished Works. Submission or distribution of documents to meet official regulatory requirements or for similar purposes in connection with the Project is not to be construed as publication in derogation of the Architect's reserved rights.

ARTICLE 3.5

COST OF THE WORK

3.5.1 DEFINITION

3.5.1.1 Total Cost. The Cost of the Work shall be the total cost or estimated cost to the Owner of all elements of the Project to be included in the Contract Documents.

3.5.1.2 Items Included. The Cost of the Work shall include the cost at current market rates of labor and materials furnished by the Owner and equipment designated, specified, selected or specially provided for by the Architect in the Contract Documents, including the cost of the Contractor's management or supervision of construction or installation, plus a reasonable allowance for the Contractor's overhead and profit. In addition, a reasonable allowance for contingencies shall be included for market conditions at the time of bidding and for changes in the Work during construction.

3.5.1.3 Items Excluded. The Cost of the Work does not include the compensation of the Architect and the Owner's or Architect's consultants, the costs of the land, rights-of-way, financing or other costs which are the responsibility of the Owner as provided in Article 3.2.

3.5.2 RESPONSIBILITY FOR COST OF THE WORK

3.5.2.1 Limitation of Responsibility. Evaluations of the Owner's Project budget, preliminary estimates of the Cost of the Work and detailed estimates of the Cost of the Work, if any, prepared by the Architect, represent the Architect's best judgment as a design professional familiar with the construction industry. It is recognized, however, that neither the Architect nor the Owner has control over the cost of labor, materials or equipment, over the Contractor's methods of determining bid prices, or over competitive bidding, market or negotiating conditions. Accordingly, the Architect cannot and does not warrant or represent that bids or negotiated prices will not vary from the Owner's Project budget or from any estimate of the Cost of the Work or evaluation prepared or agreed to by the Architect.

3.5.2.2 Fixed Limit of the Cost of the Work. If a fixed limit of the Cost of the Work has been established, the Architect shall be permitted to include contingencies for design, bidding and price escalation, to determine what materials, furniture, furnishings and equipment, component systems and types of construction are to be included in the Contract Documents, to make reasonable adjustments in the scope of the Project and to include in the Contract Documents alternate bids to adjust the Cost of the Work to the fixed limit. Fixed limits, if any, shall be increased in the amount of an increase in the Contract Sum occurring after execution of the Contract for Construction.

3.5.2.3 Adjustments. If the Bidding or Negotiation Phase has not commenced within 90 days after the Contract Documents are submitted to the Owner, the Project budget or fixed limit of the Cost of the Work shall be adjusted to reflect changes in the general level of prices in the construction industry between the date of submission of the Contract Documents to the Owner and the date on which bids or negotiated proposals are sought.

3.5.2.4 Owner's Responsibility to Meet Fixed Limit. If a fixed limit of the Cost of the Work (adjusted as provided in Subparagraph 3.5.2.3) is exceeded by the lowest bona fide bid or negotiated proposal, the Owner shall:

.1 give written approval of an increase in such fixed limit;

.2 authorize rebidding or renegotiation of the Project within a reasonable time;

.3 if the Project is abandoned, terminate in accordance with Paragraph 3.9; or

.4 cooperate in revising the Project scope and quality as required to reduce the Construction Cost.

3.5.2.5 Architect's Responsibility to Meet Fixed Limit. If the Owner chooses to proceed under Clause 3.5.2.4.4, the Architect, without additional compensation, shall modify the documents that the Architect is responsible for preparing under the Designated Services portion of this Agreement as necessary to comply with the fixed limit. The modification of such documents shall be the limit of the Architect's responsibility arising out of the establishment of a fixed limit. The Architect shall be entitled to compensation in accordance with this Agreement for all services performed whether or not the Construction Phase is commenced.

ARTICLE 3.6
PAYMENTS TO THE ARCHITECT

3.6.1 Direct Personnel Expense. Direct Personnel Expense is defined as the direct salaries of the Architect's personnel engaged on the Project and the portion of the cost of their mandatory and customary contributions and benefits related thereto, such as employment taxes and other statutory employee benefits, insurance, sick leave, holidays, vacations, pensions and similar contributions and benefits.

3.6.2 Reimbursable Expenses. Reimbursable Expenses are in addition to compensation for the Architect's services and include expenses incurred by the Architect and Architect's employees and consultants in the interest of the Project, as identified in the following Clauses:

.1 transportation in connection with the Project, authorized out-of-town travel, long-distance communications, and fees paid for securing approval of authorities having jurisdiction over the Project;

.2 reproductions, postage and handling of Drawings, Specifications and other documents;

.3 facsimile services, courier services, overnight deliveries or other similar project-related expenditures;

.4 if authorized in advance by the Owner, expense of overtime work requiring higher than regular rates;

.5 renderings, models and mock-ups requested by the Owner;

.6 additional insurance coverage or limits, including professional liability insurance, requested by the Owner in excess of that normally carried by the Architect and Architect's consultants; and

.7 Expense of computer-aided design and drafting equipment time when used in connection with the Project.

3.6.3 Payments for Contingent Additional Services and Reimbursable Expenses. Payments on account of the Architect's Contingent Additional Services and for Reimbursable Expenses shall be made monthly upon presentation of the Architect's statement of services rendered or expenses incurred.

3.6.4 Extended Time. If and to the extent that the time initially established in this Agreement is exceeded or extended through no fault of the Architect, compensation for any services rendered during the additional period of time shall be computed in the manner set forth in Article 1.6.

3.6.5 Changes Affecting Percentage Compensation Method. When compensation is based on a percentage of Construction Cost and any portions of the Project are deleted or otherwise not constructed, compensation for those portions of the Project shall be payable to the extent services are performed on those portions, in accordance with the schedule set forth in Part 1, Subparagraph 1.3.2, based on (1) the lowest bona fide bid or negotiated proposal, or (2) if no such bid or proposal is received, the most recent preliminary estimate of Construction Cost or detailed estimate of Construction Cost for such portions of the Project.

3.6.6 Payments Withheld. No deductions shall be made from the Architect's compensation on account of penalty, liquidated damages or other sums withheld from payments to contractors, or on account of the cost of changes in the Work other than those for which the Architect has been found to be liable.

3.6.7 Architect's Accounting Records. Records of Reimbursable Expenses, of expenses pertaining to Contingent Additional Services, and of services performed on the basis of a multiple of Direct Personnel Expense shall be available to the Owner or the Owner's authorized representative at mutually convenient times.

ARTICLE 3.7
DISPUTE RESOLUTION

3.7.1 Claims and Disputes. Claims, disputes or other matters in question between the parties to this Agreement arising out of or relating to this Agreement or breach thereof shall be subject to and decided by mediation and arbitration in accordance with the Construction Industry Mediation and Arbitration Rules of the American Arbitration Association currently in effect.

3.7.2 Mediation. In addition to and prior to arbitration, the parties shall endeavor to settle disputes by mediation in accordance with the Construction Industry Mediation Rules of the American Arbitration Association currently in effect. Demand for mediation shall be filed in writing with the other party to this Agreement and with the American Arbitration Association. A demand for mediation shall be made within a reasonable time after the claim, dispute or other matter in question has arisen. In no event shall the demand for mediation be made after the date when institution of legal or equitable proceedings based on such claim, dispute or other matter in question would be barred by the applicable statute of repose or limitations.

3.7.3 Arbitration. Demand for arbitration shall be filed in writing with the other party to this Agreement and with the American Arbitration Association. A demand for arbitration shall be made within a reasonable time after the claim, dispute or other matter in question has arisen. In no event shall the demand for arbitration be made after the date when institution of legal or equitable proceedings based on such claim, dispute or other matter in question would be barred by the applicable statutes of repose or limitations.

3.7.4 Consolidation and Joinder. An arbitration pursuant to this paragraph may be joined with an arbitration involving common issues of law or fact between the Architect and any person or entity with whom the Architect has a contractual obligation to arbitrate disputes. No other arbitration arising out of or relating to this Agreement shall include, by consolidation, joinder or in any other manner, an additional person or entity not a party to this Agreement, except by written consent containing a specific reference to this Agreement signed by the Owner, Architect, and any other person or entity sought to be joined. Consent to arbitration involving an additional person or entity shall not constitute consent to arbitration of any claim, dispute or other matter in question not described in the written consent or with a person or entity not named or described therein. The foregoing agreement to arbitrate and other agreements to arbitrate with an additional person or entity duly consented to by the parties to this Agreement shall be specifically enforceable in accordance with applicable law in any court having jurisdiction thereof.

3.7.5 Award. The award rendered by the arbitrator or arbitrators shall be final, and judgment may be entered upon it in accordance with applicable law in any court having jurisdiction thereof.

ARTICLE 3.8

MISCELLANEOUS PROVISIONS

3.8.1 Governing Law. This Agreement shall be governed by the law of the place of the Project.

3.8.2 Definitions. Terms in this Agreement shall have the same meaning as those in AIA Document A201, General Conditions of the Contract for Construction, and AIA Document A271, General Conditions of the Contract for Furniture, Furnishings and Equipment, current as of the date of this Agreement.

3.8.3 Statutes of Repose or Limitations. Causes of action between the parties to this Agreement pertaining to acts or failures to act shall be deemed to have accrued and the applicable statutes of repose or limitations shall commence to run not later than either the Date of Substantial Completion for acts or failures occurring prior to Substantial Completion, or the date of issuance of the final Certificate for Payment for acts or failures to act occurring after Substantial Completion.

3.8.4 Waivers of Subrogation. The Owner and the Architect waive all rights against each other and against the contractors, consultants, agents and employees of the other for damages, but only to the extent covered by property insurance during construction, except such rights as they may have to the proceeds of such insurance as set forth in the editions of AIA Document A201, General Conditions of the Contract for Construction, and AIA Document A271, General Conditions of the Contract for Furniture, Furnishings and Equipment, current as of the date of this Agreement. The Owner and Architect shall each require similar waivers from their contractors, consultants and agents.

3.8.5 Successors and Assigns. The Owner and Architect, respectively, bind themselves, their partners, successors, assigns and legal representatives to the other party to this Agreement and to the partners, successors, assigns and legal representatives of such other party with respect to all covenants of this Agreement. Neither Owner nor Architect shall assign this Agreement without the written consent of the other.

3.8.6 Titles and Headings. The titles and headings in this Agreement are for convenience and shall not be interpreted as supplementing or superseding the intent of the parties as expressed in the body of this Agreement.

3.8.7 Third Parties. Nothing contained in this Agreement shall create a contractual relationship with or a cause of action in favor of a third party against either the Owner or Architect.

3.8.8 Hazardous Materials. Unless otherwise provided in this Agreement, the Architect and Architect's consultants shall have no responsibility for the discovery, presence, handling, removal or disposal of or exposure of persons to hazardous materials or toxic substances in any form at the Project site. If the Architect is required to perform services related to hazardous materials, the Owner agrees to indemnify and hold harmless the Architect, the Architect's consultants and their agents and employees from and against any and all claims, damages, losses and expenses, including but not limited to attorneys' fees, arising out of or resulting from performance of services by the Architect, the Architect's consultants or their agents or employees related to such services, except where such liability arises from the sole negligence or willful misconduct of the person or entity seeking indemnification.

3.8.9 Publicity. The Architect shall have the right to include representations of the design of the Project, including photographs of the exterior and interior, among the Architect's promotional and professional materials. The Architect's materials shall not include the Owner's confidential or proprietary information if the Owner has previously advised the Architect in writing of the specific information considered by the Owner to be confidential or proprietary. The Owner shall provide professional credit for the Architect on the construction sign and in the promotional materials for the Project.

3.8.10 Conflict of Interest. Except with the Owner's knowledge and consent, the Architect shall not (1) accept trade discounts, (2) have a substantial direct or indirect financial interest in the Project, or (3) undertake any activity or employment or accept any contribution, if it would reasonably appear that such activity, employment, interest or contribution could compromise the Architect's professional judgment or prevent the Architect from serving the best interest of the Owner.

ARTICLE 3.9

TERMINATION, SUSPENSION OR ABANDONMENT

3.9.1 Termination for Breach. This Agreement may be terminated by either party upon not less than seven days' written notice should the other party fail substantially to perform in accordance with the terms of this Agreement, through no fault of the party initiating the termination. Failure of the Owner to make payments to the Architect in accordance with this Agreement shall be considered substantial nonperformance and cause for termination.

3.9.2 Suspension. If the Project is suspended by the Owner for more than 30 consecutive days, the Architect shall be compensated for services performed prior to notice of such suspension. When the Project is resumed, the Architect's compensation shall be equitably adjusted to provide for expenses incurred in the interruption and resumption of the Architect's services.

3.9.3 Termination on Abandonment. This Agreement may be terminated by the Owner upon not less than seven days' written notice to the Architect in the event that the Project is permanently abandoned. If the Project is abandoned by the Owner for more than 90 consecutive days, the Architect may terminate this Agreement by giving written notice to the Owner.

3.9.4 Failure of the Owner to make payments to the Architect in accordance with this Agreement shall be considered substantial nonperformance and cause for termination.

3.9.5 Suspension by Architect. If the Owner fails to make payment when due the Architect for services and expenses, the Architect may, upon seven days' written notice to the Owner, suspend performance of services under this Agreement. Unless payment in full is received by the Architect within seven days of the date of the notice, the suspension shall take effect without further notice. In the event of a suspension of services, the Architect shall have no liability to the Owner for delay or damage caused the Owner because of such suspension of services.

3.9.6 Compensation of Architect. In the event of termination not the fault of the Architect, the Architect shall be compensated for services performed prior to termination, together with Reimbursable Expenses then due and all Termination Expenses as defined in Subparagraph 3.9.7.

3.9.7 Termination Expenses. Termination expenses are in addition to compensation for the Architect's services, and include expenses which are directly attributable to termination. Termination Expenses shall be computed as a percentage of the total compensation for all services earned to the time of termination, as follows:

.1 Twenty percent of the total compensation for all services earned to date if termination occurs before or during the Predesign, Site Analysis or Schematic Design Phases; or

.2 Ten percent of the total compensation for all services earned to date if termination occurs during the Design Development Phase; or

.3 Five percent of the total compensation for all services earned to date if termination occurs during any subsequent phase.

Appendix F:

AIA Document G612--Owner's Instructions
Regarding the Construction Contract,
Insurance and Bonds, and
Bidding Procedures,
1987 Edition

Appendix

Reproduced with permission of The American Institute of Architects under license number #96051. This license expires September 30, 1997. FURTHER REPRODUCTION IS PROHIBITED. Because AIA Documents are revised from time to time, users should ascertain from the AIA the current edition of this document. Copies of the current edition of this AIA document may be purchased from The American Institute of Architects or its local distributors. The text of this document is not "model language" and is not intended for use in other documents without permission of the AIA.

OWNER'S INSTRUCTIONS REGARDING THE CONSTRUCTION CONTRACT

AIA DOCUMENT G612, PART A

PROJECT TITLE: DATE:

OWNER: PROJECT NO:

NOTATION TO OWNER—In consultation with your attorney, complete this form, which will provide your instructions regarding requirements for construction contracts for this Project. Please return the completed form, along with any other instructions concerning construction-related matters, to the Architect.

TO: (ARCHITECT)

Attention:

_____

You are hereby instructed to proceed with development of necessary construction-related documents on the basis of the advice and information given below:

1. Project title to be used (if other than above):

2. Legal name and address of the Owner for construction contract purposes:

3. The Owner is a ☐ Corporation ☐ Partnership ☐ Individual ☐ Other *(specify)*_____
   If a corporation: incorporated in the State of _____
   Qualified to do business at the Project location ☐ Yes ☐ No

4. Detailed description of building-site property, including designation of property owner if different from the Owner identified above:

5. During the Construction Phase the Owner's Representative will be_____, whose title is_____.
   On-site field representative(s) of the Owner's permanent staff ☐ will ☐ will not be utilized during the construction period.

6. The Project will be constructed utilizing:
   ☐ Single contract, stipulated sum
   ☐ Single contract, cost of the Work plus a fee
   ☐ Multiple contracts, stipulated sum
   ☐ Multiple contracts, cost of the Work plus a fee
   ☐ Phased construction (fast track)
   ☐ Portions of construction by Owner's own forces
   ☐ Other *(specify)*_____

CAUTION: You should sign an original AIA document which has this caution printed in red. An original assures that changes will not be obscured as may occur when documents are reproduced.

AIA DOCUMENT G612 • OWNER'S INSTRUCTIONS • 1987 EDITION • AIA® • ©1987 • THE AMERICAN INSTITUTE OF ARCHITECTS, 1735 NEW YORK AVENUE, N.W., WASHINGTON, D.C. 20006

WARNING: Unlicensed photocopying violates U.S. copyright laws and is subject to legal prosecution.

7. The method of selecting the Contractor(s) shall be:
   a. ☐ Bidding that is:
      ☐ Open and competitive
      ☐ By invitation only
      ☐ Other *(specify)* _____
   b. ☐ Negotiation with:
      ☐ A single Contractor
      ☐ Multiple Contractors
      ☐ Other *(specify)* _____

8. The form(s) of agreement between Owner and Contractor(s) shall be:
   ☐ Stipulated Sum, AIA Document A101
   ☐ Abbreviated Form, Stipulated Sum, AIA Document A107
   ☐ Cost of the Work Plus a Fee, AIA Document A111
   ☐ Abbreviated Form, Cost of the Work Plus a Fee, AIA Document A117
   ☐ Stipulated Sum, Furniture, Furnishings and Equipment, AIA Document A171
   ☐ Abbreviated Form, Stipulated Sum, Furniture, Furnishings and Equipment, AIA Document A177
   ☐ Other *(specify and furnish copy)* _____

9. If multiple contracts are to be utilized, the activities of the Contractors will be coordinated by the Owner:
   ☐ Through the Owner's own forces
   ☐ Through others *(specify)* _____

10. General Conditions of the Contract shall be (inapplicable if A107, A117 or A177 is marked on Item 7 above):
    ☐ AIA Document A201 (Construction)
    ☐ AIA Document A271 (Furniture, Furnishings and Equipment)
    ☐ Other *(specify and furnish copy)* _____

11. Supplementary Conditions of the Contract and General Requirements will be discussed and reviewed:
    ☐ With the Owner's Representative, who is _____
    ☐ Directly with the Owner's attorney, _____
       Whose address is _____
       Telephone ( ) _____

12. Contractor's applications for payment will be paid:
    ☐ By the _____ day of each month
    ☐ Other *(specify)* _____

13. Applications for payment must be received by the Owner _____ days before payment is due.

14. Retainage:
    a. Until Substantial Completion, retainage from progress payments to the Contractor shall be:
       ☐ _____% of each payment.
       ☐ _____% of each payment until the Work is 50% complete, after which remaining partial payments shall be paid in full without reduction of previous retainage.
       ☐ _____% of each payment (calculated separately for each Work category) until the Work is 50% complete, after which remaining partial payments shall be paid in full without reduction of previous retainage.
    b. Upon Substantial Completion, retainage shall be reduced to:
       ☐ _____%
       ☐ $_____
    c. Retained amounts will be paid into an escrow account in a financial institution chosen by the Contractor and approved by the Owner, the interest earnings from which shall accrue to the benefit of the Contractor:
       ☐ Yes ☐ No

15. Liquidated Damages:
    ☐ Required
    ☐ Not required
    If required, Liquidated Damages shall be assessed in the amount of $_____ per day for each calendar day required to achieve Substantial Completion beyond the date set forth in the Owner-Contractor Agreement, subject to agreed-on adjustments.

Appendix

(NOTE: When liquidated Damages are to be stipulated, it is important that Subcontractors be made aware of this provision of the Contract. Therefore, while space is provided in the AIA Owner-Contractor Agreement forms for this provision, it is recommended that any enacting clause be set out in the Supplementary Conditions.)

16. Special instructions are attached for items checked.
    - ☐ Equal opportunity requirements
    - ☐ Escrow of retainage
    - ☐ Extensions of time
    - ☐ Lien waivers
    - ☐ Monthly affidavits
    - ☐ Phased occupancy
    - ☐ Separate contracts
    - ☐ Special time periods during which the Contractor cannot perform construction
    - ☐ Tax exemptions
    - ☐ Wage standards to which the Contractor must conform
    - ☐ Work by the Owner's own forces
    - ☐ Other *(specify)* _____

_____
Owner

_____   _____
By   Date

OWNER'S INSTRUCTIONS FOR INSURANCE AND BONDS

AIA DOCUMENT G612, PART B

PROJECT: DATE:

OWNER: PROJECT NO:

NOTATION TO OWNER—In consultation with your insurance advisor, complete this form, which will provide your instructions regarding your requirements for bonds and insurance for this Project. Please return the completed form, along with any other instructions relating to bonds and insurance, to the Architect.

TO: (ARCHITECT)

Attention:

You are hereby instructed to include the following information and requirements in appropriate locations in the Bidding Documents and the Contract Documents. These requirements are based on Article 11 of AIA Document A201, General Conditions of the Contract for Construction, 1987 edition (copy attached hereto), and the completion of these instructions is presumed to be based thereon.

A. CONTRACTOR'S LIABILITY INSURANCE
Concerning the insurance described in Paragraph 11.1 of AIA Document A201, 1987 edition, specify the following minimum limits:

1. Workers' Compensation
   - ☐ State: Statutory
   - ☐ Voluntary Compensation (by any exempt entities): Same as State Workers Compensation
   - ☐ Applicable Federal (e.g., Longshoremen, harbor work, Work at or outside U.S. Boundaries): Statutory
   - ☐ Maritime: $ _____
   - ☐ Employer's Liability:
     $ _____ Each accident
     $ _____ Disease, Policy limit
     $ _____ Disease, Each employee
   - ☐ Benefits required by union labor contracts: As applicable

2. General Liability (including Premises-Operations; Independent Contractors' Protective; Products and Completed Operations; Broad Form Property Damage):
   (a) Bodily Injury:
       $ _____ Each Occurrence
       $ _____ Aggregate
   (b) Property Damage:
       $ _____ Each Occurence
       $ _____ Aggregate
   (c) Products and Completed Operations Insurance shall be maintained for a minimum period of ☐ 1 ☐ 2 ☐ _____ year(s) after final payment and the Contractor shall continue to provide evidence of such coverage to the Owner on an annual basis during the aforementioned period.

CAUTION: You should use an original AIA document which has this caution printed in red. An original assures that changes will not be obscured as may occur when documents are reproduced.

(d) Property Damage Liability Insurance shall include coverage for the following hazards:
- ☐ X (Explosion)
- ☐ C (Collapse)
- ☐ U (Underground)

(e) Contractual Liability (Hold Harmless Coverage):
  Bodily Injury:
  $_____ Each Occurrence
  Property Damage:
  $_____ Each Occurrence
  $_____ Aggregate

(f) Personal Injury (with Employment Exclusion deleted, if applicable):
  $_____ Aggregate

(g) If the General Liability policy includes a General Aggregate, such General Aggregate shall be not less than $_____. Policy shall be endorsed to have General Aggregate apply to this Project only:
  ☐ Yes ☐ No

3. Umbrella Excess Liability
  $_____ Over primary insurance
  $_____ Retention

4. Automobile Liability (owned, non-owned, hired):
  Bodily Injury:
  $_____ Each Person
  $_____ Each Accident
  Property Damage:
  $_____ Each Occurrence

5. Aircraft Liability (owned and non-owned) when applicable, as follows: (*Select one*)
  ☐ With limits proposed by the Contractor for the Owner's approval.
  ☐ With the following limits:
  Bodily Injury:
  $_____ Each Person
  $_____ Each Occurrence
  Property Damage:
  $_____ Each Occurrence

6. Watercraft Liability (owned and non-owned) when applicable, as follows: (*Select one*)
  ☐ With limits proposed by the Contractor for the Owner's approval.
  ☐ With the following limits:
  Bodily Injury:
  $_____ Each Person
  $_____ Each Occurrence
  Property Damage:
  $_____ Each Occurrence

7. Other Insurance:
  COVERAGE   **AMOUNT**

B. OWNER'S LIABILITY INSURANCE
Concerning the insurance described in Paragraph 11.2 of AIA Document A201, 1987 edition: (*Select one*)
☐ No modification is required.
☐ The Contractor shall provide this insurance with the following limits:

(1) Bodily Injury:
   $_____ Each Occurrence
   $_____ Aggregate
(2) Property Damage:
   $_____ Each Occurrence
   $_____ Aggregate

C. PROPERTY INSURANCE

Concerning the insurance described in Paragraph 11.3 of AIA Document A201, 1987 edition: *(Complete (a), (b) or (c), and then complete (d) and (e).)*

(a) ☐ No modification is required; the Owner will purchase (including coverage for all materials and equipment to be incorporated or used in the Project when stored off the site or when in transit).

(b) ☐ The Owner will purchase with the following modifications:

(c) ☐ The Contractor shall purchase the following: *(Select one)*
   ☐ All-Risk
   ☐ Other *(specify)* _____
   On the following form: *(Select one)*
   ☐ Completed Value
   ☐ Reporting
   In the names of the Owner, Contractor, Subcontractor and Sub-subcontractors as their interests may appear with limits as follows: *(Select one)*
   ☐ Amount equal to the Contract Sum for the Work.
   ☐ Other *(specify)* _____

(NOTE: If coverage for alterations and additions to existing structures is to be included under the Owner's existing coverage, specific instructions should be included under Item D below.)

(d) Boiler and Machinery Insurance:
Concerning the insurance described in Subparagraph 11.3.2 of AIA Document A201, 1987 edition:
(1) The Owner shall provide this insurance with a limit of $_____.
(2) Objects to be insured: *(List objects)*

(e) Loss of Use Insurance.
Concerning the insurance described in Subparagraph 11.3.3 of AIA Document A201, 1987 edition: *(Select one)*
   ☐ No modification is required.
   ☐ The Contractor shall provide this insurance with limits of $_____.

D. OTHER INSTRUCTIONS RELATED TO BONDS OR INSURANCE
(If none, please so indicate.)

Appendix

E. BONDS

(a) Performance Bond and Payment Bond as described in Paragraph 11.4 of AIA Document A201, 1987 edition, will be:
☐ Required
☐ Not required

(b) Required bonds shall be in the amount of *(Select one)*

Performance	☐ 100% of Contract Sum	☐ _____ % of Contract Sum	☐ $_____
Payment	☐ 100% of Contract Sum	☐ _____ % of Contract Sum	☐ $_____

(c) Form of bonds shall be *(Select one)*
☐ AIA Document A311   ☐ AIA Document A312
☐ Other _____
(Describe and furnish sample copy if available)

(d) Special instructions:

Owner

By _____   Date _____

OWNER'S INSTRUCTIONS REGARDING BIDDING PROCEDURES

AIA DOCUMENT G612, PART C

PROJECT:　　　　　　　　　　　　　　　　　　DATE:

OWNER:　　　　　　　　　　　　　　　　　　　PROJECT NO:

NOTATION TO OWNER—Complete this form, which will provide your instructions regarding your requirements for bidding procedures for this Project. Please return the completed form, along with any other instructions relating to bidding procedures, to the Architect.

TO: (ARCHITECT)

Attention:

The following information and data, together with that provided in Parts A and B of this AIA Document G612, are transmitted to you for use in connection with preparation of the Bidding Documents.

1. Instructions to Bidders shall be:
   ☐ AIA Document A701
   ☐ As provided by the Owner

2. Proposal Form shall be prepared by:
   ☐ The Architect
   ☐ The Owner

3. Bids shall be solicited by:
   ☐ Public Advertisement
   　Arranged by ☐ The Owner
   　　　　　　 ☐ The Architect

   ☐ Private Invitation
   　Issued by ☐ The Owner
   　　　　　　☐ The Architect

4. Instructions, if any, on the method of selection and/or qualification of Bidders:

   _____
   _____

5. Bid Security
   ☐ Is not required
   ☐ Is required, in the amount of ☐ $_____ ☐ _____% of the total Bid in the form of:
   　☐ A Bid Bond (AIA Document A310)
   　　☐ A Bid Bond (_____)
   　　　　　　　　　　(Other forms)
   　　☐ _____
   　　　　　(Other types of security)

CAUTION: You should sign an original AIA document which has this caution printed in red. An original assures that changes will not be obscured as may occur when documents are reproduced.

AIA DOCUMENT G612 • OWNER'S INSTRUCTIONS • 1987 EDITION • AIA® • ©1987 • THE AMERICAN INSTITUTE OF ARCHITECTS, 1735 NEW YORK AVENUE, N.W., WASHINGTON, D.C. 20006

WARNING: Unlicensed photocopying violates U.S. copyright laws and is subject to legal prosecution.

Appendix

6. Copies of the Bidding Documents may be made available in plan rooms: ☐ Yes ☐ No
   If yes, plan rooms will be:
   ☐ Designated by the Owner
   ☐ Selected by the Architect

7. Date and time for receipt of Bids:
   ☐ *(Specify)* _____
   ☐ Will be determined later by the Owner
   ☐ Shall be determined by the Architect

8. Bids shall be received:
   ☐ At the Architect's office
   ☐ Other *(Specify)* _____

9. Bid tabulation forms will be prepared by:
   ☐ the Owner
   ☐ the Architect

10. Bids shall be publicly opened and read aloud: ☐ Yes ☐ No
    If opened in private:
    Bid tabulation may be furnished to Bidders: ☐ Yes ☐ No
    Bids may be made public: ☐ Yes ☐ No
    Special instructions: _____

11. In addition to the usual original signed Bid, the Owner requires the following: _____

12. Bids shall not be withdrawn by Bidders for _____ days after the receipt of Bids.

13. If a Contract is awarded, construction at the site may commence:
    ☐ Upon execution of the Agreement
    ☐ Upon, but not before, receipt of a Notice to Proceed
    ☐ Not earlier than _____ days after award of the Contract
    ☐ Other *(Specify)* _____

14. Work shall be substantially complete:
    ☐ _____ calendar days after the Date of Commencement
    ☐ By _____, 19____
    ☐ In the number of calendar days stipulated by the Bidder in the bid form

15. Special instructions:

_____
Owner

_____   _____
By   Date

Appendix G:

Synopses of AIA Standard Form Documents, 1994 Edition

A-SERIES

The documents in the A-Series relate to various forms of agreement between an owner and a contractor.

AIA DOCUMENT A101, OWNER-CONTRACTOR AGREEMENT FORM—STIPULATED SUM

This is a standard form of agreement between owner and contractor for use where the basis of payment is a stipulated sum (fixed price). The A101 document adopts by reference and is designed for use with AIA Document A201, General Conditions of the Contract for Construction, thus providing an integrated pair of legal documents. When used together, they are appropriate for most projects. For projects of limited scope, however, use of AIA Document A107 might be considered.

AIA DOCUMENT A101/CMa, OWNER-CONTRACTOR AGREEMENT FORM—STIPULATED SUM—CONSTRUCTION MANAGER-ADVISER EDITION

A101/CMa is a standard form of agreement between owner and contractor for use on projects where the basis of payment is a stipulated sum (fixed price), and where, in addition to the contractor and the architect, a construction manager assists the owner in an advisory capacity during design and construction. The document has been prepared for use with AIA document A201/CMa, General Conditions of the Contract for Construction—Construction Manager-Adviser Edition. This integrated set of documents is appropriate for use on projects where the Construction Manager serves only in the capacity of an adviser to the owner, rather than as constructor, the latter relationship being represented in AIA Documents A121/CMc and A131/CMc). A101/CMa is suitable for projects where the cost of construction has been predetermined, either by bidding or by negotiation.

AIA DOCUMENT A105, STANDARD FORM OF AGREEMENT BETWEEN OWNER AND CONTRACTOR FOR A SMALL PROJECT

AIA DOCUMENT A205, GENERAL CONDITIONS OF THE CONTRACT FOR CONSTRUCTION OF A SMALL PROJECT

AIA Documents A105 and A205 are intended to be used in conjunction with one another. The two documents are only sold as a set, and they share a common Instruction Sheet. They have been developed for use where payment to the Contractor is based on a stipulated sum (fixed price) and where the project is modest in size and brief in duration.

A105 and A205 are two of the three documents that comprise the Small Projects family of documents. They have been developed for use with AIA Document B155, Standard Form of Agreement Between Owner and Architect for a Small Project.

These documents are specifically coordinated for use as a set. Although A105, A205, and B155 may share some similarities with other AIA documents, the Small Projects documents should NOT be used in tandem with other AIA document families without careful side-by-side comparison of contents.

A205 is considered to be the keystone document of the Small Projects family, since it is specifically adopted by separate reference into both A105 and B155. A205 is a vital document, in that it is used to allocate proper legal responsibilities among the parties, while providing both a common ground and a means of coordination within the Small Projects family. In order to maintain the condensed nature of this document, arbitration and other ADR provisions have been omitted. ADR provisions may be included in A105 under Article 6.

AIA DOCUMENT A107, ABBREVIATED OWNER-CONTRACTOR AGREEMENT FORM—STIPULATED SUM—FOR CONSTRUCTION PROJECTS OF LIMITED SCOPE

As an abbreviated form of agreement between owner and contractor, this document is intended for use where the basis of payment is a stipulated sum (fixed price). It is appropriate for construction projects of limited scope not requiring the complexity and length of the combination of AIA Documents A101 and A201. The document contains abbreviated general conditions based on AIA Document A201. It may be used when the owner and contractor have established a prior working relationship (e.g., a previous project of like or similar nature) or where the project is relatively simple in detail or short in duration.

AIA DOCUMENT A111, OWNER-CONTRACTOR AGREEMENT FORM—COST OF THE WORK PLUS A FEE, WITH OR WITHOUT A GUARANTEED MAXIMUM PRICE

This standard form of agreement between owner and contractor is appropriate for use when the basis of payment to the contractor is the cost of the work plus a fee, which in turn may be either a stipulated amount or a percentage of the construction cost. A guaranteed maximum price may be designated. A111 adopts by reference and is intended for use with AIA Document A201, General Conditions of the Contract for Construction, thus providing an integrated pair of legal documents. These documents are appropriate for most projects. For projects of limited scope, however, use of AIA Document A117 might also be considered.

AIA DOCUMENT A117, ABBREVIATED OWNER-CONTRACTOR AGREEMENT FORM—COST OF THE WORK PLUS A FEE—FOR CONSTRUCTION PROJECTS OF LIMITED SCOPE, WITH OR WITHOUT A GUARANTEED MAXIMUM PRICE

AIA Document A117 is an abbreviated form of agreement between owner and contractor for use where the basis of payment to the contractor is the cost of the work plus a fee. It may be used for construction projects of limited scope that do not require the complexity and length of the combination of AIA Documents A111 and A201, as A117 incorporates abbreviated general conditions based on A201. A guaranteed maximum price may be designated. A117 may be used where the owner and the contractor have established a prior working relationship (e.g., a previous project of like or similar nature), or where the project is relatively simple in detail or short in duration.

Appendix

AIA DOCUMENT A121/CMc—AGC DOCUMENT 565, OWNER-CONSTRUCTION MANAGER AGREEMENT WHERE THE CONSTRUCTION MANAGER IS ALSO THE CONSTRUCTOR

This document represents the collaborative efforts of The American Institute of Architects and The Associated General Contractors of America. AIA designates this document as A121/CMc and AGC designates it as AGC 565. A121/CMc is intended for use on projects where a construction manager, in addition to serving as adviser to the owner, assumes financial responsibility for construction of the project. The construction manager provides the owner with a guaranteed maximum price proposal, which the owner may accept, reject, or negotiate. Upon the owner's acceptance of the proposal by execution of an amendment, the construction manager becomes contractually bound to provide labor and materials for the project. The document divides the construction manger's services into two phases: the preconstruction phase and the construction phase, portions of which may proceed concurrently in order to fast track the process. A121/CMc is coordinated for use with AIA Document A201, General Conditions of the Contract for Construction, and B141, Standard Form of Agreement Between Owner and Architect. Check Article 5 of B511 for guidance in this regard. *Caution: to avoid confusion and ambiguity, do not use this construction management document with any other AIA or AGC construction management document.*

AIA DOCUMENT A131/CMc—AGC DOCUMENT 566, OWNER-CONSTRUCTION MANAGER AGREEMENT WHERE THE CONSTRUCTION MANAGER IS ALSO THE CONSTRUCTOR— COST PLUS A FEE, NO GUARANTEE OF COST

Similar to A121/CMc, the new CM-constructor agreement is also intended for use when the owner seeks a constructor who will take on responsibility for providing the means and methods of construction. However, the method of determining cost of the work diverges sharply in the two documents, with A121/CMc allowing for a Guaranteed Maximum Price (GMP) while A131/CMc uses a Control Estimate. A131/CMc employs the cost-plus-a-fee method, wherein the owner can monitor cost through periodic review of the Control Estimate, which is revised as the project proceeds. It is important to note that, while the CM-constructor may be assuming varied responsibilities, there are still just three primary players on the project: the owner, architect, and CMc. The A201 *General Conditions* continues to apply, although it is modified (in part) by the A131/CMc agreement. *Caution: to avoid confusion and ambiguity, do not use this construction management document with any other AIA or AGC construction management document.*

AIA DOCUMENT A171, OWNER-CONTRACTOR AGREEMENT FORM—STIPULATED SUM—FOR FURNITURE, FURNISHINGS, AND EQUIPMENT

This is a standard form of agreement between owner and contractor for furniture, furnishings, and equipment (FF&E) where the basis of payment is a stipulated sum (fixed price). A171 adopts by reference and is intended for use with AIA Document A271, General Conditions of the Contract for Furniture, Furnishings, and Equipment. It may be used in any arrangement between the owner and the contractor where the cost of FF&E has been determined in advance, either through bidding or negotiation.

AIA DOCUMENT A177, ABBREVIATED OWNER-CONTRACTOR AGREEMENT FORM—STIPULATED SUM—FOR FURNITURE, FURNISHINGS, AND EQUIPMENT

A177 is an abbreviated document that philosophically derives much of its content from a combination of the more complex and lengthy A171 and A271 documents. Its abbreviated terms and conditions may be used on projects where the contractor for furniture, furnishings, and equipment (FF&E) has a prior working relationship with the owner, or where the project is relatively simple in detail or short in duration. *Caution: this document is not intended for use on major construction work that may involve life safety systems or structural components.*

AIA DOCUMENT A191, OWNER-DESIGN/BUILDER AGREEMENTS

This document contains two agreements to be used in sequence by an owner contracting with one entity serving as a single point of responsibility for both design and construction services. Design/build entities may be architects, contractors, or even businesspersons, so long as they comply with governing laws; especially those pertaining to licensing and public procurement regulations. The first agreement covers preliminary design and budgeting services, while the second deals with final design and construction. Although it is anticipated that an owner and a design/builder entering into the first agreement will later enter into the second, the parties are not obligated to do so and may conclude their relationship after the terms of the first agreement have been fulfilled.

AIA DOCUMENT A201, GENERAL CONDITIONS OF THE CONTRACT FOR CONSTRUCTION

The General Conditions are an integral part of the contract for construction, in that they set forth the rights, responsibilities, and relationships of the owner, contractor, and architect. While not a party to the contract for construction between owner and contractor, the architect does participate in the preparation of the contract documents and performs certain duties and responsibilities described in detail in the general conditions. This document is typically adopted by reference into certain other AIA documents, such as owner-architect agreements, owner-contractor agreements, and contractor-subcontractor agreements. Thus, it is often called the "keystone" document.

Since conditions vary by locality and by project, supplementary conditions are usually added to amend or supplement portions of the General Conditions as required by the individual project. Review the model language provided in A511 as a guide in creating supplementary conditions for A201.

AIA DOCUMENT A201/CMa, GENERAL CONDITIONS OF THE CONTRACT FOR CONSTRUCTION—CONSTRUCTION MANAGER-ADVISER EDITION

A201/CMa is an adaptation of AIA Document A201 and has been developed for construction management projects where a fourth player—a construction manager—has been added to the team of owner, architect, and contractor. Under A201/CMa, the construction manager has the role of an independent adviser to the owner. Thus, the document carries the CMa suffix. A major difference between A201 and A201/CMa occurs in Article 2, Administration of the Contract, which deals with the duties and responsibilities of both the architect and the construction manager-adviser. Another major difference implicit in A201/CMa is the use of multiple construction contracts directly with trade contractors. *Caution: it is vital that A201/CMa not be used in combination with documents where it is assumed that the construction manager takes on the role of constructor, gives the owner a guaranteed maximum price, or contracts directly with those who supply labor and materials for the project.*

AIA DOCUMENT A201/SC, FEDERAL SUPPLEMENTARY CONDITIONS OF THE CONTRACT FOR CONSTRUCTION

A201/SC is intended for use on certain federally assisted construction projects. For such projects, A201/SC adapts A201 by providing (1) necessary modifications of the General Conditions, (2) additional conditions, and (3) insurance requirements for federally assisted construction projects.

AIA DOCUMENT A271, GENERAL CONDITIONS OF THE CONTRACT FOR FURNITURE, FURNISHINGS, AND EQUIPMENT

When the scope of a contract is limited to furniture, furnishings, and equipment (FF&E), A271 is intended for use in a manner similar to the way in which A201 is used for construction projects. The document was jointly developed by the AIA and the American Society of Interior Designers (ASID). Because the Uniform Commercial Code (UCC) has been adopted in virtually every jurisdiction, A271 has been drafted to recognize the commercial standards set forth in Article 2 of the UCC, and uses certain standard UCC terminology. Except for minor works, A271 should not be used for construction involving life safety systems or structural components.

AIA DOCUMENT A305, CONTRACTOR'S QUALIFICATION STATEMENT

An owner preparing to request bids or to award a contract for a construction project often requires a means of verifying the background, history, references, and financial stability of any contractor being considered. The time frame for construction and the contractor's performance history, previous experience, and financial stability are important factors for an owner to investigate. This form provides a sworn, notarized statement with appropriate attachments to elaborate on important aspects of the contractor's qualifications.

AIA DOCUMENT A310, BID BOND

This simple one-page form was drafted with input from the major surety companies to ensure its legality and acceptability. A bid bond establishes the maximum penal amount that may be due the owner if the selected bidder fails to execute the contract and provide any required performance and payment bonds.

AIA DOCUMENT A312, PERFORMANCE BOND AND PAYMENT BOND

This form incorporates two bonds covering first, the contractor's performance, and second, the contractor's obligations to pay subcontractors and others for material and labor. In addition, the A312 document obligates the surety to act responsively to the owner's requests for discussions aimed at anticipating or preventing a contractor's default.

AIA DOCUMENT A401, STANDARD FORM OF AGREEMENT BETWEEN CONTRACTOR AND SUBCONTRACTOR

This document is intended for use in establishing the contractual relationship between the contractor and subcontractor. It spells out the responsibilities of both parties and lists their respective obligations, which are written to parallel AIA Document A201, General Conditions of the Contract for Construction. Blank spaces are provided where the parties can supplement the details of their agreement. A401 may be modified for use as a subcontractor- sub-subcontractor agreement.

AIA DOCUMENT A491, DESIGN/BUILDER-CONTRACTOR AGREEMENTS

A491 contains two agreements to be used in sequence by a design/builder and a construction contractor. The first agreement covers management consulting services to be provided during the preliminary design and budgeting phase of the project, while the second covers construction. It is presumed that the design/builder has contracted with an owner to provide design and construction services under the agreements contained in AIA Document A191.

Although it is anticipated that a design/builder and a contractor entering into the first agreement will later enter into the second, the parties are not obligated to do so, and may conclude their relationship after the terms of the first agreement have been fulfilled. It is also possible that the parties may forgo entering into the first agreement and proceed directly to the second.

AIA DOCUMENT A501, RECOMMENDED GUIDE FOR COMPETITIVE BIDDING PROCEDURES AND CONTRACT AWARDS FOR BUILDING CONSTRUCTION

This guide outlines appropriate procedures in the bidding and award of contracts when competitive lump sum bids are requested in connection with building and related construction. The Guide is a joint publication of the AIA and the Associated General Contractors of America (AGC).

AIA DOCUMENT A511, GUIDE FOR SUPPLEMENTARY CONDITIONS

A511 is a guide for modifying and supplementing A201, the General Conditions of the Contract for Construction. It provides model language with explanatory notes to assist users in adapting A201 to local circumstances. Although A201 is considered the keystone in the legal framework of the construction contract, because it is a standard document, it cannot cover all the particulars of a specific project. Thus, A511 is intended as an aid to users of A201 in developing supplementary conditions.

This document is printed with a column containing the model text and an adjacent column providing explanatory notes to the user. Excerpting of the model text is permitted by the AIA under a limited license for reproduction granted for drafting the supplementary conditions of a particular project.

AIA DOCUMENT A511/CMa, GUIDE FOR SUPPLEMENTARY CONDITIONS—CONSTRUCTION MANAGER-ADVISER EDITION

Similar to A511, the A511/CMa document is a guide to model provisions for supplementing the general conditions of the contract for construction, construction manager-adviser edition (AIA Document A201/CMa). A511/CMa should only be employed—as should A201/CMa—on projects where the construction manager is serving in the capacity of *adviser* to the owner (as represented by the CMa document designation), and not in situations where the Construction Manager is also the constructor (CMc document-based relationships). *Caution: CMc documents are based on utilization of the A201 document, which in turn should be modified using A511 as a guide.*

Like A511, this document contains suggested language for supplementary conditions, along with notes on appropriate usage. However, many important distinctions are made to ensure consistency with other construction manager-adviser documents.

AIA DOCUMENT A521, UNIFORM LOCATION OF SUBJECT MATTER

A521 is a joint publication of the AIA and the Engineers Joint Contract Documents Committee (EJCDC), which is composed of the National Society of Engineers, American Consulting Engineers Council, and American Society of Civil Engineers. By consensus of these organizations, the AIA and EJCDC documents follow A521's tabular guide with regard to the placement of subject matter among the various contract and bidding documents. A521 is a tabulation to guide the user in determining the proper placement and phrasing of information customarily used on a construction project. This document shows the importance of maintaining uniformity in location and language from document to document with respect to subject matter. Inconsistencies in either area may cause confusion, delay, or unanticipated legal problems.

AIA DOCUMENT A571, GUIDE FOR INTERIORS SUPPLEMENTARY CONDITIONS

Similar to A511, AIA Document A571 is intended as an aid to practitioners in preparing supplementary conditions on interiors projects.

AIA Document A571 provides additional information to address local variations in project requirements where A271, General Conditions of the Contract for Furniture, Furnishings, and Equipment, is used.

AIA DOCUMENT A701, INSTRUCTIONS TO BIDDERS

This document is used when competitive bids are to be solicited for construction of the project. Coordinated with A201 and its related documents, A701 contains instructions on procedures to be followed by bidders in preparing and submitting their bids, including bonding. Specific instructions or special requirements, such as the amount and type of bonding, are to be attached to A701 as supplementary conditions.

AIA DOCUMENT A771, INSTRUCTIONS TO INTERIORS BIDDERS

Similar to A701, A771 is used for projects dealing with furniture, furnishings, and equipment (FF&E). It parallels A701, but contains minor changes to maintain consistency with A271 and its related FF&E documents.

B-SERIES

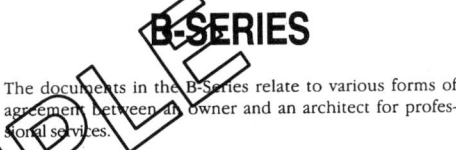

The documents in the B-Series relate to various forms of agreement between an owner and an architect for professional services.

AIA DOCUMENT B141, STANDARD FORM OF AGREEMENT BETWEEN OWNER AND ARCHITECT

This document is the first and foremost standard form among the various owner and architect agreements in the B-Series. Its five phases represent an architect's traditional package of professional services that commence with schematic design and continue through construction contract administration. A variety of more comprehensive and specialized services may be provided through B141 in the form of additional services or, alternately, through use of other B-Series documents such as B163 (a designated services agreement). B141's construction phase services are coordinated with and referenced to the architect's duties and responsibilities as set forth in AIA Document A201, General Conditions of the Contract for Construction.

AIA DOCUMENT B141/CMa, STANDARD FORM OF AGREEMENT BETWEEN OWNER AND ARCHITECT, CONSTRUCTION MANAGER-ADVISER EDITION

B141/CMa is a standard form of agreement between owner and architect for use on building projects where construction management services are to be provided under separate contract with the owner. It is coordinated with AIA Document B801/CMa, an owner-construction manager-adviser agreement where the construction manager is an independent, professional adviser to the owner throughout the course of the project. Both B141/CMa and B801/CMa are based on the premise that a separate construction contractor will also contract with the owner. The owner-contractor agreement is jointly administered by the architect and the construction manager under AIA Document A201/CMa, General Conditions of the Contract for Construction—Construction Manager-Adviser Edition.

Caution: B141/CMa is not coordinated with and should not be used with documents where the construction manager acts as the constructor (i.e., contractor) as depicted in AIA Documents A121/CMc and A131/CMc.

AIA DOCUMENT B144/ARCH-CM, STANDARD FORM OF AMENDMENT FOR THE AGREEMENT BETWEEN THE OWNER AND ARCHITECT WHERE THE ARCHITECT PROVIDES CONSTRUCTION MANAGEMENT SERVICES AS AN ADVISER TO THE OWNER

B144/ARCH-CM is an amendment for use in circumstances where the architect agrees to provide the owner with a package of construction management services to expand upon, blend with, and supplement the architect's design and other construction administration services as described in AIA Document B141, Standard Form of Agreement Between Owner and Architect. This amendment maintains and elaborates upon the architect's normal role as an adviser to the owner as set forth in both B141 and AIA Document A201, General Conditions of the Contract for Construction. Under this amendment, it is not intended that the architect provide the owner with a Guaranteed Maximum Price (GMP) or dictate the means and methods of, or safety requirements for, the construction by contracting with subcontractors. Although this document is philosophically similar to the AIA's construction manager-adviser (CMa) documents, it should not be used in conjunction with any CMa documents, because the underlying premise of B144/ARCH-CM is that there are only three primary players on the project: the owner, the architect (who is also providing construction management services), and the contractor. In contrast, the CMa documents envision a fourth player: an independent construction manager-adviser.

AIA DOCUMENT B151, ABBREVIATED OWNER-ARCHITECT AGREEMENT

This abbreviated owner-architect agreement is intended for use on projects of limited scope where a concise, readable contract is needed, but where the services and detail of B141 or B163 may not be required. B151 provides for three phases of service, while B141 provides for five and B163 allows for nine. B151 is also appropriate where the owner and architect have established a prior working relationship (e.g., a previous project of like or similar nature), or where the project is relatively simple in detail or short in duration.

AIA DOCUMENT B155, STANDARD FORM OF AGREEMENT BETWEEN OWNER AND ARCHITECT FOR A SMALL PROJECT

AIA Document B155 is a standard form of agreement between Owner and Architect intended for use on a Small Project; one that is modest in size and brief in duration. B155 is one of three documents that comprise the Small Projects family of documents. It has been developed for use with AIA Document A105, Standard Form of Agreement Between Owner and Contractor for a Small Project, and A205, General Conditions of the Contract for Construction of a Small Project. These documents are specifically coordinated for use as a set. Although A105, A205, and B155 may share some similarities with other AIA documents, the Small Project documents should NOT be used in tandem with other AIA document families without careful side-by-side comparison of contents.

In addition, B155 adopts the A205 document by reference as it pertains to the Architect's responsibilities in administration of the construction contract between Owner and Contractor.

AIA DOCUMENT B161, STANDARD FORM OF AGREEMENT BETWEEN OWNER AND ARCHITECT FOR DESIGNATED SERVICES

THIS DOCUMENT HAS BEEN REPLACED BY
AIA DOCUMENT B163.

AIA DOCUMENT B162, SCOPE OF DESIGNATED SERVICES

THIS DOCUMENT HAS BEEN REPLACED BY
AIA DOCMENT B163.

AIA DOCUMENT B163, STANDARD FORM OF AGREEMENT BETWEEN OWNER AND ARCHITECT FOR DESIGNATED SERVICES

B163 is the most comprehensive AIA owner-architect agreement. This three-part document contains, among other things, a thorough list of 83 possible services divided among nine phases, covering pre-design through supplemental services. This detailed classification allows the architect to more accurately estimate the time and personnel costs required for a particular project. Both owner and architect benefit from the ability to clearly establish the scope of services required for the project, as responsibilities and compensation issues are negotiated and defined. The architect's compensation may be calculated on a time/cost basis through use of the worksheet provided in the instructions to B163.

Part One of the document deals with variables typical of many owner-architect agreements, such as compensation and scope of services. The scope of services is delimited through use of a matrix that allows the parties to designate their agreed-upon services and responsibilities. Part Two contains detailed descriptions of the specific services found in Part One's matrix. Part Three contains general descriptions of the parties' duties and responsibilities. B163's list of services has been expanded beyond those of any of its predecessor documents through inclusion of construction management and interiors services.

AIA DOCUMENT B171, STANDARD FORM OF AGREEMENT BETWEEN OWNER AND ARCHITECT FOR INTERIOR DESIGN SERVICES

B171 is intended for use when the architect agrees to provide an owner with design and administrative services for the procurement of interior furniture, furnishings, and equipment (FF&E). Unlike B141, which is used for building design, this document includes programming of the interior spaces and requirements as part of the overall package of basic services. The authority to reject goods is left in the hands of the owner rather than the architect, since the procurement of goods is governed by the Uniform Commercial Code (UCC), which would in turn make the architect's mistaken rejection or acceptance of goods binding upon the owner. B171 is coordinated with and adopts by reference AIA Document A271, General Conditions of the Contract for Furniture, Furnishings, and Equipment. When B171 is used, it is anticipated that A271 will form part of the contract between the owner and the contractor for FF&E.

AIA DOCUMENT B177, ABBREVIATED FORM OF AGREEMENT FOR INTERIOR DESIGN SERVICES

B177 is an abbreviated document that is similar to B171, but with less complexity and detail. This document may be used where the owner and architect have a continuing relationship from previous work together or where the project is relatively simple in detail or short in duration.

AIA DOCUMENT B181, STANDARD FORM OF AGREEMENT BETWEEN OWNER AND ARCHITECT FOR HOUSING SERVICES

This document has been developed with the assistance of the U.S. Department of Housing and Urban Development and other federal housing agencies, and is primarily intended for use in multi-unit housing design. B181 requires that the owner, (and not the architect), furnish cost-estimating services. B181 is coordinated with and adopts by reference AIA Document A201, General Conditions of the Contract for Construction.

DOCUMENT B352, DUTIES, RESPONSIBILITIES AND LIMITATIONS OF AUTHORITY OF THE ARCHITECT'S PROJECT REPRESENTATIVE

When and if the owner wants additional project representation at the construction site on a full- or part-time basis, B141 and other AIA owner-architect agreements reference B352 to establish the project representative's duties, responsibilities, and limitations of authority.

The project representative is employed and supervised by the architect. In contrast, up until the early 1950s, B352 predecessor documents called the representative "the Clerk of the Works", because such persons were hired by the owner but supervised by the architect. The split between hiring and supervision caused numerous problems, which have been resolved under B352 by designating the architect as both employer and supervisor. B352 is coordinated for use with both B141 and B163, as well as A201.

AIA DOCUMENT B431, ARCHITECT'S QUALIFICATION STATEMENT

The Architect's Qualification Statement is a standardized outline of information that a client may wish to review prior to selecting an architect for a particular project. It may be used as part of a request for proposals (RFP) or as a final check on the credentials of an architect. Under some circumstances, B431 may also be attached to the owner-architect agreement to show, for example, the team of professionals and consultants expected to be employed on the owner's project.

AIA DOCUMENT B727, STANDARD FORM OF AGREEMENT BETWEEN OWNER AND ARCHITECT FOR SPECIAL SERVICES

B727 is the most flexible of the AIA owner-architect agreements, in that the description of services is left entirely up to the ingenuity of the parties. Otherwise, many of the terms and conditions are very similar to those found in AIA Document B141. B727 is often used for planning, feasibility studies, and other services (such as construction administration), which do not follow the complete phasing sequence of services set forth in B141 and other AIA documents. If construction administration services are to be provided, care must be taken to coordinate B727 with the appropriate general conditions of the contract for construction.

AIA DOCUMENT B801/CMa, STANDARD FORM OF AGREEMENT BETWEEN OWNER AND CONSTRUCTION MANAGER WHERE THE CONSTRUCTION MANAGER IS NOT A CONSTRUCTOR

This standard form of agreement is intended for use on projects where construction management services are assumed by a single entity who is separate and independent from the architect and the contractor, and who acts solely as an adviser (CMa) to the owner throughout the course of the project.

B801/CMa is coordinated for use with AIA Document B141/CMa, Standard Form of Agreement Between Owner and Architect—Construction Manager-Adviser Edition. Both B801/CMa and B141/CMa are based on the premise that there will be a separate, and possibly multiple, construction contractor(s) whose contracts with the owner are jointly administered by the architect and the construction manager under AIA Document A201/CMa, General Conditions of the Contract for Construction—Construction Manager-Adviser Edition. *Caution: B801/CMa is not coordinated with and should not be used with documents where the construction manager acts as the constructor (i.e., contractor) for the project, such as AIA Documents A121/CMc or A131/CMc.*

AIA DOCUMENT B901, STANDARD FORM OF AGREEMENT BETWEEN DESIGN/BUILDER AND ARCHITECT

This document contains two agreements to be used in sequence by a design/builder and an architect, the first covering preliminary design and the second covering final design. It is presumed that the design/builder has previously contracted with an owner to provide design and construction services under the agreements contained in AIA Document A191. Although it is anticipated that a design/builder and an architect entering into the first agreement will later enter into the second, the parties are not obligated to do so and may conclude their relationship after the terms of the first agreement have been fulfilled.

Design/build entities may be architects, contractors, or even businesspersons, so long as they comply with the governing laws; especially those pertaining to licensing and public procurement regulations. Prior to proceeding in this fashion or entering into either agreement contained in this document with any other entity, architects are advised to contact their legal, insurance, and management advisers.

C-SERIES

The documents in the C-Series relate to various forms of agreement between an architect and other professionals, including engineers, consultants, and other architects.

AIA DOCUMENT C141, STANDARD FORM OF AGREEMENT BETWEEN ARCHITECT AND CONSULTANT

This is a standard form of agreement between architect and consultant, establishing their respective responsibilities and mutual rights. C141 is most applicable to engineers, but may also be used by consultants in other disciplines providing services to scope of basic services, incurring reimbursable expenses, or proceeding with certain additional services. The document should only be used in conjunction with an earlier agreement for professional services to provide a written record of such authorizations, giving particulars of activities, time spans, and compensation involved.

AIA DOCUMENT G612, OWNER'S INSTRUCTIONS REGARDING THE CONSTRUCTION CONTRACT, INSURANCE AND BONDS, AND BIDDING PROCEDURES

This document is formatted as a questionnaire and is divided into three parts. Part A relates to the contracts, Part B covers insurance and bonds, and Part C deals with bidding procedures. The sections follow a project's normal chronological sequence to provide information when it will be needed. Because many of the items relating to the contract will have some bearing on the development of construction documents, it is important to place Part A in the owner's hands at the earliest possible phase of the project. The owner's responses to Part A will lead to a selection of the appropriate delivery method and contract forms, including the general conditions. Part B naturally follows after the selection of the general conditions, because insurance and bonding information is dependent upon the type of general conditions chosen. Thereafter, answers to Part C will follow as the contract documents are further developed.

AIA DOCUMENT G701, CHANGE ORDER

G701 may be used as written documentation of changes in the work, contract sum, or contract time that are mutually agreed to by the owner and contractor. G701 provides space for the signatures of the owner, architect, and contractor, and for a complete description of the change.

AIA DOCUMENT G701/CMa, CHANGE ORDER—CONSTRUCTION MANAGER-ADVISER EDITION

The purpose of this document is essentially the same as that of G701. The major difference is that the signature of the construction manager-adviser, along with those of the owner, architect, and contractor, is required to validate the change order.

AIA DOCUMENT G702, APPLICATION AND CERTIFICATE FOR PAYMENT; AIA DOCUMENT G703, CONTINUATION SHEET

These documents provide convenient and complete forms on which the contractor can apply for payment and the architect can certify that payment is due. The forms require the contractor to show the status of the contract sum to date, including the total dollar amount of the work completed and stored to date, the amount of retainage (if any), the total of previous payments, a summary of change orders, and the amount of current payment requested. G703, Continuation Sheet, breaks the contract sum into portions of the work in accordance with a schedule of values required by the general conditions. The form serves as both the contractor's application and the architect's certification. Its use can expedite payment and reduce the possibility of error. If the application is properly completed and acceptable to the architect, the architect's signature certifies to the owner that a payment in the amount indicated is due to the contractor. The form also allows the architect to certify an amount different than the amount applied for, with explanation provided by the architect.

AIA DOCUMENT G702/CMa, APPLICATION AND CERTIFICATE FOR PAYMENT—CONSTRUCTION MANAGER-ADVISER EDITION; AIA DOCUMENT G703, CONTINUATION SHEET

Though the use and purpose of G702/CMa remains substantially similar to that of G702, the construction manager-adviser edition expands responsibility for certification of payment to include both architect and construction manager. Similarly, both architect and construction manager may certify a different amount than that applied for, with each initialing the figures that have been changed and providing written explanation(s) accordingly. The standard G703 Continuation Sheet is appropriate for use with G702/CMa.

AIA DOCUMENT G704, CERTIFICATE OF SUBSTANTIAL COMPLETION

G704 is a standard form for recording the date of substantial completion of the work of a designated portion thereof. The contractor prepares a list of items to be completed or corrected, and the architect verifies and amends this list. If the architect finds that the work is substantially complete, the form is prepared for acceptance by the contractor and the owner. Appended thereto is the list of items to be completed or corrected. The form provides for agreement as to the time allowed for completion or correction of the items, the date when the owner will occupy the work or designated portion thereof, and a description of responsibilities for maintenance, heat, utilities, and insurance.

AIA DOCUMENT G705, CERTIFICATE OF INSURANCE

THIS DOCUMENT HAS BEEN REPLACED BY
AIA DOCUMENT G715.

AIA DOCUMENT G706, CONTRACTOR'S AFFIDAVIT OF PAYMENT OF DEBTS AND CLAIMS

The contractor submits this affidavit with the final request for payment, stating that all payrolls, bills for materials and equipment, and other indebtedness connected with the work for which the owner might be responsible has been paid or otherwise satisfied. G706 requires the contractor to list any indebtedness or known claims in connection with the construction contract that have not been paid or otherwise satisfied. The contractor may also be required to furnish a lien bond or indemnity bond to protect the owner with respect to each exception.

AIA DOCUMENT G706A, CONTRACTOR'S AFFIDAVIT OF RELEASE OF LIENS

G706A supports AIA Document G706 in the event that the owner requires a sworn statement of the contractor stating that all releases or waivers of liens have been received. In such event, it is normal for the contractor to submit G706 and G706A, along with attached releases or waivers of liens for the contractor, all subcontractors, and others who may have lien rights against the owner's property. The contractor is required to list any exceptions to the sworn statement provided in G706A, and may be required to furnish to the owner a lien bond or indemnity bond to protect the owner with respect to such exceptions.

AIA DOCUMENT G707, CONSENT OF SURETY TO FINAL PAYMENT

By obtaining the surety's approval of final payment to the contractor and its agreement that final payment will not relieve the surety of any of its obligations, the owner may preserve its rights under the bonds.

AIA DOCUMENT G707A, CONSENT OF SURETY TO REDUCTION IN OR PARTIAL RELEASE OF RETAINAGE

This is a standard form for use when a surety company is involved and the owner-contractor agreement contains a clause whereby retainage is reduced during the course of the construction project. When duly executed, G707A assures the owner that such reduction or partial release of retainage does not relieve the surety of its obligations.

AIA DOCUMENT G709, PROPOSAL REQUEST

This form is used to obtain price quotations required in the negotiation of change orders. G709 is not a change order or a direction to proceed with the work; it is simply a request to the contractor for information related to a proposed change in the construction contract.

AIA DOCUMENT G710, ARCHITECT'S SUPPLEMENTAL INSTRUCTIONS

Architect's supplemental instructions are used by the architect to issue additional instructions or interpretations or to order minor changes in the work. The form is intended to assist the architect in performing obligations as interpreter of the contract document requirements in accordance with the owner-architect agreement and the general conditions. This form should not be used to change the contract sum or contract time. If the contractor believes that a change in the contract sum or contract time is involved, other G-Series documents must be used.

AIA DOCUMENT G711, ARCHITECT'S FIELD REPORT

The Architect's Field Report is a standard form for the architect's project representative to use in maintaining a concise record of site visits or, in the case of a full-time project representative, a daily log of construction activities.

AIA DOCUMENT G712, SHOP DRAWING AND SAMPLE RECORD

This is a standard form by which the architect can schedule and monitor shop drawings and samples. Since this process tends to be complex, the schedule provided in G712 shows the progress of a submittal, which in turn contributes to the orderly processing of work. G712 can also serve as a permanent record of the chronology of the submittal process.

AIA DOCUMENT G714, CONSTRUCTION CHANGE DIRECTIVE

This document replaces former AIA Document G713, Construction Change Authorization. G714 was developed as a directive for changes in the work which, if not expeditiously implemented, might delay the project. In contrast to a Change Order (AIA Document G701), G714 is to be used where the owner and contractor, for whatever reason, have not reached agreement on proposed changes in the contract sum or contract time. Upon receipt of a completed G714, the contractor *must* promptly proceed with the change in the work described therein.

AIA DOCUMENT G714/CMa, CONSTRUCTION CHANGE DIRECTIVE, CONSTRUCTION MANAGER-ADVISER EDITION

G714/CMa is designed to effect the same type of substantive changes in the work described in the synopsis of G714, above. The difference between the two lies not in purpose, but in execution: whereas the owner and architect must both sign the G714 in order for the directive to become a valid contractual instrument, G714/CMa requires execution by owner, architect, *and construction manager-adviser.*

AIA DOCUMENT G722/CMa, PROJECT APPLICATION AND PROJECT CERTIFICATE FOR PAYMENT; AIA DOCUMENT G723/CMa, PROJECT APPLICATION SUMMARY

These documents are similar in purpose to the combination of G702 and G703, but are for use on construction management projects where the CM serves as an adviser to the owner.

Each contractor submits separate G702/CMa and G703/CMa documents to the construction manager-adviser, who collects and compiles them to complete G723/CMa. G723/CMa then serves as a summary of the contractors' applications, with project totals being transferred to a G722/CMa. The construction manager-adviser can then sign the form, have it notarized, and submit it along with the G723/CMa (which has all of the separate contractors' G702/CMa forms attached) to the architect for review and appropriate action.

Index

A

AAA	266
Abbreviated Documents	310
Form	40
Accident Prevention	101, 187
Accord and Satisfaction	160
Additional Agreements	345
Time	203
ADR	263
Adverse Weather	272
Advising Owner	125
Advisor CM	72
Advisors	107
Aesthetic Decisions	250
Affidavit of Payment	227
AGC	71
AGC Document 566	72
AGC Form 565	41
Agreement Forms	49
AIA Document A101	5, 14, 39, 89, 184, 210
AIA Document A101/CMa	41
AIA Document A105	14, 25, 40
AIA Document A107	14, 40
AIA Document A111	14, 39, 77, 89, 184, 218
AIA Document A117	40
AIA Document A121/CMc	41, 72
AIA Document A131/CMc	41, 72
AIA Document A191	41
AIA Document A201	5, 11, 103, 118, 133, 142, 148, 155, 162, 173, 184, 185, 210, 248
AIA Document A201/CMa	41
AIA Document A205	40
AIA Document A305	84
AIA Document A310	88
AIA Document A312	137
AIA Document A401	210
AIA Document A491	42
AIA Document A511	17
AIA Document A701	84
AIA Document B141	5, 24, 28, 110, 114, 118, 142, 160, 168, 210
AIA Document B141/CMa	26, 41
AIA Document B151	24
AIA Document B155	25, 40
AIA Document B163	26
AIA Document B171	26
AIA Document B177	27
AIA Document B181	27
AIA Document B188	27, 311
AIA Document B352	130
AIA Document B801	72
AIA Document B801/CMa	41
AIA Document B901	27, 42
AIA Document C141	110, 142, 149
AIA Document C142	149
AIA Document G612	82, 114, 321
AIA Document G701	163
AIA Document G702	156, 174, 227, 336
AIA Document G703	156, 174
AIA Document G704	219
AIA Document G705	114
AIA Document G706	160, 227
AIA Document G706A	160, 227
AIA Document G707	139, 160
AIA Document G707A	139, 160, 230
AIA Document G709	163
AIA Document G711	123
AIA Document G712	120, 150
AIA Document G714	163
AIA General Conditions	11, 39, 98, 104, 170, 260
AIA Standard Contracts	42
Allowances	60, 103
Alternate Bids	87
Alternates	87
Alternative Dispute Resolution	263
American Arbitration Association	266
Amount of Retainage	51
Appeals	250
Application for Payment	155
Approval, Architect	152
Approvals	171, 346
Approve Drawings	350
APR	129, 131
Arbitration	265
Arbitration Demand	298
Fees	266
Hearings	307
Rules	268
Architect Removal	192
Suspension	35
Architect's Approval	240
Authority	128
Certificate	220, 244

Architect's Certificates 339
    Certifications 336
    Consultants 37
    Decisions 106, 247
    Interpretations 18, 296
    Rejection 252
    Representative 129
    Research 284
    Responsibility 3
    Review 152
Architectural Fee 30, 121
    Scale 353
    Services 308, 330
    Services Agreement 23, 118
Asbestos 206
    Removal 189
Assisting Lawyer 300
Associated General Contractors 213
Associations, Professional 279
ASTM 252
Attorney Selection 300
Authorizations 346

B

Bankruptcy 246
Base Bid 87
Bid Bond 87, 88
    Package 85
Bidding 24, 83, 90, 332
    Conditions 82
    Documents 85, 87, 93
    Period 86
    Practices 82
    Procedure 83, 323
    Systems 93
Billing Changes 164
Blueprints 349
Bond Bid 88
    Claims 139
    Construction 133
    Cost 134
    Form 243
    Surety 47
Bonding Capacity 91
    Company 88
Bonds 105, 115
Boundary Survey 98
Breach Contract 194
Budget Overruns 292
    Project 329
Building Codes 200
    Inspector 224
Building Permits 171

C

CADD 22
Calender Days 270
Capital Requirements 92
Care, Standard 288
Cash Allowances 62, 230
Cause, Sufficient 339
Certificate, Architectural 336
    of Completion 219
    of Insurance 114
    Payment 227, 336
Certificates of Occupancy 223
Certification, Payment 120, 155
Changing Architect's 192
Change Costs 164
    Order Form 163
    Orders 46, 121, 161, 167, 180, 204
Changed Conditions 269
Changes 20, 204
Chemical Toilets 97
Claims 123, 203, 261
    for Changes 167
    Owner's 176
    Third-Party 308
Clarifying Documents 255
Clean Site 175
Clerk of Works 129
Closing Project 217
CM 68, 69
    Contractor 72
    Contracts 71
CMAA 71
Commence Job 102
Communications 100, 188, 211
Competitive Bidding 81
    Environment 92
Completion 217
    Date 56, 58
    Formalities 181
    Job 232
    of Construction 225
    Substantial 274
Components 144
Computer Aided Design 22
Concealed Conditions 203
Conciliation 265
Conciliator 265
Condemning Work 252
Conference, Preconstruction 96
Conflicting Viewpoints 257
Conflicts, Drawings 111
Confusion, Contract 19
Consent of Surety 138, 228
Consequential Damages 254

Index

Construction Administrator 23
    Bond 133
    Budget 330
    Defects 317
    Documents 24, 331
    Funding 92
    Insurance 113
    Lenders 55, 224
    Management 68
    Management (CM) Agreements 41
    Manager 70
    Phase 118
    Progress 99, 105
    Quality 186
    Schedule 272
    Time 56, 58, 205
Consultant's Review 149
Consultants 107
Contactor's Rights 212
Continuing Performance 267
Contract Administration 4, 203
    Administrators 2
    Bond 134
    Breach 194
    Clauses 56
    Documents 5, 11, 151, 171
    Drawings 185
    Errors 97
    Performance 203
    Precedence 16
    Price 43
    Provisions 58
    Signing 171, 184
    Sum 61, 242
    Termination 138, 174, 233, 236, 238
    Time 218
Contractor CM 72
    Prequalification 135
    Responsibilities 183
    Retainage 52
Contractor's Fee .... 48 - 49
    Indemnication 192
    License 195
    Payment 337
    Qualification Statement 84
    Responsibility 3
    Superintendent 197
Contracts 39
Contractural Relationships 6, 100
Controversial Drawings 19
Converting Scales 354
Coordination 104
Correcting Deficiencies 338
    Work 207
Cost Estimating .... 332 - 333

Cost Plus Contracts 59
Counsel, Legal 299
Court Decisions 288
CPM Schedule 274
Critical Path 274
Custom Construction 178
Customer Service 191

D

Daily Log 99, 200
Dated Drawing 20
Debris Storage 97
Defective Documents 193
    Work 128, 157, 177, 207, 251, 253
Defects .... 317 - 319
Delay Claim 275
    Damages 274
Delays 99, 237, 269
    Job 180
Demand, Arbitration 265
Design Budget 325
    Build 67
    Concept 98, 249
    Development 24, 331
    Documents 331
    Flaws 319
    Process 110
Design/Builder-Architect 42
Designated Services 26
Designer Liability 278
Development Documents 331
Differing Conditions 271
Difficult Contracts 181
Directives, Change 162
Disclosure 289
Dispute Resolution 259
Disputes 121, 261
Document Review 14
Documents, Construction 1
Drafting 20
    Conventions 351
Drawing Approvals 350
    Dated 20
    Reproduction 21
    Revision 21
    Submittals 141
Drawings 185, 349, 355
    Controversial 19
    Shop 141

E

Early Completion 60, 219
Easements 172
Ejecting Contractor 238
Eliminated Bidder 85
Eliminating Errors 292
Engineering Fee 32
    Scale 353
Entry Log 201
Error Correction 194
Errors .... 286, 291 - 294
    Contract 97
    of Inconsistencies 15
Estimating Changes 164
Expert Witness 284
Expertise Contractor 194
Explosives 206
Extension Time 166
Extras Unexpected 180

F

Fabricator Drawings 147
Failure to Pay 174
Fair Bidding 92
Fast Track 73, 75
Faster Track 76
Faulty Directions 195
Fee Architectural 28
Fees 121, 315
Fiduciary Duty 126
Field Conditions 199
    Professionals 123
    Report 122, 126
Figure 1-1 7
Final Completion 219, 226, 227
    Payment 159, 231, 338
    Reconciliation 230
    Submittals 225
Financial Budget 325
    Incentives 59
Finding Errors 292
Fixed Costs 332
Form of Agreement 26
Forman 198

G

General Conditions 253
    Contractor 183

GMP 44, 76
Governmental Regulation 172
Graphic Communication 349
Guaranteed Maximum Fee 31
    Maximum Price 44, 76, 218

H

Half Size Drawing 21
Hazardous Materials 206
    Waste 189
Hearings Arbitration 307
Hourly Rate 30

I

Impact Claims 166
Inadvertence 291
Inclement Weather 45
Incomplete Documents 77
Inconsistencies 15
Indemnification 192
Indemnity 311
Informing Client 150
    Owner 124
Innovation 283
    Product 280
    Responsible 282
Inspections 102, 117, 208
Instruction to Bidder 84, 85
Insurance 322
    Certificate 104, 114, 173
    Construction 113
    Liability 295
    Owner's 105
Intent of Documents 16, 248
Interest Past Due 47
Interior Design Services 26
Interrelationships 210
Introduction 1
Invitation Bid 84

J

Job Delays 180
    Log 200
    Safety 187
    Visit 100, 199
Jobsite Conference 200
    Office 97

Index

K

Keeping Records.. 305

L

Late Completion 245
Lawsuits......................... 297, 303, 304, 315
Lawyers 267, 299
Layout, Project 98
Leased Premises.. 224
Legal Claims... 297
    Counsel.. 299
    Formbooks 57
    Process 311
    Representation................................ 267
    Response 298
    Responsibilities 193
Legends.. 352
Letter, Agreement 28
Liability 295
    Architects 287
    Claim 312
    Insurance 299, 313
License Law... 195
Licenses 187
Lien Law.. 220
Limitation Statute 306
Limiting Liability 108
Line of Communication 129
Liquidated Damages 45, 55, 106, 230, 251, 270
List of Bidders 83
Litigation, Surety 136
Local Practices.. 278
Log Book 200
Long Form 25
Low Bidder 88, 89, 93, 94
Lump Sum 48
    Sum Fee.. 31

M

Management, Construction 70
Material Technology................................. 279
Materials Storage.. 97
Mechanics' Lien 172, 221
Mediation.. 264, 320
Metrics... 352
Minor Changes 255
Mistakes... 290
Misuse of Drawings.................................. 153
Mobilization 52
Monitoring Submissions........................... 150
Monthly Payments 155

N

Negligence 277
Negotiated Compromise........................... 287
    Contract 42, 81
Negotiation............................. 24, 43, 88, 264
    Contract 88
    Phase.. 332
New Materials 280
    Systems... 66
Nonconforming Work 128, 168
Notice of Cessation 223
    of Completion................................. 220
    to Proceed 102
Notices 340
NSPE.. 71

O

Objections 346
Office Standards....................................... 315
Omissions.. 286
Opening Bid.. 88
Oral Agreements 13, 27
Orders.. 346
    Change 161
Original Drawing 21
OSHA... 187
Outside Consultants 107
Overall Budget.. 327
Overcertification 158
Overhead... 28
Overhead and Profit 54, 104
Overpaying Contractor............................. 139
Overruns.. 292
Overselling.. 312
Overshooting... 292
Overtime 166
Owner Relationships 177
    Suspension.. 35
    Design/Builder................................. 41
Owner's Budget 325, 331
    Changes 162
    Claims... 176
    Death... 36
    Information 172
    Instructions 321
    Program 327
    Responsibilities.............................. 169
    Rights...................................... 175, 241

P

Package Bid 85
    Deal 67
Partial Occupancy 229
Past Due Interest 47
Payments 120, 125, 155, 158, 173, 215, 315
    Application 155, 174
    Bond 173
    Certificate 336
    Final 159
    Progress 47
PCBs 189, 206
Pending Claims 123
Performance Bond 173
Permits 171
Personality Conflict 36
Personell Availability 91
Plan Conflicts 111
    Services 86
Post-Contract 26
Precedence, Contract 16
Pre-construction 95, 104, 200, 205
    Meeting 102
Pre-Design Phase 26
Primary Visit 199
Priority of Documents 17
Product Data 142, 202, 212
Professional Liability 190
    Negligence 277
    Responsibility 277
    Standard 278
Profit 104
Program Design 325
Progress Payments 47
    Schedules 99, 105, 202
Project Bond 133
    Budget 329
    Delays 99
    Delivery Systems 65
    Manual 13
    Proximity 91
    Size 91
    Type 91
Property Protection 206
Prospective Bidders 92
Protection, Property 206
Provisions Termination 234
Public Work Projects 224
Published Standards 252

Q

Qualified Personnel 151
Quality 330
    Construction 186
Quotations Changes 165

R

Reading Drawings 349
Record Drawings 225
    Jobsite 99
    Keeping 123, 305
Rejecting Work 179, 215, 251, 254
Related Costs 62
Release of Retainage 50
Replacing Superintendents 208
Reporting 120
Representations 156
Reproduction, Drawing 21
Request for Payment 101
Resolution, Dispute 37
Resolving Disputes 121, 260, 261
Responsibilities, Contractor 183
Responsibility, Professional 277
Responsible Innovation 282
Retainage 46, 50, 229
Review Document 14, 151
    Submittals 120
Revision Drawing 21

S

Safety 187
    Procedures 101
Samples 202
    Product 142
Scaling Drawings 351
Schedule Changes 165
    Construction 272
    of Values 53, 105
    Progress 202
Scheduling Shop Drawings 147
Schematic Design 24, 330, 332
Scope of Work 43
Sechedule Fee 28
Selected Contractor 81
Selecting Attorney 299
    Contractor 77, 81
    of Bidders 83
Selling Services 285
Separate Contractors 176, 187, 204
Settlement 306
Shared Savings 45

Index

Shop Drawing Review 148
    Drawings 141, 142, 147, 202, 212
Short Form Agreement 25
Short Form of Agreement 24, 40
Signing Contract 171, 184
Site Analysis 26
    Visit 100, 119
Size of Project 91
Small Project Editions 25
Special Forms 26
Specifications 185, 355
Stamping Drawings 151
Standard Documents 1
    of Care 288
    Professional 277
Statute of Limitations 305
Stipulated Sum 89
Stop Work 175
Subcontract Agreement 213
    List 213
    Payments 157
    Changes 165
Subcontractors 104, 188, 209
Submission Shop Drawings 147
Submittal Schedules 105
Submittals 99, 104, 142, 212, 343, 344
Substantial Completion 205, 218, 219, 227, 274, 338
    Performance 253
Substantiating Payments 160
Substitutions 98
Successful Project 95
Sufficient Cause 339
Superintendent 198
Superintendent's Responsibility 100
Supervision 201
Supplementary Drawings 163
Suppliers 188
Surety 134
    Approval 136
    Bond 47, 105
    Claims 136
    Obligations 134
Surveys 98
Suspending Contract 242
Systems Project Delivery 65

T

Taxes 187
Temporary Fences 97
    Services 98
Terminate Contracts 32, 138, 174, 216
    for Cause 240

Termination 231, 245
    Contract .... 233 - 238
    Expense 34
    of Contract 138, 174
Terms and Conditions 43
Testimony Legal 307
Testing 208
    Laboratory 172
    Materials 102
Third-Party Claims 52, 308
Time Changes 165
    Compression 76
    Construction 205
    Delay 275
    of Completion 45
    Schedule 328
Transmitting Bond 137
Turnkey 67

U

Unforeseen Conditions 271
Unilateral Terminations 231
Unit Prices 49, 87, 103, 164
Unknown Conditions 203
Unsettled Claims 231
Unspecified Drawings 144
Unsuccessful Innovations 281
Unwritten Expectations 214
Urealistical Expectations 285
Use Partial 229

W

Warranty 190
    Period 48, 190
Waste 189
Weather 99
    Delays 272
Willful Breach 194
Withholding Payment 157
Withold Certificate 228
Witness Appearances 306
    Expert 284
Work Defective 177
    Rejection 179
Working Days 270
Written Communication 127, 100, 335, 348
    Journal 200
    Warranties 225